Molecular Mechanisms of Cellular Growth

Developments in Molecular and Cellular Biochemistry

Series Editor: Naranjan S. Dhalla, Ph.D., FACC

1. Victor A. Najjar (ed.): *Biological Effects of Glutamic Acid and Its Derivatives.* 1981 ISBN 90-6193-841-4
2. Victor A. Najjar (ed.): *Immunologically Active Peptides.* 1981 ISBN 90-6193-842-2
3. Victor A. Najjar (ed.): *Enzyme Induction and Modulation.* 1983 ISBN 0-89838-583-0
4. Victor A. Najjar and Laszlo Lorand (eds.): *Transglutaminase.* 1984 ISBN 0-89838-593-8
5. Ger J. van der Vusse (ed.): *Lipid Metabolism in Normoxic and Ischemic Heart.* 1989 ISBN 0-7923-0479-9
6. Jan F.C. Glatz & Ger J. van der Vusse (eds.): *Cellular Fatty Acid-Binding Proteins.* 1990 ISBN 0-7923-0896-4
7. H.E. Morgan (ed.): *Molecular Mechanisms of Cellular Growth.* 1991 ISBN 0-7923-1183-3

Molecular Mechanisms of Cellular Growth

edited by

HOWARD E. MORGAN

Senior Vice President
Geisinger Clinic
Weis Center for Research
Danville, Pennsylvania 178222-2601, USA

Reprinted from *Molecular and Cellular Biochemistry*, Volume 104, Nos. 1 & 2 (1991)

Springer-Science+Business Media, B.V.

Library of Congress Cataloging-in Publication Data

Molecular mechanisms of cellular growth / edited by Howard E. Morgan
... [et al.].
p. cm. -- (Developments in molecular and cellular
biochemistry ; v. 7)
"Reprinted from Molecular and cellular biochemistry, volume 104."
ISBN 978-0-7923-1183-6 ISBN 978-1-4615-3886-8 (eBook)
DOI 10.1007/978-1-4615-3886-8
1. Cells--Growth--Regulation--Congresses. 2. Paracrine
mechanisms--Congresses. 3. Autocrine mechanisms--Congresses.
I. Morgan, H. E. (Howard E.), 1927- . II. Molecular and cellular
biochemistry. III. Series.
QH604.M66 1991
574.87'61--dc20 91-8478
CIP

CONTENTS

MOLECULAR MECHANISMS OF CELLULAR GROWTH SYMPOSIUM

A SCIENTIFIC SYMPOSIUM WAS HELD AT GEISINGER ON JUNE 17–21, 1990 TO CELEBRATE THE 75TH ANNIVERSARY OF THE GEISINGER HEALTH CARE SYSTEM

Molecular and Cellular Biochemistry **104**: 1–3, 1991.

Foreword: Geisinger Clinic – 75 years of commitment to patient care, education and research

In 1990 the Geisinger Health Care System celebrated its 75th anniversary. Since its founding, Geisinger has grown from a single 63-bed hospital in Danville into a multi-institutional system of health care that delivers services in 45 locations throughout central and northeastern Pennsylvania.

The system includes:

- Geisinger Foundation – parent organization.
- Geisinger Clinic – the multi-specialty, salaried physician group, which includes more than 460 physicians practicing in 65 specialty areas. The clinic administers the research program and conducts continuing medical education programs. Geisinger Clinic physicians recorded more than 1 million patient visits last year.
- Geisinger Medical Center – a 577-bed regional tertiary referral center located in Danville, PA.
- Geisinger Wyoming Valley Medical Center – a 230-bed secondary referral center in Wilkes-Barre.
- Marworth – a nationally-recognized alcohol and chemical dependency treatment center that operates two facilities – one in Waverly, near Scranton, and one in Shawnee on Delaware, near the Delaware Water Gap.
- Geisinger Health Plan – a 88,000 member health maintenance organization licensed to deliver services in 17 Pennsylvania counties.
- Geisinger Medical Management Corporation – a for-profit corporation that provides contract management, consulting, and other medical services to health care providers inside and outside the Geisinger system.

The Sigfried and Janet Weis Center for Research stands as tangible evidence of the Geisinger organization's commitment to fulfilling its triple mission of patient care, education, and research.

Dedicated in September 1987, the center includes 65,000 square feet of laboratories, conference areas, offices, and informal meeting rooms. Initially opened with a shelled-in third floor, the center has ten laboratories for its ten staff scientists, eight associate scientists, seven postdoctoral fellows, and 25 research technicians.

Focusing at the cellular and molecular level of cardiovascular function, work in progress is funded this year at more than $1.3 million by outside agencies, including the National Institutes of Health, the American Heart Association, Mars Foundation, National Science Foundation, and the Arthritis Foundation.

Current research topics include studies of extracellular matrix proteins and control of cellular growth, the role of renin angiotensin systems in cardiac cell growth, second messengers in physical signal transduction, the expression of G-proteins in myocytes, post-translational modification of proteins by isoprenoids.

Other topics include the control of cell cycling, transacting factors in rDNA expressions, effects of contraction on extracellular matrix protein genes, and control of contraction of vascular smooth muscles.

The Sigfried and Janet Weis Center for Research has been named to honor the commitment of the Weis family to the Geisinger system. The support of Sigfried and Janet Weis has been key in initiation of the Basic Research Program at Geisinger and in its continued development. To recognize these contributions, the publication of this scientific symposium entitled 'Molecular Mechanisms of Cellular Growth' is dedicated to Sigfried and Janet Weis. The symposium was held at Geisinger on June 17–21, 1990 to celebrate the 75th Anniversary of the Geisinger Health Care System.

Sigfried and Janet Weis

Sigfried Weis became chairman of the Geisinger Medical Center Board of Directors in 1978. Since 1981, when he became the first chairman of the Geisinger Foundation, Mr. Weis has been a source of inspiration and wisdom for Geisinger's transformation into a multi-institutional system of health care. The Geisinger system today is one of the largest not-for-profit health systems in the eastern United States.

Janet Weis, a published author and civic leader, has served for five years as co-chairperson of the Children's Miracle Network Telethon at Geisinger Medical Center. She has contributed her energy to help the Geisinger Telethon raise more than $3.8 million to 'help make miracles happen'.

The Weis family has a long history of active community involvement. The Sigmund and Claire G. Weis Center for the Performing Arts at Bucknell University is named in honor of Mr. Weis's parents. Brother-in-law Charles Degenstein has established the Degenstein Fund for Distinguished Scientists in the Weis Center for Research at Geisinger. The lobby of the research center has been named in his honor.

On behalf of the Geisinger Health Care System, I want to thank the participants of the Symposium and those who submitted papers for this focussed issue of Molecular and Cellular Biochemistry. I am especially appreciative for the support of Naranjan S. Dhalla, Editor who invited publication of these papers.

H.E. Morgan, M.D., Guest Editor
Senior Vice President for Research
Geisinger Clinic, Sigfried and Janet Weis Center, for Research,
Danville, PA 17822-2601, USA

SIGNAL GENERATION

Molecular and Cellular Biochemistry **104**: 7–13, 1991.

Molecular control of myogenesis: antagonism between growth and differentiation

Eric N. Olson, Thomas J. Brennan, Tushar Chakraborty, Tse-Chang Cheng, Peter Cserjesi, Diane Edmondson, Guy James and Li Li
Department of Biochemistry and Molecular Biology, The University of Texas M.D. Anderson Cancer Center, 1515 Holcombe Blvd., Box 117, Houston, Texas 77030, USA

Key words: myogenesis, helix-loop-helix proteins, growth factors

Abstract

Insight into the molecular mechanisms that control establishment of the skeletal muscle phenotype has recently been obtained through cloning of a family of muscle-specific regulatory factors that can activate myogenesis when transfected into non-muscle cells. This family of factors, which includes MyoD, myogenin, myf-5, and MRF4, can bind DNA and transactivate muscle-specific genes in collaboration with ubiquitous cellular factors. Growth factors play an antagonistic role in myogenesis by suppressing the actions of the myogenic regulatory factor family. This review will focus on the regulation and mechanism of action of this family of myogenic regulatory factors and on the central role of peptide growth factors in modulating their expression and biological activities.

Understanding the mechanisms that determine cell fates and control the expression of specific gene sets in specialized cell types during development is a major question in molecular biology. Skeletal muscle has become one of the most well-defined systems for dissecting these mechanisms due to the recent discovery of a family of skeletal muscle-specific factors that can activate the entire myogenic phenotype. Figure 1 shows a hypothetical representation of the events involved in determination and differentiation within the myogenic lineage. Early in embryogenesis, totipotent stem cells become restricted in their potential fate and may give rise to multipotential mesodermal stem cells through mechanisms that remain undefined. Mesodermal stem cells are the precursors of several different cell types, including adipoblasts, chondroblasts, and myoblasts. Following generation of these stable lineages, the events of terminal differentiation can be triggered, often in response to extracellular cues. In the case of the myogenic lineage, myoblasts which arise from mesoderm have two options. They can remain in a determined, but undifferentiated state, or they can enter the differentiation pathway. Activation of the differentiation program is contingent on cessation of cell division and is accompanied by the expression of an array of unlinked muscle-specific genes and fusion to form multinucleate myotubes. Formation of a mature muscle fiber has been shown to involve precise control at the level of transcription, mRNA splicing, translation and protein assembly. The decision to differentiate is controlled through a repression-type mechanism in which mitogens in serum or specific peptide growth factors such as fibroblast growth factor (FGF) and transforming growth factor type-β (TGF-β) block expression of muscle-specific genes [1–3]. Depletion of growth factors below a critical level results in activation of the differentiation program. Although it is well

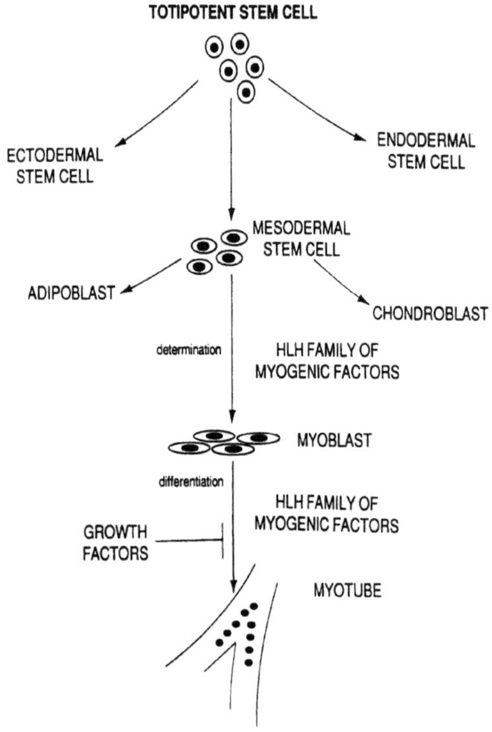

Fig. 1. Hypothetical model for myogenic lineage determination. In early embryogenesis, totipotent stem cells give rise to stem cells of the three germ layers (endoderm, ectoderm and mesoderm). Mesodermal stem cells have the potential to generate adipoblasts, chondroblasts and myoblasts. Myoblasts are triggered to differentiate when the extracellular concentration of growth factors drops below a critical threshold. The HLH family of myogenic regulatory factors may participate in myogenic lineage determination and differentiation.

established that myogenesis is controlled through a growth factor-dependent mechanism, how growth factors, which act at the cell surface, coordinately suppress an array of more than 20 unlinked muscle-specific genes remains unclear.

A family of myogenic regulatory factors determines the muscle phenotype

Studies from several laboratories over many years supported the notion that establishment of the muscle phenotype was controlled by a single myogenic master gene that controlled all subsequent events. The first such regulatory gene to be isolated was MyoD, which was identified by subtraction-hybridization of cDNA from myoblasts against cDNA from 10T1/2 fibroblasts [4]. MyoD is expressed exclusively in skeletal muscle and can activate myogenesis when transfected into 10T1/2 cells and several other cell types. Cells of mesodermal origin appear to be the most susceptible to myogenic conversion by MyoD, whereas cells from the other germ layers show less, or no, responsiveness to MyoD [5–6]. Soon after the identification of MyoD, three related factors, myogenin [7, 8], myf-5 [9] and MRF4/herculin/myf-6 [10–12] were independently isolated using different strategies. Like MyoD, these factors are restricted to skeletal muscle and can activate myogenesis in transfected cells.

Each of the myogenic regulatory factors are localized to the nucleus and share greater than 80% homology within a 70 amino acid segment that encompasses a basic region responsible for DNA binding and a helix-loop-helix (HLH) motif that mediates dimerization [13–15]. In the case of MyoD, this 70 amino acid segment has been shown to be sufficient to activate the myogenic program in stably transfected 10T1/2 cells [16]. This basic-HLH region also shares significant homology with several proteins implicated in transcription control, including the immunoglobulin enhancer-binding factors E12, E47, E2-2 and TFE3, the *Drosophila* genes *daughterless, achaete scute* T3 and T4, *enhancer-of-split, hairy, twist,* and *extramacrochaetae,* which regulate lineage decisions in early development, and members of the c-*myc* family of oncogenes. Oligomerization appears to be a prerequisite for binding of HLH proteins to DNA and seems to occur preferentially between ubiquitously expressed HLH proteins such as E12, E47 and the *daughterless* gene product and cell type-specific HLH proteins such as the myogenic regulatory factors and the products of the *achaete scute* complex (AS-C) [13, 14]. Combinatorial interactions among different HLH proteins dramatically expands the regulatory potential of these factors and provides opportunities for positive and negative interactions.

Most of the HLH proteins studied thus far bind to variations of the DNA consensus sequence

CANNTG, referred to as an E-box [13]. This motif was first identified within the immunoglobulin enhancers as a nuclear factor binding site and was subsequently found within numerous cell type-specific enhancers. Individual HLH proteins exhibit considerable selectivity with respect to the E-boxes they recognize, the specificity being determined by the central nucleotides of the concensus, as well as by nucleotides flanking the concensus sequence. Myogenin : E12 and MyoD : E12 hetero-oligomers, for example, bind with high affinity to the E-box concensus, CCACCTGC [14, 15, 17–19], referred to as a κE-2 site, but do not show significant binding to the E-box, TCATGTGG, referred to as a μE-3 site. Conversely, homo-oligomers of the HLH protein TFE3 bind preferentially to the μ-E3 site and do not recognize the κE-2 site [20]. In addition to the regulatory diversity imparted by hetero-oligomerization, the ability of different combinations of HLH hetero-oligomers to bind similar or identical sequences also allows for negative regulation through competition for DNA binding.

The myogenic regulatory factors bind DNA and activate muscle-specific genes

Although several cell type-specific transcription factors have been described, the myogenic regulatory factors are the only mammalian factors identified thus far that can generate a complete differentiated phenotype requiring activation of a large array of unlinked genes. How do the myogenic regulatory factors activate muscle-specific genes? The exact mechanism through which these factors activate their target genes is unknown, but there is considerable evidence to indicate that they activate at least some genes through direct interactions with muscle-specific control regions. The *muscle creatine kinase, myosin light chain-1/3,* and *acetylcholine receptor α-subunit* genes, for example, have been shown to contain multiple E-boxes in their regulatory regions that bind the myogenic regulatory factors [17–21]. Mutagenesis of these E-boxes abolishes muscle-specific expression and abrogates responsiveness to myogenin and MyoD. Cis-acting sequences that confer muscle-specific expression are generally quite complex and contain binding sites for numerous transcription factors. Thus, it is likely that the myogenic regulatory factors collaborate with other transcription factors to generate muscle-specific transcription. Indeed, in the case of the MCK enhancer, there are two E-boxes that are necessary, but not sufficient for full muscle-specific enhancer activity [22].

There are several muscle-specific genes, such as *myosin heavy chain* and *α-skeletal actin* that are activated in response to the myogenic regulatory factors, but which do not contain obvious binding sites for these factors. How then are these genes controlled? One possibility is that this subset of muscle-specific genes is controlled indirectly through intermediate muscle-specific transcription factors that are induced in response to the myogenic factors. Two such factors have been identified, MEF-2 [22], MCBF [23], each of which are expressed in muscle and bind to cis-acting sequences in different muscle-specific genes.

Growth factors regulate the expression of the myogenic regulatory factors

Growth factors have long been known to inhibit differentiation of determined myoblast cell lines. Recent studies suggest that the myogenic regulatory factors may serve as the targets for growth factor-dependent repression. This repression is exerted at multiple levels; through suppression of the transcription of the myogenic regulatory factor genes and through inhibition of the activities of these factors in cells that express them constitutively [4, 8, 24].

Depending on the cell line, either serum, FGF or TGF-β can partially or fully block the expression of the myogenic regulatory factors [4, 8, 24]. Thus, repression of the differentiation program can be at least partially explained by growth factor-dependent suppression of the factors responsible for activation of 'downstream' muscle specific genes. How might growth factor signals suppress expression of the myogenic regulatory factor genes? During commitment of mesodermal stem cells to the myogenic lineage, the myogenic regulatory factor genes acquire the potential to be expressed (Fig. 1). Since myoblasts exposed to high concentrations of growth factors maintain their position in the lineage, but often do not express members of the MyoD family at detectable levels, there must be additional factors that participate in myogenic lineage commitment. Perhaps the activities of these additional factors are modulated by growth factors such that they can only activate the expression of the MyoD family when growth signals are diminished. Such regulation could be mediated by cis-acting DNA sequences that act in a negative manner and override the positive actions of lineage-specific activators or they could act indirectly by interfering with the activity of the activators. When the control regions of the myogenic regulatory factor genes are defined it should be possible to distinguish among these potential modes of regulation.

Growth factors regulate the activities of the myogenic regulatory factors

In addition to inhibiting the expression of myogenin and MyoD in myoblasts, growth factors have the ability to inhibit the activities of these factors in cells that express them constitutively [4, 8, 24]. Several potential pathways through which growth factors might suppress the actions of the myogenic regulatory factors can be envisioned. Benezra *et al.* have described an inhibitory HLH protein Id, which is expressed in a wide range of proliferating cell types and is downregulated upon withdrawal of mitogens [25]. Id can heterodimerize with E12 and to a lesser extent with MyoD. However, it lacks a basic domain and therefore generates heterodimers that cannot bind DNA. The ability of Id to attenuate the DNA binding activity of MyoD could account for the observation that MyoD is expressed in some proliferating myoblasts, but does not activate myogenesis until mitogens have been depleted from the medium.

Because Id is down-regulated in quiescent cells, it is likely to account for only certain aspects of the myogenic program. Inhibitory signals that do not require cell proliferation, such as TGF-β [2] and ras [26, 27], are likely to act through a mechanism that does not involve Id. Indeed, TGF-β, the most potent growth factor inhibitor of myogenesis yet identified, can suppress the ability of myogenin to activate muscle target genes through a mechanism distal to DNA sequence recognition and independent of Id [27a].

Another mechanism whereby growth factor signals could regulate the activity of the myogenic regulatory factors would be by changes in protein phosphorylation. MyoD and myogenin have been shown to be phosphorylated. Activation of intracellular growth factor cascades could therefore interfere with the activities of these proteins through phosphorylation of sites involved in DNA binding, transcriptional activation, or oligomerization. Indeed, there are numerous examples in which transcription factors are regulated by phosphorylation. At present, the sites in these factors that are phosphorylated and the kinases for which they are substrates are unknown.

An additional level of complexity is added by the finding that, like many nuclear regulatory factors, members of the MyoD family positively autoregulate their own expression and cross-activate each others' expression [28, 29]. These regulatory interactions may confer stability to the myogenic phenotype and reinforce the decision to differentiate. The autoregulatory loops among the myogenic regulatory factors might function as intranuclear sensors of external growth factor signals. When growth factors are high, these interactions are dampened, whereas when growth factors are low, they are amplified leading to terminal differentiation. It is conceivable that suppression of myogenic regulatory factor expression in response to growth factors is the result of a decrease in the transcription activating potential of the myogenic regulatory factors themselves.

It is also possible that the myogenic factors are not the targets of intracellular growth factor cascades, but rather another cellular component that these factors require for activation of muscle-specific genes. If this is the case, this component would appear to be quite specific since growth factors do not lead to generalized transcriptional repression.

The myogenic regulatory factors inhibit cell proliferation

Initial transfection experiments showed a 10-fold reduction in stable colonies following transfection with a MyoD expression vector relative to a selectable marker alone, suggesting that forced expression of MyoD at high levels was incompatible with cell proliferation [4]. The growth inhibitory actions of MyoD are influenced by the concentration of extracellular mitogens and can be overcome by high concentrations of serum or FGF. The ability of mitogens to overcome the growth inhibitory effects of MyoD supports the notion that the actions of MyoD are governed in a negative manner by incoming growth factor signals and explains the observation that several types of myoblasts express MyoD prior to differentiation.

How do members of the MyoD family inhibit cell proliferation? Transfection experiments using MyoD deletion mutants indicate that the growth inhibitory actions of MyoD map to the basic region and are dependent on DNA binding [30]. Interestingly, mutants in which the basic domain of MyoD has been replaced with that of E12 also inhibit proliferation, but are unable to activate myogenesis. These results indicate that inhibition of cell proliferation is independent of differentiation. Similarly, MyoD can suppress cell growth in cell lines that are refractory to myogenic conversion by MyoD.

One mechanism whereby members of the MyoD family could inhibit cell growth would be by binding and repression of the control regions of genes that are required for cell cycle progression. Members of the MyoD family also could interact directly with gene products that regulate cell proliferation such as *fos, myc,* or *jun*. Finally, excessive levels of the myogenic regulatory factors could inhibit cell proliferation by squelching or transcriptional interference, as a consequence of their interaction with limiting transcription factors required for cell division.

Activated oncogenes can mimic the effects of growth factors on the differentiation program

One approach that has been taken to dissect the intracellular pathways through which growth factors inhibit myogenesis has involved modification of myoblasts with activated oncogenes whose products transduce growth factor signals from the cell membrane to the nucleus. These studies have shown that members of the *ras* family of guanyl nucleotide regulatory proteins can, when activated by missense mutations, block activation of the muscle differentiation program [26, 27]. Like exogenous growth factors, *ras* proteins appear to exert their effects at multiple levels. *ras*-transformed myoblasts do not express MyoD or myogenin [31]. These results suggest that *ras* inhibits myogenesis by extinguishing expression of MyoD (and myogenin). As with growth factors, the actions of *ras* and MyoD are dependent on the relative levels of each. Forced expression of MyoD is sufficient to activate muscle-specific genes in cells that express *ras* whereas high levels of *ras* expression can block myogenesis in cells that express MyoD constitutively [31].

The nuclear factors *fos* and *jun* bind the DNA consensus sequence TGACTCA, referred to as an AP-1 site, and have been proposed to function as intranuclear mediators of growth factor and *ras*-dependent pathways. Indeed, transformation of myoblasts by v-*fos* [31] or v-*jun* (P. Vogt, personal communication) is sufficient to block differentiation. Similarly, c-*fos* and *jun*B can block trans-activation of the MCK enhancer by myogenin and MyoD [31; L. Li and E. Olson, in preparation]. Anti-sense *fos* also has been shown to rescue the ability of MyoD to trans-activate the MCK enhancer in HeLa cells, which are normally refractory to myogenic conversion [5]. It remains to be determined whether myogenic regulatory factor genes contain AP-1 sites in the regulatory regions that may mediate growth factor-dependent repression.

The precise role of *myc* in myogenic differentiation has been examined intensely, but remains unclear. Whereas transformation of primary chick myoblasts with avian myocytomatosis virus MC29 encoding v-*myc* is sufficient to prevent differentiation [32], deregulated expression of c-*myc* in established muscle cell lines does not necessarily extinguish the myogenic phenotype [33]. Inhibition of myogenesis by v-*myc* appears to involve continued stimulation of cell division which prevents differentiation indirectly [32, 34]. In contrast to v-*myc*, deregulated expression of c-*myc* in the BC_3H1 muscle cell line is insufficient to support cell proliferation and does not fully extinguish muscle-specific gene expression [33].

The observation that c-*myc* shares homology with members of the myogenic regulatory factor family within the basic-HLH region suggested the attractive hypothesis that *myc* proteins could disrupt myogenesis through formation of functionally inactive oligomeric complexes with the myogenic factors or E12, or could possibly compete for binding to conserved DNA sequences. However, analysis of the oligomerization [14, 15] and DNA-binding properties of *myc* have failed to provide evidence to support this notion, leaving the exact mechanism of action of *myc* unclear.

Summary

The identification and characterization of the myogenic regulatory gene family has yielded significant insight into the mechanisms responsible for generation of the skeletal muscle phenotype and has begun to shed light on the molecular basis for the antagonism between cell growth and differentiation. In the future, it will be important to determine whether regulatory mechanisms operative in skeletal muscle have parallels in other specialized cell types.

Acknowledgements

The secretarial assistance of E. Madson is appreciated. Work in the authors' laboratory is supported by grants from the National Institutes of Health and American Cancer Society. ENO is an Established Investigator of the American Heart Association.

References

1. Spizz G, Roman D, Strauss A, Olson EN: Serum and fibroblast growth factor inhibit myogenic differentiation through a mechanism dependent on protein synthesis and independent of cell proliferation. J Biol Chem 261: 9483–9488, 1986
2. Olson EN, Sternberg E, Hu JS, Spizz G, Wilcox C: Regulation of myogenic differentiation by type beta transforming growth factor. J Cell Biol 103: 1799–1805, 1986
3. Clegg CH, Linkhart TA, Olwin BB, Hauschka SD: Growth factor control of skeletal muscle differentiation occurs in G1-phase and is repressed by fibroblast growth factor. J Cell Biol 105: 949–956, 1987
4. Davis RL, Weintraub H, Lassar AB: Expression of a single transfected cDNA converts fibroblasts to myoblasts. Cell 51: 987–1000, 1987
5. Weintraub H, Tapscott SJ, Davis RL, Thayer MJ, Adam MA, Lassar AB, Miller AD: Activation of muscle specific genes in pigment, nerve, fat, liver and fibroblast cell lines by forced expression of MyoD. Proc Natl Acad Sci 86: 5434–5438, 1989
6. Schafer BW, Blakely BT, Darlington GJ, Blau HM: Effect of cell history on response to helix-loop-helix family of myogenic regulators. Nature 344: 454–458, 1990
7. Wright WE, Sassoon DA, Lin VK: Myogenin, a factor regulating myogenesis, has a domain homologous to MyoD1. Cell 56: 607–617, 1989
8. Edmondson DG, Olson EN: A gene with homology to the *myc* similarity region of MyoD1 is expressed during myogenesis and is sufficient to activate the muscle differentiation program. Genes Dev 3: 628–640, 1989
9. Braun T, Buschhausen-Denker G, Bober E, Tannich E, Arnold HH: A novel human muscle factor related to but distinct from MyoD1 induces myogenic conversion in 10T1/2 fibroblasts. EMBO J 8: 701–709, 1989
10. Rhodes SJ, Konieczny SF: Identification of MRF4: A new member of the muscle regulatory factor gene family. Genes Dev 3: 2050–2061, 1989
11. Miner JH, Wold B: Herculin, a fourth member of the MyoD family of myogenic regulatory genes. Proc Natl Acad Sci USA 87: 1089–1093, 1990
12. Braun T, Bober E, Winter B, Rosenthal N, Arnold HH: *Myf-6*, a new member of the human gene family of myogenic determination factors: evidence for a gene cluster on chromosome 12. EMBO J 9: 821–831, 1990
13. Murre C, McCaw PS, Baltimore D: A new DNA binding and dimerization motif in immunoglobulin enhancer binding, *daughterless, MyoD,* and *myc* proteins. Cell 56: 777–783, 1989a
14. Murre C, McCaw PS, Vaessin H, Caudy M, Jan LY, Jan YN, Cabrera CV, Buskin JN, Hauschka SD, Lassar AB, Weintraub H, Baltimore D: Interactions between heterologous helix-loop-helix proteins generate complexes that bind specifically to a common DNA sequence. Cell 58: 537–544, 1989b

15. Davis RL, Cheng P-F, Lassar AB, Weintraub H: The MyoD DNA binding domain contains a recognition code for muscle-specific gene activation. Cell 60: 733–746, 1990
16. Tapscott SJ, Davis RL, Thayer MJ, Cheng P-F, Weintraub H, Lassar AB: MyoD1: A nuclear phosphoprotein requiring a *myc* homology region to convert fibroblasts by myoblasts. Science 242: 405–411, 1988
17. Brennan TJ, Olson EN: Myogenin resides in the nucleus and acquires high affinity for a conserved enhancer element on heterodimerization. Genes Dev 4: 582–595, 1990
18. Buskin JN, Hauschka SD: Identification of a myocyte-specific nuclear factor which binds to the muscle-specific enhancer of the mouse muscle creatine kinase gene. Mol Cell Biol 9: 2627–2640, 1989
19. Lassar AB, Buskin JN, Lockshon D, Davis RL, Apone S, Hauschka SD, Weintraub H: MyoD is a sequence-specific DNA binding protein requiring a region of *myc* homology to bind to the muscle creatine kinase enhancer. Cell 58: 823–831, 1989a
20. Beckmann H, Su L-K, Kadesch T: TFE3: A helix-loop-helix protein that activates transcription through the immunoglobulin enhancer μE3 motif. Genes Dev 4: 167–179, 1990
21. Piette J, Bessereau J-L, Huchet M, Changeux J-P: Two adjacent MyoD1-binding sites regulate expression of the acetylcholine receptor α-subunit gene. Nature 345: 353–355, 1990
22. Gossett L, Kelvin D, Sternberg E, Olson EN: A new myocyte-specific enhancer binding factor that recognizes a conserved element associated with multiple muscle-specific genes. Mol Cell Biol 9: 5022–5033, 1989
23. Mar JH, Ordahl CP: M-CAT binding factor, a novel trans-acting factor governing muscle-specific transcription. Mol Cell Biol, in press, 1990
24. Vaidya TB, Rhodes SJ, Taparowsky EJ, Konieczny SF: Fibroblast growth factor and transforming growth factor β repress transcription of the myogenic regulatory gene MyoD1. Mol Cell Biol 9: 3576–3579, 1989
25. Benezra R, Davis RL, Lockshon D, Turner DL, Weintraub H: The protein Id: a negative regulator of helix-loop-helix DNA binding proteins. Cell 61: 49–59, 1990
26. Olson EN, Spizz G, Tainsky MA: The oncogenic form of N-*ras* prevents skeletal myoblast differentiation. Mol Cell Biol 7: 2104–2111, 1987
27. Gossett LA, Zhang W, Olson EN: Dexamethasone-dependent inhibition of differentiation of C2 myoblasts bearing steroid-inducible N-*ras* oncogenes. J Cell Biol 106: 2127–2137, 1988
27a. Brennan T, Chakraborty T, Olson EN: TGFB inhibits the actions of myogenin through a mechanism independent of DNA binding. Roc Natl Acad Sci USA, in press, 1991.
28. Thayer MJ, Tapscott SJ, Davis RL, Wright WE, Lassar AB, Weintraub H: Positive autoregulation of the myogenic determination gene MyoD1. Cell 58: 241–248, 1989
29. Braun T, Bober E, Buschhausen-Denker G, Kotz S, Grzeschik K, Arnold HH: Differential expression of myogenic determination genes in muscle cells: possible autoactivation by the *Myf* gene products. EMBO J 8: 3617–3625; 1989b
30. Sorrentino V, Pepperkok R, Davis RL, Ansorge W, Philipson L: Cell proliferation inhibited by MyoD1 independently of myogenic differentiation. Nature 345: 813–815, 1990
31. Lassar AB, Thayer MJ, Overell RW, Weintraub H: Transformation by activated RAS or FOS prevents myogenesis by inhibiting expression of MyoD1. Cell 5: 659–667, 1989b
32. Falcone G, Tato F, Alema S: Distinctive effects of the viral oncogenes *myc, erb, fps,* and *src* on the differentiation program of quail myogenic cells. Proc Natl Acad Sci 82: 426–430, 1985
33. Schneider MD, Perryman MB, Payne PA, Spizz G, Roberts R, Olson EN: Autonomous expression of c-*myc* in BC_3H1 cells partially inhibits but does not prevent myogenic differentiation. Mol Cell Biol 7: 1973–1977, 1987
34. LaRocca SA, Grossi M, Falcone G, Alema S, Tato F: Interaction with normal cells suppresses the transformed phenotype of v-*myc* transformed quail muscle cells. Cell 58: 123–131, 1989

Address for offprints: E.N. Olson, Department of Biochemistry and Molecular Biology, The University of Texas M.D. Anderson Cancer Center, 1515 Holcombe Blvd., Box 117, Houston, TX 77030, USA

Molecular and Cellular Biochemistry **104**: 15–19, 1991.

Transgenic animals as a tool for studying the effect of the c-*myc* proto-oncogene on cardiac development

Twila Jackson[1], Michael F. Allard[2], Catherine M. Sreenan[2], Lisa K. Doss[2], Sanford P. Bishop[2] and Judith L. Swain[1]
[1] *Departments of Medicine and Microbiology, Duke University Medical Center, Durham, North Carolina 27710, USA;* [2] *Department of Pathology, University of Alabama at Birmingham, Birmingham, Alabama 35294, USA*

Key words: cardiac growth, cardiac myocytes, c-*myc* mRNA, heart development

Abstract

Transgenic animals provide a model system to elucidate the role of specific proteins in development. This model is now being used increasingly in the cardiovascular system to study cardiac growth and differentiation. During cardiac myocyte development a transition occurs from hyperplastic to hypertrophic growth. In the heart the switch from myocyte proliferation to terminal differentiation is synchronous with a decrease in c-*myc* mRNA abundance. To determine whether c-*myc* functions to regulate myocyte proliferation and/or differentiation, we examined the *in vivo* effect of increasing c-*myc* expression during fetal development and of preventing the decrease in c-*myc* mRNA expression that normally occurs during myocyte development. The model system used was a strain of transgenic mice exhibiting constitutive expression of c-*myc* mRNA in cardiac myocytes throughout development. Increased c-*myc* mRNA expression is associated with both atrial and ventricular enlargement in the transgenic mice. This increase in cardiac mass is secondary to myocyte hyperplasia, with the transgenic hearts containing greater than twice as many myocytes as nontransgenic hearts. The results of this study indicate that constitutive expression of c-*myc* mRNA in the heart during development results in enhanced hyperplastic growth, and suggest a regulatory role for the c-*myc* protooncogene in cardiac myogenesis.

Introduction

Cardiac development has been extensively studied *in vivo*. Myocytes proliferate throughout fetal and early postnatal development, and proliferation ceases at approximately 2 weeks of age. Cardiac growth after 2 weeks of age primarily occurs through an increase in myocyte size rather than number [1] and by proliferation of nonmuscle cells. The factors which control this developmental progression are unknown.

The c-*myc* protooncogene has been implicated in controlling both proliferation and differentiation in various cell types [2]. Increased expression of c-*myc* increases proliferative rate in some cell types, possibly through a decrease in the G1 phase of the cell cycle [3]. Evidence exists for a role for c-*myc* in skeletal myocyte development. Differentiation of skeletal myoblasts is accompanied by a decrease in c-*myc* expression [4, 5], and overexpression of c-*myc* inhibits differentiation [6, 7].

The role of c-*myc* in the control of proliferation and differentiation of cardiac myocytes is unknown. We and others [8] have demonstrated an association between decreased c-*myc* mRNA expression and the transition from hyperplastic to hypertrophic cardiac growth during cardiac development. To examine the effects of preventing c-*myc* downregulation during development, a strain of transgenic mouse was developed in which c-*myc* mRNA is constitutively expressed at high levels in the heart throughout development. The results of this study demonstrate that increased c-*myc* mRNA expression induces marked cardiac hyperplasia during fetal development, with the transgenic hearts containing greater than two-fold more myocytes than nontransgenic hearts. These results indicate that increased expression of c-*myc* can augment the proliferative capacity of cardiac myocytes, and suggests a regulatory role for this nuclear localized protooncogene product in cardiac development.

Materials and methods

Production and analysis of transgenic mouse strain

The strain of transgenic mouse used in these studies has been described in detail elsewhere [9]. Mice were genotyped by extracting DNA from either a short section of tail or from an organ, and Southern analysis performed after digesting the DNA with either Bam H I or Hpa II as previously described [9]. RNA was obtained for analysis by a modification of the method of Chirgwin *et al.* [10] and expression of the c-*myc* transgene was assessed by RNAase protection as previously described [9].

Isolated myocyte preparation and morphometry

For the studies examining transgene expression *in vitro,* hearts were obtained from fetuses at 12 to 15 days of gestation, and primary myocyte cultures prepared by a modification of the method of Simpson [11]. For the morphologic studies of adult hearts, myocytes were isolated according to previously described methods [12, 13] after perfusion with 0.1% collagenase. The percentage of myocytes with single or multiple nuclei was determined by counting 200 H & E stained cells for each preparation. Cell length was determined by direct microscopic measurement of 50 glutaraldehyde-fixed cells in suspension using an inverted microscope. Cell volume was determined on suspended glutaraldehyde-fixed cells with a Coulter Channelyzer system [14]. The median value for $2\text{–}4 \times 10^4$ cells was determined for each preparation. Myocyte number in the total ventricle was calculated by an extension of a method previously described [13, 15]. Total calculated ventricular cardiac myocyte volume was divided by mean isolated myocyte volume to obtain number of cells in the ventricles.

Results

Transgene expression during cardiac development

RNAase protection analysis was performed to determine if cardiac transgene expression is developmentally regulated, and to identify the cell type responsible for transgene expression. RNA was prepared from hearts obtained from embryos at 15 days of gestation, neonates at 1, 8, 11, and 14 days of age, and adults at 6 months of age. Primary myocyte cultures were prepared from 15 day embryonic hearts, and RNA extracted from these cells after 4 days in culture. The probe used has been previously described [9] and was derived from the murine c-*myc* cDNA.

The results of RNAase protection analysis demonstrate that the transgene is expressed in cardiac myocytes in primary culture, and in embryonic, neonatal and adult heart, and c-*myc* mRNA abundance is increased approximately 10 fold in the transgenic compared to nontransgenic hearts. In addition, transgene expression also was demonstrated in both atria and both ventricles of the heart. The expression of the transgene in primary cultures of cardiac myocytes, and the absence of transgene expression in nonmyocyte companion cultures, indicates that transgene expression occurs in the myocyte rather than other cell types such as fibroblasts, endothelial cells, and smooth muscle cells.

Effects of increased cardiac c-myc *mRNA expression*

To determine the effect of increased c-*myc* expression on the rate of cardiac growth during development mice were sacrificed at 1, 15, or 60 days of age, and heart weights and body weights measured (Table 1). Cardiac enlargment in the transgenic animals was apparent by 1 day of age and persists throughout development. The percent increase in cardiac mass of the transgenic compared to nontransgenic hearts is relatively constant during postnatal development, indicating that the rate of postnatal growth of transgenic hearts is similar to that of nontransgenic hearts. This result suggests that increased c-*myc* expression accelerates cardiac growth during fetal development, but although transgene mRNA continues to be expressed during postnatal development, further acceleration of postnatal growth does not occur. The heart weight to body weight ratio in the c-*myc* expressing animals is greater than that in the nontransgenic animals at each stage of development, further evidence that a significant increase in cardiac mass is present.

The possibility that the increase in cardiac size noted in the c-*myc* transgenic mice is secondary to proliferation of connective tissue was examined. Transverse sections were taken from adult transgenic and nontransgenic hearts and stained with Masson trichrome to identify fibrotic tissue. Histologic examination revealed no evidence of increased fibrosis in the c-*myc* expressing transgenic hearts. Therefore, the increased mass of the transgenic hearts is not secondary to changes in the connective tissue content of the heart. In addition, no abnormalities in cardiac architecture were observed in the transgenic mice, and no cardiac tumors were detected.

Constitutive cardiac c-myc *expression results in myocyte hyperplasia*

The increased cardiac mass noted as a consequence of increased c-*myc* expression could be secondary to myocyte enlargement (i.e. hypertrophy) and/or to myocyte proliferation (hyperplasia) during fetal development. Therefore isolated myocytes were subjected to morphometric analysis and the results are shown in Table 2. The transgenic hearts contain more than twice as many myocytes compared to the nontransgenic hearts. In addition, transgenic myocytes are significantly (40%) smaller than the nontransgenic myocytes. Taken together, these data indicate that increased expression of c-*myc* in the heart during development leads to myocyte hyperplasia, and that by adulthood the resultant myocytes are smaller in size than comparable myocytes from nontransgenic mice.

Discussion

An activated c-*myc* gene was introduced into the germline of transgenic mice to test the hypothesis that increased c-*myc* expression increases proliferation of cardiac myocytes during development. The strain of transgenic mice developed displays cardiac-specific c-*myc* transgene expression [9],

Table 1. Heart weight and body weight in nontransgenic versus c-*myc* expressing transgenic mice

	1 day of age		15 days of age		60 days of age	
	non TG (n = 43)	TG (n = 20)	non TG (n = 109)	TG (n = 63)	non TG (n = 46)	TG (n = 35)
Ht wt (mg)	10.6 ± 2.8	15.3 ± 3.0*	45.6 ± 13.9	66.6 ± 26.0*	131 ± 23	176 ± 50*
Body wt (gm)	1.72 ± 0.029	1.56 ± 0.09	7.69 ± 2.58	7.25 ± 2.40	31.1 ± 4.7	28.8 ± 3.1*
Ht/bd wt (× 10^3)	6.22 ± 1.56	9.93 ± 2.83*	6.49 ± 5.65	9.29 ± 3.14*	4.25 ± 0.44	6.11 ± 1.61*

Non TG, nontransgenic; TG, transgenic; Ht wt, heart weight; Body wt, body weight; Ht/bd wt, heart weight to body weight ratio. All values are mean ± SD. * $p < 0.05$ TG vs non TG.

and a greater than 10-fold increase in cardiac c-*myc* mRNA expression throughout development. This study demonstrates that increased expression of c-*myc* during cardiac development results in myocyte hyperplasia. At six weeks of age transgenic hearts contain greater than two-fold more myocytes than nontransgenic hearts. This finding together with the significant increase in heart size noted by 1 day of age, suggests that the transgenic myocytes have increased proliferative capacity. The fact that the rate of postnatal cardiac growth in the transgenic and nontransgenic animals is comparable suggests that continued myocyte proliferation does not occur during postnatal development even though elevated c-*myc* mRNA expression persists. Thus the data support a role for c-*myc* in cardiac myocyte development *in vivo*.

Although terminally differentiated cardiac myocytes do not undergo karyokinesis and cytokinesis, these cells do exhibit increases in protein content and size during development [1]. It has been postulated that c-*myc* participates in regulating this growth process [16, 17]. The fact that myocytes in the c-*myc* expressing transgenic hearts are smaller than those in nontransgenic hearts does not eliminate a role for this protooncogene product in regulating post-differentiated myocyte growth. Since more than twice as many myocytes are present in the transgenic hearts, yet the work of the heart is similar in transgenic and nontransgenic hearts (i.e. body size and presumably cardiac demand are similar), it is anticipated that the hypertrophic growth that normally occurs in post-differentiated myocytes would be less in the transgenic hearts. Thus the results of our study do not eliminate a role for c-*myc* in the hypertrophic growth of differentiated cardiac myocytes. Further studies in these mice are in progress to address the role of c-*myc* in modulating the effects of agents that induce hypertrophic growth in terminally differentiated cardiac myocytes.

Table 2. Myocyte number and size in six to eight week old nontransgenic and c-*myc* expressing transgenic mice

	Non TG (n = 20)	TG (n = 9)
Myocyte#/heart ($\times 10^{-6}$)	4.9 ± 1.4	11.1 ± 1.9*
Median myocyte volume (fl $\times$ 10^{-4})	2.1 ± 0.5	1.3 ± 0.2*
Myocyte length (μm)	95 ± 10	88 ± 9
Cross sectional area (μm^2)	200 ± 42	148 ± 22*

Non TG, nontransgenic; TG, transgenic. All values mean ± SD. * $p < 0.05$ TG vs non TG.

Acknowledgements

The authors wish to thank Dr. Philip Leder for many helpful discussions, and for providing the strain of transgenic mice used in this study. Dr. Robert Robbins assisted with the histologic examination of the hearts. This work was supported in part by a Research Award from the American Heart Association and through Public Health Service Awards HL26831 (JLS) and HL36892 (SPB). The data in this manuscript has been previously published in an expanded form [18].

References

1. Bugaisky L, Zak R: Biological mechanisms of hypertrophy. In: H Fozzard, E Haber, R Jennings, A Katz, H Morgan (eds) The Heart and Cardiovascular System – Scientific Foundations. Raven Press, New York, 1986, pp 1491–1506
2. Marcu KB: Regulation of expression of the c-*myc* protooncogene. Bioessays 6: 28–32, 1987
3. Ramsay GM, Moscovici G, Moscovici C, Bishop JM: Neoplastic transformation and tumorigenesis by the human protooncogene MYC. Proc Natl Acad Sci USA 87: 2102–2106, 1990
4. Endo T, Nadal-Ginard B: Transcriptional and posttranscriptional control of c-*myc* during myogenesis: its mRNA remains inducible in differentiated cells and does not suppress the differentiated phenotype. Mol Cell Biol 6: 1412–1421, 1986
5. Sejersen T, Sumegi J, Ringertz NR: Density-dependent arrest of DNA replication is accompanied by decreased levels of c-*myc* mRNA in myogenic but not in differentiation-defective myoblasts. J Cell Physiol 125: 465–470, 1985
6. Denis N, Blanc S, Leibovitch MP, Nicolaiew N, Dautry F, Raymondjean M, Kruh J, Kitzia A: c-*Myc* oncogene ex-

pression inhibits the initiation of myogenic differentiation. Exper Cell Res 172: 212–217, 1987
7. Schneider MD, Perryman MB, Payne PA, Spizz G, Roberts R, Olson E: Autonomous expression of c-*myc* in BC3H1 cells partially inhibits but does not prevent myogenic differentiation. Mol Cell Biol 7: 1973–1977, 1987
8. Schneider MD, Payne PA, Ueno H, Perryman MB, Roberts R: Dissociated expression of c-*myc* and a *fos*-related competence gene during cardiac myogenesis. Mol Cell Biol 6: 4140–4143, 1986
9. Swain JL, Stewart TA, Leder P: Parental legacy determines methylation and expression of an autosomal transgene: A molecular mechanism for parental imprinting. Cell 50: 719–727, 1987
10. Chirgwin JM, Przybyla AE, MacDonald RJ, Rutter WJ: Isolation of biologically active ribonucleic acid from sources enriched in ribonuclease. Biochemistry 18: 5294–5299, 1979
11. Simpson P, Savion S: Differentiation of rat myocytes in single cell culture with and without proliferating nonmyocardial cells: Cross-striations, ultrastructure, and chronotropic response to isoproterenol. Circ Res 50: 101–116, 1982
12. Bishop SP, Drummond JL: Surface morphology and cell size measurement of isolated rat cardiac myocytes. J Mol Cell Cardiol 11: 423–433, 1979
13. Clubb FJ, Bell PD, Kriseman JD, Bishop SP: Myocardial cell growth and blood pressure development in neonatal spontaneously hypertensive rats. Lab Invest 56: 189–197, 1987
14. Nash GB, Tatham PER, Powell T, Twist VW, Speller RD, Loverock LT: Size measurements on isolated rat heart cells using Couldter analysis and light scatter flow cytometry. Biochem Biophys Acta 587: 99–111, 1979
15. Gerdes AM, Kriseman JD, Bishop SP: Changes in myocardial cell size and number during the development and reversal of hyperthyroidism in neonatal rats. Lab Invest 48: 598–602, 1983
16. Izumo S, Nadal-Ginard B, Mahdavi V: Protooncogene induction and reprogramming of cardiac gene expression produced by pressure overload. Proc Natl Acad Sci USA 85: 339–343, 1988
17. Starksen NF, Simpson PC, Bishopric N, Coughlin SR, Lee WM, Escobedo JA, Williams LT: Cardiac myocyte hypertrophy is associated with c-*myc* protooncogene expression. Proc Natl Acad Sci USA 83: 8348–8350, 1986
18. Jackson T, Allard M, Sreenan C, Doss L, Bishop S, Swain JL: The c-*myc* protooncogene regulates cardiac development in transgenic mice. Molec Cell Biol 10: 3709–3716, 1990

Address for offprints: J.L. Swain, Box 3828, Duke University Medical Center, Durham, NC 27710, USA

Molecular and Cellular Biochemistry **104**: 21–28, 1991.

Biological functions of proteoglycans: use of specific inhibitors of proteoglycan synthesis

David J. Carey
Weis Center for Research, Geisinger Clinic, North Academy Avenue, Danville, PA 17822, USA

Key words: extracellular matrix, glycosaminoglycan

Abstract

The use of specific inhibitors of proteoglycan synthesis have demonstrated essential functions for these molecules. Proteoglycans do not appear to be essential for cell viability or proliferation but are necessary for stable assembly of ECM and functional cell-ECM interaction. Because of their ability to nearly completely abolish ECM assembly in cell culture systems, proteoglycan synthesis inhibitors are useful tools to examine effects of ECM on the phenotypic behavior of cells.

Despite its usefulness, the use of proteoglycan synthesis inhibitors has some drawbacks. The most significant of these is the fact that synthesis of all proteoglycans is inhibited, making it difficult to assign a particular function to a specific proteoglycan type. For this reason, inhibition studies need to be done in conjunction with additional biochemical, immunological or molecular biological studies.

Introduction

Proteoglycans are ubiquitous constituents of extracellular matrix (ECM) and cell surfaces. As illustrated in Fig. 1a, proteoglycans consist of a protein core with one or more covalently attached unbranched chains of carbohydrate called glycosaminoglycan [1, 2]. Several structurally distinct types of proteoglycans have been described. Proteoglycans have been suggested to be involved in many important and diverse cell and tissue functions, including macromolecular filtration, cell-cell and cell-matrix adhesion, regulation of cell growth, and modulation of lipid metabolism [3]. Direct evidence for the participation of proteoglycans in these processes is still lacking in most cases, however. This is due, in part, to the difficulty of obtaining purified preparation of proteoglycans for *in vitro* functional studies.

Here I review studies from my laboratory that utilize an alternative approach to studying proteoglycan function, i.e. the use of specific metabolic inhibitors of proteoglycan synthesis. By administering these inhibitors to primary cultures of differentiating cells and observing the functional consequences that result, insights into the biological functions of proteoglycans have been gained.

Effects of β-D-xyloside on proteoglycan synthesis

As shown in Fig. 1b, the glycosaminoglycan-core protein linkage structure consists of a D-xylose moiety attached by a β-O-glycosidic bond to a serine residue of the core protein [1]. The xylose residue is added to newly synthesized core proteins in the Golgi apparatus by the enzyme xylosyltransferase [4]. The xylosylated core proteins then serve as the primers for glycosaminoglycan chain synthesis. The proteoglycan synthesis inhibitors that are most commonly used are derivatives of D-xylose in which the sugar is attached by a β-glycosidic

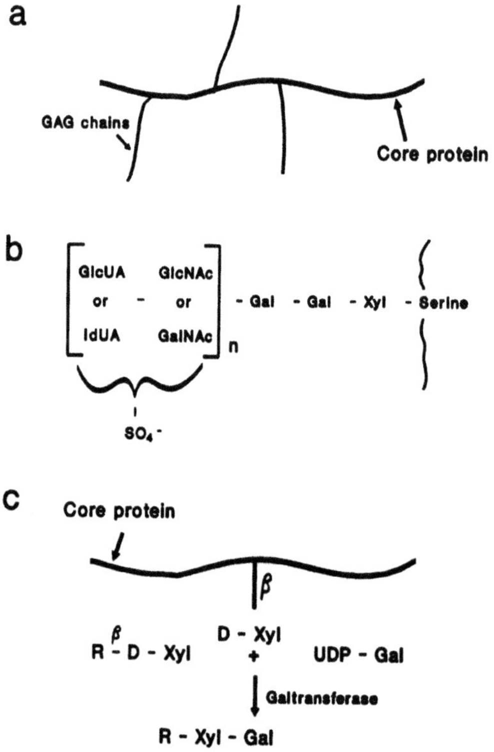

Fig. 1. Proteoglycan structure and mechanism of action of proteoglycan synthesis inhibitors. a: general proteoglycan structure; b: glycosaminoglycan – core protein linkage structure; c: inhibition of addition of galactose residue to xylose on core protein by exogenous β-D-xylose derivatives (R-D-xyl).

bond to a bulky organic moiety. Such molecules are similar in structure to the primer for glycosaminoglycan chain synthesis on nascent proteoglycan molecules. As shown in Fig. 1c, the exogenous β-D-xyloside derivatives accumulate intracellularly in sufficient amounts to compete with the endogenous xylosylated core proteins for the enzyme that adds the second sugar of the initiated glycosaminoglycan chain, i.e. galactosyltransferase I. Subsequent to galactose addition to the exogenous primers, glycosaminoglycan chain elongation then proceeds on the exogenous acceptor molecules.

Support for this mechanism comes from several observations. First, of the monosaccharides tested only xylose derivatives specifically inhibit proteoglycan synthesis and these are effective only when they are present in a β- and not α-conformation [5, 6]. Second, administration of β-D-xyloside derivatives to most cultured cells inhibits proteoglycan synthesis but stimulates synthesis of free glycosaminoglycan chains [5–8], by providing additional primers for chain initiation. This result also suggests that in most cells under normal conditions core protein synthesis is rate limiting for glycosaminoglycan chain synthesis. The free glycosaminoglycan chains synthesized in response to β-D-xyloside derivatives are secreted into and accumulate in the culture medium [6–8].

Biological effects of inhibition of proteoglycan synthesis

In my laboratory β-D-xyloside derivatives have been utilized to investigate the role of proteoglycans in the development of 2 types of primary cell cultures: co-cultures of embryonic rat peripheral nerve tissue (dorsal root ganglia) containing sensory neurons and Schwann cells, and vascular smooth muscle cells derived from thoracic aortas of adult rats.

Schwann cell-neuron co-cultures

Primary cultures of rat embryo Schwann cells and neurons undergo normal peripheral nerve development, including formation of axons by the neurons and ensheathment and myelination of these axons by the Schwann cells [9]. Prior to and during the period of myelination Schwann cells synthesize and deposit a specialized sheet of ECM (basement membrane) on their outer surfaces. The major biochemical components of the basement membrane are laminin [10], a large multi-domain adhesive glycoprotein, collagen type IV [11], and a large heparan sulfate proteoglycan [12]. Schwann cells also synthesize at least 2 types of cell associated heparan sulfate proteoglycans that differ in their mechanisms of attachment to the cells: one by a phosphatidylinositol lipid anchor [13, 14] and the other by a transmembrane hydrophobic domain of the protein core (Carey and Evans, unpublished observations). Previously the functions of these proteoglycans were not known.

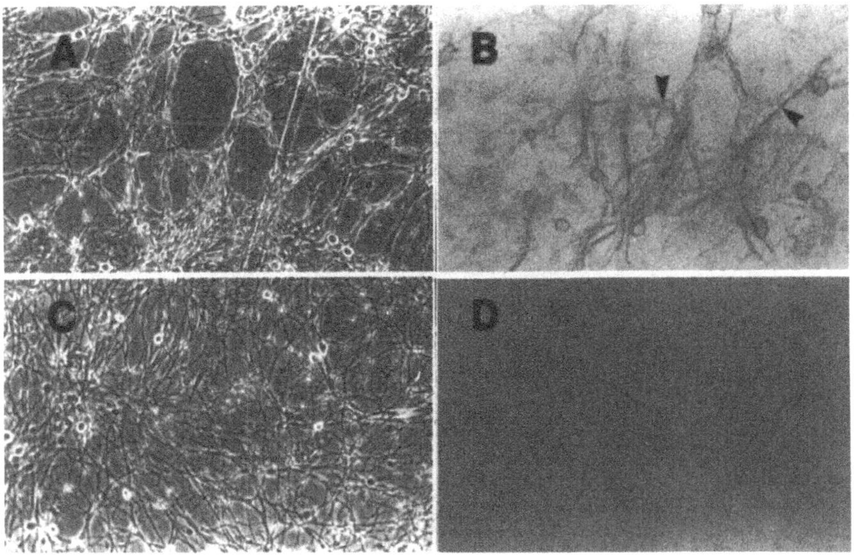

Fig. 2. Inhibition of Schwann cell myelin formation by proteoglycan synthesis inhibitor. Rat embryo Schwann cell-neuron co-cultures grown in the absence (A, B) or presence (C, D) of 1 mM 4-methylumbelliferyl-β-D-xylose. Phase contrast (A, C) and bright field (B, D) light microscopic images are shown. Myelin segments are visible with bright field optics (arrowheads) (magnification 150 ×).

Administration to Schwann cell-neuron cultures of 4-methylumbelliferyl-β-D-xyloside results in a dose dependent inhibition of synthesis of both the basement membrane and cell associated Schwann cell proteoglycans with maximal inhibition observed at a xyloside concentration of 1 mM [6]. In addition, there is an approximately ten-fold increase in the synthesis of free glycosaminoglycan chains.

Immunofluorescent staining of the cultures with antibodies against laminin and collagen type IV revealed that inhibition of proteoglycan synthesis causes a substantial decrease in the deposition of these basement membrane proteins onto the surface of the Schwann cells [6]. This is not caused by an inhibition of the synthesis or secretion of these molecules, since it can be shown by immunoblotting experiments that similar amounts of these proteins are present in the media of control and xyloside treated cultures. From these data we conclude that proteoglycans are required for the stable assembly of basement membrane on the Schwann cell surface.

In addition, in xyloside treated cultures there is a near total inhibition of myelin formation by the Schwann cells as monitored by light microscopic examination of the cultures (Fig. 2). This effect can be shown to be due to the lack of basement membrane in xyloside treated cultures, since addition of exogenous basement membrane-like extracellular matrix to xyloside treated cultures restores their ability to produce myelin [6].

Vascular smooth muscle cells

In large elastic arteries vascular smooth muscle (VSM) cells occupy the medial layer where they are arranged as concentric layers of cells around the circumference of the vessel. In the vessel wall the VSM cells are connected to and embedded within a dense network of ECM consisting of several types of glycoproteins, collagens and proteoglycans [15]. The vascular ECM (or specific ECM components) have been suggested to modulate a variety of functions including mechanical properties of the vessel wall, proliferation and migration of VSM cells and lipid and cholesterol metabolism [15–17].

Early passage cultures of VSM cells re-establish

Fig. 3. Vascular smooth muscle extracellular matrix. Post-confluent VSM cell clture viewed by transmission electron microscopy after ruthenium red staining. The culture was sectioned at a right angle to the culture dish surface. Portions of 2 VSM cells are visible. The space between the cells is filled with fibrillar and granular ECM material. By this method, proteoglycans appear as dark granules (arrowheads) (magnification 52,500 ×).

to a large extent the general tissue organization seen *in vivo,* and form multi-layered cellular arrangements with numerous ECM structures visible between the cells (Fig. 3). We have used 4-methylumbelliferyl-β-D-xyloside to inhibit proteoglycan synthesis in these cultures and observed the functional consequences on VSM cell phenotype. Incubation of VSM cells in medium containing 1 mM β-D-xyloside significantly inhibits proteoglycan synthesis, but different classes of proteoglycans are affected to different extents [7]. For example, synthesis of proteoglycans secreted into the culture medium, mostly chondroitin sulfate and dermatan sulfate proteoglycans destined for incorporation into the VSM ECM, is inhibited by approximately 75%, whereas synthesis of cell associated proteoglycans, mostly heparan sulfate proteoglycans, is inhibited by approximately 65%.

Similar to the effect of the proteoglycan synthesis inhibitor on Schwann cell-neuron cultures, in the VSM cell cultures the deposition of ECM is significantly inhibited [7]. This affects not only proteoglycan deposition but also glycoprotein deposition into the ECM. Immunofluorescence light microscopy with specific antibodies demonstrates that deposition of laminin, a basement membrane specific glycoprotein, as well as fibronectin (Fig. 4) and thrombospondin, interstitial ECM glycoprotein, are significantly diminished in β-D-xyloside treated cultures [7]. Thus, in VSM culture assembly of both basement membrane and interstitial ECM is inhibited in the absence of proteoglycans.

Measurement of cellular proliferation rates in control and β-xyloside treated cultures revealed that subconfluent growth rates are not affected by the inhibitor, suggesting that proteoglycans do not play an essential role in proliferation per se [7]. Under normal culture conditions, VSM cells continue to proliferate at a significant rate subsequent to reaching confluency, forming multi-layers of cells. In contrast to this behavior, the β-xyloside treated cultures do not proliferate beyond the confluent density. Electron microscopic examination of the cultures reveals the nature of the difference in proliferation patterns: whereas the control cells form the typical multi-layered arrangement, the β-xyloside treated cells do not form multi-layers but only a single confluent monolayer [17]. Thus,

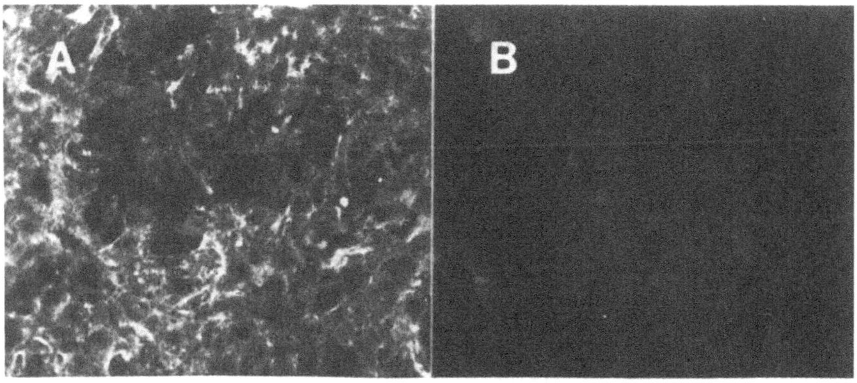

Fig. 4. Inhibition of fibronectin deposition in VSM cell cultures by proteoglycan synthesis inhibitor. Rat thoracic aorta VSM cells were cultured in the absence (A) or presence (B) of 1 mM 4-methylumbelliferyl-β-D-xylose and stained with anti-fibronectin antibodies. Bound antibodies were visualized by means of fluorescein-conjugated secondary antibodies. (magnification 300 ×).

inhibition of proteoglycan synthesis in these cultures abolishes the ability of the cells to arrange themselves into a 3-dimensional organization that is similar to structures observed *in vivo*.

Whereas these results demonstrate that proteoglycans are necessary for ECM assembly and multicellular organization of the cells, other experiments have demonstrated that organization at the level of the individual cells is also affected. Differentiated VSM cells synthesize and utilize a specific actin sub-type, α-smooth muscle actin. Immunofluorescent staining of control post-confluent VSM cell cultures reveals a brightly stained fibrillar pattern within the cells. In contrast to this, cells from β-D-xyloside treated cultures show no fibrillar staining with anti-α-smooth muscle actin antibodies (Fig. 5). When the inhibitor is removed from the cultures, a fibrillar pattern of staining reappears within several days. Surprisingly, when the α-smooth muscle actin content of the cells is determined by immunoblotting, it is found that the control and β-D-xyloside treated cultures do not differ significantly [7]. Thus, inhibition of proteoglycan synthesis alters the utilization (organization of α-smooth muscle actin into filaments) but not the synthesis of α-smooth muscle actin. This probably results from the effects of the inhibitor on ECM assembly (see below).

General mechanisms

Extracellular matrix assembly

The experiments with β-xyloside derivatives demonstrate that in the absence of proteoglycans the assembly of ECM is significantly diminished. While the mechanisms underlying ECM assembly are not fully understood, ECM assembly is thought to occur via the spontaneous self-assembly of molecular components interacting initially through specific non-covalent associations [18, 19]. In some cases covalent bonds between matrix components are later formed to further stabilize the matrix structure. Proteoglycans, especially heparan sulfate proteoglycans, probably play a critical role in this process by virtue of their ability to bind to several of the major molecular constituents of the ECM via their glycosaminoglycan chains. The large heparan sulfate proteoglycan of basement membranes, for example, binds to collagen, laminin [20], and fibronectin [21]. Since this proteoglycan contains several glycosaminoglycan chains it can function as a multivalent cross-linker of matrix components (Fig. 6). Cell surface proteoglycans also appear to participate in this process. A lipid-anchored heparan sulfate proteoglycan isolated from Schwann cells binds to laminin and fibronectin through its glycosaminoglycan chains [22] and

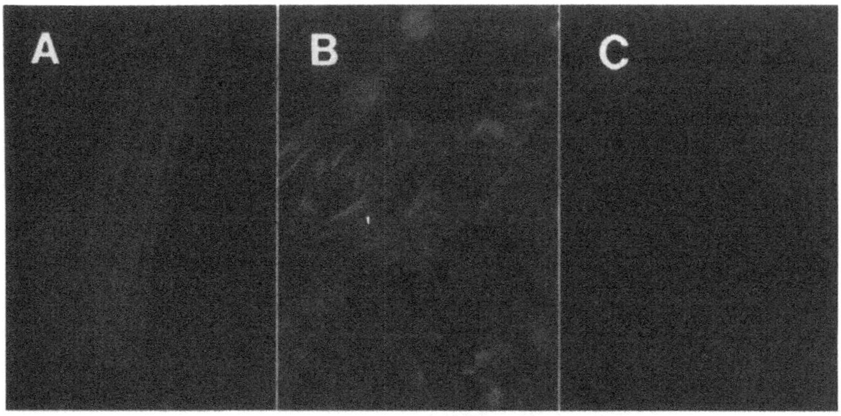

Fig. 5. Proteoglycan synthesis inhibitor alters α-smooth muscle actin assembly in cultured VSM cells. Rat VSM cell cultures were grown for a: 9 days in normal medium; b: 6 days in medium containing the inhibitor followed by 3 days in normal medium; c: 9 days in medium containing 1 mM 4-methylumbelliferyl-β-D-xylose. The cultures were stained with a monoclonal anti-smooth muscle α-actin antibody and fluorescent secondary antibody.

may function to bind ECM molecules to the cell surface (Fig. 6).

Extracellular matrix-cytoskeleton interactions

ECM contact exerts a wide range of effects on the phenotypic behavior of cells. Figure 6 illustrates a model that explains many of these effects. The model proposes that there are direct physical links

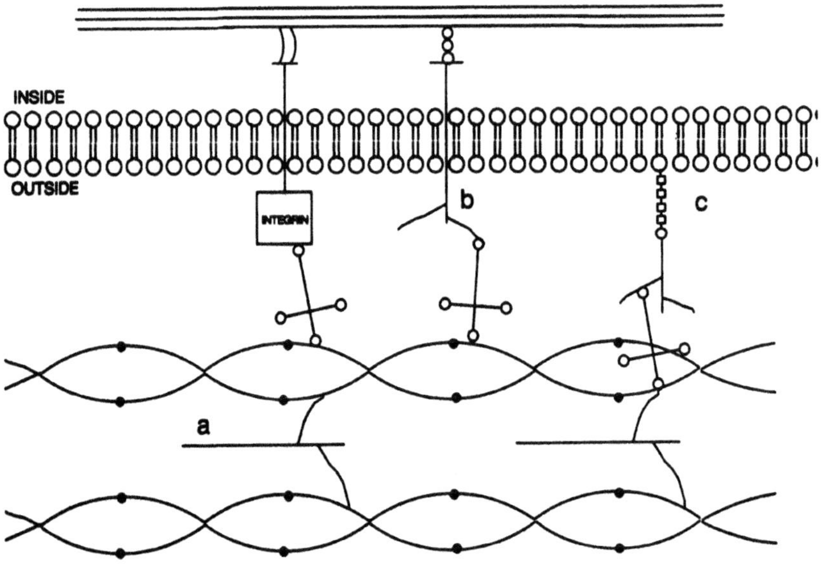

Fig. 6. Schematic model of proposed proteoglycan functions. a: matrix proteoglycan cross-linking ECM molecules; b: transmembrane proteoglycan linking ECM and cytoskeletal structures; c: lipid-anchored proteoglycan binding ECM to the cell surface. Cross-shaped structures represent laminin molecules; helical structures represent collagen molecules.

between the ECM outside the cells and cytoskeletal structures inside the cell. The linkage is mediated by cell surface transmembrane receptors that interact with both structures. Recently, a family of transmembrane ECM receptor proteins, termed the integrin superfamily, has been identified [23]. These proteins appear to fulfill the criteria predicted for such transmembrane linkers between the ECM and cytoskeleton [24, 25]. Additional evidence suggests that in addition to the integrins other proteins are also important in this process. Among these are membrane intercalated cell surface proteoglycans.

Work in several laboratories has shown that various cells express on their surface heparan sulfate proteoglycans that possess transmembrane core proteins that are responsible for anchoring them to the cell surface [26, 27]. The precise function of these proteoglycans is not known but they bind to ECM proteins in *in vitro* assays [28–30]. Preliminary evidence suggests these proteoglycans comprise a family of related proteoglycans that are expressed in a tissue specific manner (26; Carey and Garbes, unpublished observations). Evidence for an important function for the cytoplasmic domain of these proteoglycan core proteins is suggested by the high degree of sequence conservation among the proteoglycan expressed in different tissues and different species (26, 27, 31; Carey and Evans, unpublished observations). We suggest that the function of these cytoplasmic extensions is to interact with specific cytoskeletal components, thereby providing, like the integrins, a direct link between the ECM and the cytoskeleton. In support of this, biochemical and immunological studies have shown that membrane proteoglycans from Schwann cells [32], liver tissue [33], and cultured fibroblasts [34] are bound tightly but non-covalently to the cytoskeleton. The structures responsible for this binding remain to be determined.

Acknowledgements

Work in the author's laboratory is supported by grants from the National Institutes of Health (NS 21925) and The American Heart Association, Pennsylvania Affiliate. Brian Tucker produced the electron micrograph in Fig. 4. We thank Brian Shoop for graphics and Kathy Knarr for secretarial assistance.

References

1. Hascall VC, Hascall GK: Proteoglycans. In: ED Hay (ed.) Cell Biology of Extracellular Matrix. Plenum Press, New York, 1981, pp 39–63
2. Hook M, Kjellen L, Johansson S, Robinson J: Cell surface glycosaminoglycans. Ann Rev Biochem 53: 847–869, 1984
3. Ruoslahti E: Proteoglycans in cell regulation. J Biol Chem 264: 13369–13372, 1989
4. Nuwayhid N, Glaser JH, Johnson JC, Conrad HE, Hauser SC, Hirschberg CB: Xylosylation and glucuronosylation reactions in rat liver golgi apparatus and endoplasmic reticulum. J Biol Chem 261: 12936–12941, 1986
5. Schwartz NB, Galligani L, Ho P-L, Dorfman A: Stimulation of synthesis of free chondroitin sulfate chains by β-D-xylosides in cultured cells. Proc Natl Acad Sci USA 71: 4047–4051, 1974
6. Carey DJ, Rafferty CM, Todd MS: Effects of inhibition of proteoglycan synthesis on the differentiation of cultured rat Schwann cells. J Cell Biol 105: 1013–1021, 1987
7. Hamati HF, Britton EL, Carey DJ: Inhibition of proteoglycan synthesis alters extracellular matrix deposition, proliferation, and cytoskeletal organization of rat aortic smooth muscle cells in culture. J Cell Biol 108: 2495–2505, 1989
8. Thompson HA, Spooner BS: Inhibition of branching morphogenesis and alteration of glycosaminoglycan biosynthesis in salivary glands treated with β-D-xyloside. Dev Biol 89: 417–424, 1982
9. Bunge RP, Bunge MB, Eldridge CF: Linkage between axonal ensheathment and basal lamina production by Schwann cells. Ann Rev Neurosci 9: 305–328, 1986
10. Cornbrooks CJ, Carey DJ, McDonald JA, Timpl R, Bunge RP: *In vivo* and *in vitro* observations on laminin production by Schwann cells. Proc Natl Acad Sci USA 80: 3850–3854, 1983
11. Carey DJ, Eldridge CF, Cornbrooks CJ, Timpl R, Bunge RP: Biosynthesis of type IV collagen by cultured rat Schwann cells. J Cell Biol 97: 473–479, 1983
12. Mehta H, Orphe C, Todd MS, Cornbrooks CJ, Carey DJ: Synthesis by Schwann cells of basal lamina and membrane associated heparan sulfate proteoglycans. J Cell Biol 101: 660–666, 1985
13. Carey DJ, Evans DM: Membrane anchoring of heparan sulfate proteoglycans by phosphatidylinositol and kinetics of synthesis of peripheral and detergent-solubilized proteoglycans in Schwann cells. J Cell Biol 108: 1891–1897, 1989
14. Carey DJ, Stahl RC: Identification of a lipid-anchored

heparan sulfate proteoglycan in Schwann cells. J Cell Biol in press, 1990
15. Wight TN: Cell biology of arterial proteoglycans. Arteriosclerosis 9: 1–20, 1989
16. Schwartz SM, Campbell GR, Campbell JH: Replication of smooth muscle cells in vascular disease. Circulation Research 58: 427–444, 1986
17. Carey DJ: Control of growth and differentiation of vascular cells by extracellular matrix proteins. Ann Rev Physiol in press, 1991
18. Trelstad RL, Silver FH: Matrix Assembly. In: ED Hay (ed.) Cell Biology of Extracellular Matrix. Plenum Press, New York, 1981, pp 179–215
19. Yurchenko PD, Schittny JC: Molecular architecture of basement membranes. FASEB J 4: 1577–1590, 1990
20. Timpl R: Structure and biological activity of basement membrane proteins. Eur J Biochem 180: 487–502, 1989
21. Heremans A, DeCock B, Cassiman JJ, Van den Berghe H, David G: The core protein of the matrix-associated heparan sulfate proteoglycan binds to fibronectin. J Biol Chem 265: 8716–8724, 1990
22. Carey DJ, Crumbling DM, Stahl RC, Evans DM: Association of cell surface heparan sulfate proteoglycans of Schwann cells with extracellular matrix proteins. J Biol Chem in press, 1990
23. Ruoslahti E: Fibronectin and its receptors. Ann Rev Biochem 57: 375–413, 1988
24. Horwitz A, Duggan K, Buck C, Beckerle MC, Burridge K: Interaction of plasma membrane fibronectin receptor with talin-a transmembrane linkage. Nature 320: 531–533, 1986
25. Otey CA, Pavalko FM, Burridge K: An interaction between α-actinin and the β1 integrin subunit *in vitro*. J Cell Biol 111: 721–729, 1990
26. Saunders S, Jalkanen M, O'Farrell S, Bernfield M: Molecular cloning of syndecan, an integral membrane proteoglycan. J Cell Biol 108: 1547–1556, 1989
27. Marynen P, Zhang J, Cassiman J-J, Van den Berghe H, David G: Partial primary structure of 48- and 90-kilodalton core proteins of the cell surface-associated heparan sulfate proteoglycans of lung fibroblasts. J Biol Chem 264: 7017–7024, 1989
28. Koda JE, Rapraeger A, Bernfield M: Heparan sulfate proteoglycans from mouse mammary epithelial cells: cell surface proteoglycan as a receptor for interstitial collagens. J Biol Chem 260: 8157–8162, 1985
29. Saunders S, Bernfield M: Cell surface proteoglycan binds mouse mammary epithelial cells to fibronectin and behaves as a receptor for interstitial matrix. J Cell Biol 106: 423–430, 1988
30. Sun X, Mosher DF, Rapraeger A: Heparan sulfate-mediated binding of epithelial cell surface proteoglycan to thrombospondin. J Biol Chem 264: 2885–2889, 1989
31. Mali M, Jaakkola P, Arvilommi A-M, Jalkanen M: Sequence of human syndecan indicates a novel gene family of integral membrane proteoglycans. J Biol Chem 265: 6884–6889, 1990
32. Carey D, Todd M: A cytoskeleton-associated plasma membrane heparan sulfate proteoglycan in Schwann cells. J Biol Chem 261: 7518–7525, 1986
33. Carey DJ, Rafferty CM, Schramm MS: Association of heparan sulfate proteoglycan and laminin with the cytoskeleton in rat liver. J Biol Chem 262: 3376–3381, 1987
34. Woods A, Couchman JR, Hook M: Heparan sulfate proteoglycans of rat embryo fibroblasts: a hydrophobic form may link cytoskeleton and matrix components. J Biol Chem 260: 10872–10879, 1985

Address for offprints: D.J. Carey, Weis Center for Research, Geisinger Clinic, North Academy Avenue, Danville, PA 17822, USA

Molecular and Cellular Biochemistry **104**: 29–34, 1991.

Expression of a neuromodulin-β-galactosidase fusion protein in primary cultured neurons and its accumulation in growth cones

Yuechueng Liu and Daniel R. Storm
Department of Pharmacology, SJ-30, University of Washington School of Medicine, Seattle, WA 98195, USA

Key words: neuromodulin, GAP-43, lipofection

Abstract

Cultured embryonic neurons share a number of characteristic morphological and physiological properties with their counterparts *in vivo*. For example, differentiating hippocampal neurons in culture develop two distinct classes of processes that serve as dendrites and axons. It has also been shown that the microtubule organization and composition in axons differs from those in dendrites, which may contribute to differential transport of macromolecules into axons or dendrites. We have expressed a neuromodulin-β-galactosidase fusion gene in cultured mesencephalic neurons in order to study the transport of the neurospecific protein neuromodulin into neurite growth cones. When β-galactosidase alone was expressed in neurons, it was found in the cell bodies with diffuse neurite staining. In marked contrast, the neuromodulin-β-galactosidase fusion protein was rapidly transported into neurites and was concentrated in the growth cones. This system may provide a useful model for studying the structural domain(s) of neuromodulin that are required for transport and accumulation of neuromodulin in the growth cones of neurons.

Introduction

Neuromodulin (also known as GAP-43, P-57, B-50, F1) is a neurospecific phosphoprotein which was first isolated from bovine cerebral cortex membranes as a calmodulin binding protein [1]. It has been proposed that neuromodulin may play important neuromodulatory roles in the nervous system (for review, see 2, 3) including regulation of axonal growth and regeneration [4, 5], calmodulin binding [6], long term potentiation [7] and phosphatidylinositide metabolism [8]. The amino acid sequence of bovine neuromodulin [9] and the deduced amino acid sequences from cDNAs encoding murine [10], rat [11], human [12], chicken [13], and gold fish neuromodulin [14] indicate that the protein is a highly charged, elongated molecule with no apparent transmembrane domain. The N-terminal region of neuromodulin, including the calmodulin binding domain [15], the membrane attachment site [16] and the protein kinase C phosphorylation site [17], are highly conserved from chicken to human.

During neuronal development and regeneration, neuromodulin is rapidly transported down axons, presumably via transporting membrane vesicles, and is concentrated in the growth cones [18]. A recent report has indicated that the 10 amino acid N-terminus of neuromodulin may be sufficient for the transport of neuromodulin and its accumulation in the growth cones of PC12 cells [16]. PC12 cells can be induced by NGF or cAMP to differentiate into sympathetic neuron-like cells with many characteristic properties of neurons including extended processes, electrical excitability and neurotransmitter release [19]. However, PC12 cells may not be suitable for studies involving specific neuronal functions. For example, unlike primary neurons in culture, PC12 cells do not produce distinctive axons and dendrites, and the neurite growth rate for PC12 cells is much slower than that in neurons. In this report, we studied the transport of neuromodulin into neurites by expressing a neuromodulin/β-galactosidase fusion gene in primary neuron cultures. It is demonstrated that the fusion protein is rapidly transported into neurites and concentrated in the growth cones.

Materials and methods

Materials

The CDM8 expression plasmid was a generous gift from Dr. B. Seed [21], Massachusetts General Hospital, Boston, MA. Rabbit anti-β-galactosidase antibody was purchased from Cappel. Rhodamine conjugated goat anti-rabbit IgG was purchased from Boehringer Mannheim Biochemicals and rabbit anti-goat IgG were purchased from Pierce Chemical. Monoclonal anti-microtubule-associated protein (MAP)-2 antibody, rabbit anti-tubulin IgG and all other analytical grade chemicals were obtained from Sigma.

Methods

Plasmids
The N1a III restriction fragment of the neuromodulin cDNA coding sequence was subcloned into the Sph 1 site of pUC18 [10], and the BamH1 and EcoR1 fragment of *Escherichia coli* β-galactosidase from plasmid pMC 1871 [20] was fused inframe to the 3′ end of the neuromodulin at the BamH1 and EcoR1 multiple cloning sites of pUC18-neuromodulin plasmid. The fusion gene was then subcloned into the pCDM8 expression vector [21]. The β-galactosidase vector, used as a control, was constructed by subcloning the full length β-galactosidase cDNA from pCH110 (Pharmacia) into pCDM8.

Cell culture and transfection
Mesencephalic neurons were prepared from 17 day old rat embryo brain by dissociation in 20 mM Hepes, 150 mM NaCl, pH 7.4 (Hepes-saline buffer) by trituration through a fire polished glass pipette. Approximately 2×10^6 cells were plated onto poly-L-lysine (Sigma, 50 μg/ml) coated 100 mm plates. The cells were maintained in chemically defined DMEM medium [22] supplied with 5 μg/ml insulin, 100 μg/ml transferrin, 20 nM progesterone, 100 μM putrescine, 1 mM sodium pyruvate, 30 nM sodium selenite, and 0.1% ovalbumin. After attachment of the cells to the plate during an initial 1–2 hr incubation, transfection was performed using the Lipofectin reagent (BRL), with 20 μg of plasmid and 70 μg Lipofectin reagent in 4 ml Hepes-saline per 100 mm dish. After 5 hr incubation in a humidified incubator supplied with 5% CO_2 at 37° C, medium was added to the plates and the incubation was continued for 12 hr before replacing with fresh medium.

Immunochemistry
Cultured cells were fixed with 3% paraformaldehyde in 25 mM KH_2PO_4, 25 mM Na_2HPO_4, 100 mM NaCl, 1 mM $MgCl_2$, and 0.01% NaN_3, pH 7.4 (PBS) for 15 min at room temperature and permeabilized with − 20° C ethanol for 5 min. Nonspecific binding sites were blocked with 4% BSA and 10% fetal calf serum in PBS for 30 min at room temperature. Indirect immunofluorescence staining was carried out by incubation with rabbit anti-β-galactosidase (Cappel, 3 μg/ml) overnight at 4° C, followed by incubation with rhodamine conjugated goat anti-rabbit IgG (1 : 1500) for 30 min at room temperature. After washing twice with PBS, rabbit anti-goat IgG (1: 1500) was added and incubated for an additional 30 min at room temperature. The cells were washed with PBS and examined by using a Leitz Dialux 20 fluorescence microscope equipped with a Leitz Vario Orthomat camera system.

Results

Culture of mesencephalic neurons
Most neurons became attached to the culture dish within 30 min of plating and began to extend pro-

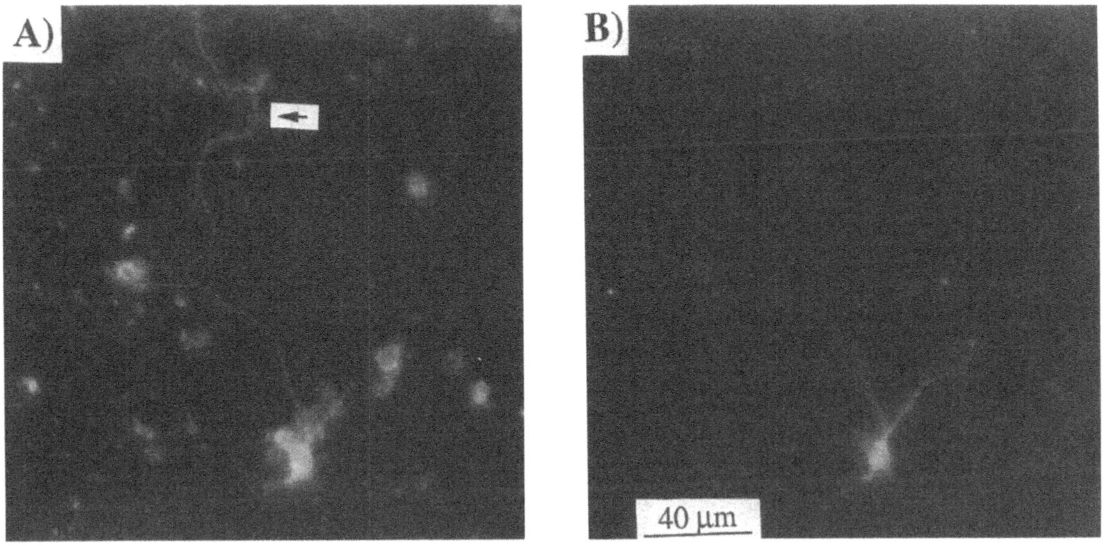

Fig. 1. Anti-tubulin and anti-MAP2 immunostaining of primary mesencephalic neuron cultures from 17 day rat embryo brain. A) Fluorescent micrograph of anti-tubulin (1 : 250) staining with rhodamine conjugated goat anti-rabbit IgG (1 : 1500, Boehringer Mannheim Biochemicals) and rabbit anti-goat IgG (1 : 1500, Pierce) as second and third antibodies. B) The same field showing anti-MAP2 (1 : 500) staining with FITC-conjugated horse anti-mouse IgG (1 : 1500, Pierce) and rabbit anti-horse IgG (1 : 1500) as second and third antibodies. The MAP2 immunoreactivity is mostly restricted to the cell bodies and dendrites, whereas anti-tubulin staining revealed intense immunoreactivity with both dendrites and axons (arrow). Cell culture conditions and immunofluorescence was carried out as described in Methods.

cesses 2 hr after plating. After 24 hr, the cultured neurons elaborated extensive processes and by 96 hr, the neurons extended characteristic dendrites and axons. This is illustrated in Fig. 1 using double immunofluorescence staining with anti-MAP-2 and anti-tubulin antibodies. As previously reported for cultured hippocampal neurons [23], the MAP2 staining was mostly restricted to the cell bodies and short dendritic processes, whereas the anti-tubulin immunoreactivity was present in both dendrites and axons.

Transfection of primary neuron cultures

Introduction of foreign genes into primary neuron cultures has proven to be very difficult. Conventional transfection procedures such as the calcium-phosphate, DEAE-dextran and electroporation methods are not useful for primary neuron cultures due to their toxicity for cell survival. Recently, retrovirus mediated gene transfer has been used to express a foreign gene in primary neuron cultures [24]. We employed a simple procedure which utilizes a synthetic cationic lipid to introduce DNA into cells. This transfection method is less toxic to neurons in culture and it has been used for transfection of a number of cell lines [25] and *Xenopus* embryonic brain tissue [26]. By using this method and plating at a relatively high cell density, we were able to express the β-galactosidase gene in cultured neurons. Although the efficiency was relatively low (about 0.02–0.1%) compared to transfections with established cell lines like COS or CHO cells, the number of transfected cells were sufficient to examine the transport of the expressed protein into neurites. *Escherichia coli* β-galactosidase is a cytosolic protein when expressed in COS-7 cells [27]. As illustrated in Fig. 2, expression of β-galactosidase alone showed accumulation of the enzyme in the cell bodies with diffuse staining in the neurites. When the β-galactosidase gene was fused to the C-terminus of the neuromodulin cDNA and expressed in neurons, the β-galactosidase fluores-

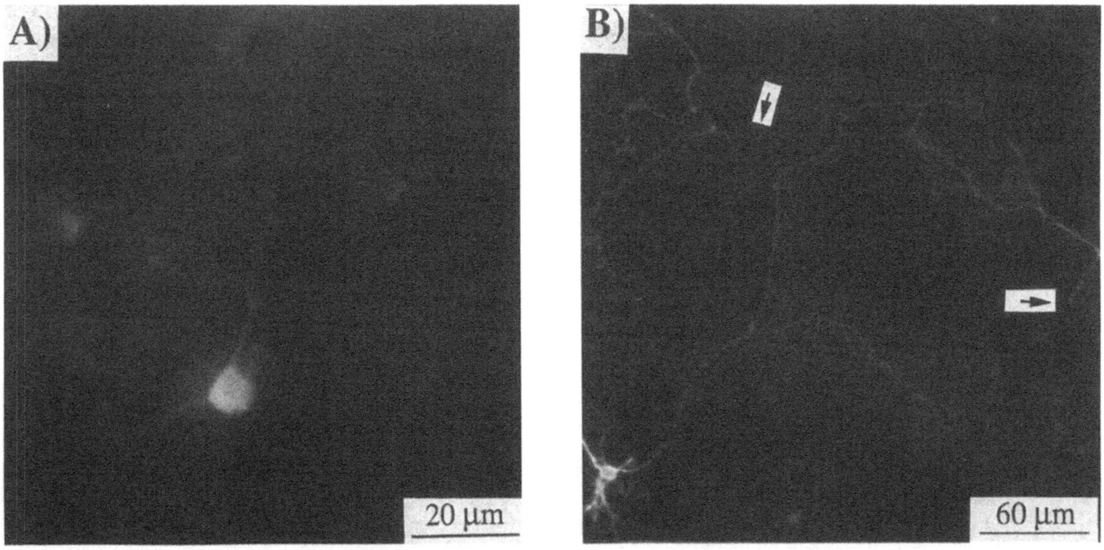

Fig. 2. Immunofluorescence of mesencephalic neurons transfected with a β-galactosidase plasmid or a plasmid encoding the neuromodulin-β-galactosidase fusion protein after 4 days in culture. A) Neurons transfected with the control β-galactosidase plasmid and stained with an anti-β-galactosidase antibody. Anti-β-galactosidase immunoreactivity was mostly concentrated in the cell bodies. B) Anti-β-galactosidase immunostaining of the cells transfected with the neuromodulin-β-galactosidase fusion construct. Note the intense staining in the growth cones. Cell culture conditions and immunofluorescence was carried out as described in Methods.

cence showed an intense punctuate staining pattern, presumably reflecting association of the fusion protein with transporting membrane vesicles. In most cases, the axonal growth cones were strongly stained, which is in agreement with the report that neuromodulin accumulates in the axonal growth cones of hippocampal neurons and regenerating sympathetic neurons [18]. We also examined the time course for the transport of the neuromodulin-β-galactosidase fusion protein into growth cones. The fusion protein appeared in the growth cones just a few hours after it was detected in the cell bodies, about 38 hr after transfection (Fig. 3).

Discussion

In this report, we have shown that a neuromodulin-β-galactosidase fusion protein was rapidly transported into neurites and was concentrated in growth cones. Although β-galactosidase is a relatively large protein (116 kD monomer) compared to neuromodulin which has a Mr of 24,000, it did not appear to interfere with the transport and localization of the fusion protein. Moreover, the fusion of β-galactosidase to neuromodulin did not affect the membrane binding ability of neuromodulin (Liu and Storm, unpublished observation), which suggests that neuromodulin may contain a 'signal sequence' for its rapid transport to the growth cones.

Although the physiological functions of neuromodulin are still being defined, its structural domains including its calmodulin binding domain [6], palmitoylation sites [28], and protein kinase C phosphorylation site [17] have been identified. It has been reported that replacement of the N-terminal cysteines of neuromodulin by threonine or glycine residues abolished its membrane association [16]. Furthermore, Zuber *et al.* have reported that the N-terminal 10 amino acid sequence of neuromodulin, which includes cysteines 3 and 4, is sufficient for its targeting of chloramphenicol acetyltransferase into the neurites and the growth cones of PC12 cells [16]. However, we have observed that

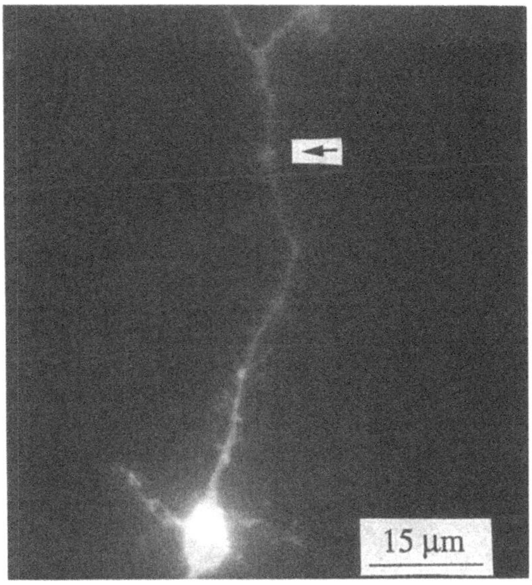

Fig. 3. Immunofluorescence of mesencephalic neurons transfected with a plasmid encoding the neuromodulin-β-galactosidase fusion protein 38 hr after transfection. Cells were immunostained with anti-β-galactosidase antibody 38 hr after transfection as described in Methods.

β-galactosidase alone expressed in PC12 cells was also accumulated in the growth cones (Liu and Storm, unpublished observation). Since the processes of PC12 cells which were involved in the studies were significantly shorter compared to those of the neuron cultured for 4 days, it is possible that passive diffusion may occur and thus result in the accumulation of the protein in the growth cones. Regardless, it is not clear that the mechanism for neuromodulin transport into PC12 cell growth cones is analogous to that for neurons.

It has been reported that neuromodulin is exclusively located in axons and absent from dendrites in cultured hippocampal neurons [29]. However, we have not observed specific axonal vs dendritic accumulation of the neuromodulin-β-galactosidase fusion protein expressed in neurons. This may reflect differing targeting mechanisms for subtypes of neurons or saturation of anchoring sites for neuromodulin in the axons which resulted in overflow of neuromodulin into dendritic processes. In conclusion, we feel that the system described in this study should provide a useful tool for the identification of structural elements of neuromodulin required for its transport and growth cone accumulation in neurons.

Acknowledgements

We thank Dr. B. Seed for the gift of pCDM8 plasmid and Ed Chapman for providing the neuromodulin-CDM8 plasmid from which the fusion construct was built. This work was supported by NIH grant GM 33708.

References

1. Andreasen TJ, Luetje CW, Heideman W, Storm DR: Purification of a novel calmodulin binding protein from bovine cerebral cortex membranes. Biochem 22: 4615–4618, 1983
2. Skene JHP: Axonal growth-associated proteins. Ann Rev Neurosci 12: 127–156, 1989
3. Liu Y, Storm DR: Regulation of free calmodulin levels by neuromodulin: neuron growth and regeneration. Trends in Pharm Sci 11: 107–111, 1990
4. Skene JHP, Willard M: Axonally transported proteins associated with axon growth in rabbit central and peripheral nervous systems. J Cell Biol 89: 96–103, 1981
5. Benowitz LI, Lewis ER: Increased transport of 44,000 to 49,000-dalton acidic proteins during regeneration of the goldfish optic nerve: A two-dimensional gel analysis. J Neurosci 3: 2153–2163, 1983
6. Alexander KA, Cimler BM, Meier KE, Storm DR: Regulation of calmodulin binding to P-57. J Biol Chem 262: 6108–6113, 1987
7. Akers RF, Routtenberg: Calcium-promoted translocation of protein kinase C to synaptic membranes: relation to the phosphorylation of an endogenous substrate (protein F1) involved in synaptic activity. J Neurosci 7: 3976–3983, 1987
8. Oestreicher AB, Van Dongen CJ, Zwiers H, Gispen WH: Affinity-purified anti-B-50 protein antibody: interference with the function of the phosphoprotein B-50 in synaptic plasma membrane. J Neurochem 41: 331–340, 1983
9. Wakim BT, Alexander KA, Masure HB, Cimler BM, Storm DR, Walsh KA: Amino acid sequence of P-57, a neurospecific calmodulin binding protein. Biochem 26: 7466–7470, 1987
10. Cimler BM, Giebelhaus DH, Wakim BT, Storm DR, Moon RT: Characterization of murine cDNAs encoding P-57, a neurospecific calmodulin-binding protein. J Biol Chem 262: 12158–12163, 1987
11. Basi GS, Jacobson RD, Virag I, Schilling J, Skene JHP: Primary structure and transcriptional regulation of

GAP-43, a protein associated with nerve growth. Cell 49: 785–791, 1987
12. Kosik KS, Orecchio LD, Bruns GAP, MacDonald GP, Cox DR, Neve RL: Human GAP-43: its deduced amino acid sequence and chromosomal localization in mouse and human. Neuron 1: 127–132, 1988
13. Baizer L, Alkan S, Stocker K, Ciment G: Chicken growth-associated protein (GAP)-43: primary structure and regulated expression of mRNA during embryogenesis. Mol Brain Res 7: 61–68, 1990
14. LaBate ME, Skene JHP: Selective conservation of GAP-43 structure in vertebrate evolution. Neuron 3: 299–310, 1989
15. Alexander KA, Wakim BT, Doyle GS, Walsh KA, Storm DR: Identification and characterization of the calmodulin binding domain of neuromodulin, a neurospecific calmodulin binding protein. J Biol Chem 263: 7544–7549, 1988
16. Zuber MX, Stittmatter SM, Fishman MC: A membrane-targeting signal in the amino terminus of the neuronal protein GAP-43. Nature 341: 345–348, 1989
17. Apel ED, Byford MF, Au D, Walsh KA, Storm DR: Identification of the protein kinase C phosphorylation site in neuromodulin. Biochem 29: 2330–2335, 1990
18. Meiri KF, Johnson MI, Willard M: Distribution and phosphorylation of the growth associated protein GAP-43 in regenerating sympathetic neurons in culture. J Neurosci 8: 2571–2581, 1988
19. Greene LA, Tischler AS: PC12 pheochromocytoma cultures in neurobiological research. Adv Cell Neurobiol 3: 373–415, 1982
20. Casadaban MJ, Martinez-Arias A, Shapira SK, Chou J: β-galactosidase gene fusion for analyzing gene expression in *Escherichia coli* and yeast. Methods Enzymol 100: 293–308, 1983
21. Seed B: An LFA-3 cDNA encodes a phospholipid-linked membrane protein homologous to its receptor CD2. Nature 329: 840, 1987
22. Bottenstein JE, Sato GH: Growth of a rat neuroblastoma cell line in serum-free supplemented medium. Proc Natl Acad Sci USA 70: 514–519, 1979
23. Caceres A, Banker GA, Binder L: Immunocytochemical localization of tubulin and microtubule-associated protein 2 during the development of hippocampal neurons in culture. J Neurosci 6: 714–722, 1986
24. Geller A, Freese A: Infection of cultured central nervous system neurons with a defective herpes simplex virus 1 vector results in stable expression of *Escherichia coli* β-galactosidase. Proc Natl Acad Sci USA 87: 1149–1153, 1990
25. Felgner PL, Gadek TR, Holm M, Roman R, Chan HW, Wenz M, Northrop JP, Ringold GM, Danielsen M: Lipofection: a highly efficient, lipid-mediated DNA-transfection procedure. Proc Natl Acad Sci USA 84: 7413–7417, 1987
26. Holt CE, Garlick N, Cornel E: Lipofection of cDNAs in the embryonic vertebrate central nervous system. Neuron 4: 203–214, 1990
27. Picard D, Yamamoto K: Two signals mediate hormone-dependent nuclear localization of the glucocorticoid receptor. EMBO J 6: 3333–3340, 1987
28. Skene JHP, Virag I: Post-translational membrane attachment and dynamic fatty acylation of neuronal growth cone protein, GAP-43. J Cell Biol 108: 613–625, 1989
29. Goslin K, Schreyer DJ, Skene JHP, Banker GA: Development of neuronal polarity: GAP-43 distinguishes axonal from dendritic growth cones. Nature 336: 672–674, 1988

Address for offprints: D.R. Storm, Department of Pharmacology, SJ-30, University of Washington, Seattle, WA 98195, USA

Molecular and Cellular Biochemistry **104**: 35–43, 1991.

Adrenergic hormones and control of cardiac myocyte growth

Paul C. Simpson, Ken-ichi Kariya, Larry R. Karns, Carlin S. Long and Joel S. Karliner
Cardiology Section and Research Service, Veterans Administration Medical Center, San Francisco, California, USA and Cardiovascular Research Institute and Department of Medicine, the University of California, San Francisco, USA

Key words: α_1-adrenergic receptors, β-adrenergic receptors, cardiac muscle, cell culture, gene expression, protein kinase C

Abstract

The molecular mechanisms of cardiac myocyte growth are relevant to important problems in cardiovascular disease. A cell culture model has been developed to explore the role of adrenergic hormones in cardiac myocyte growth and gene expression. Activation of a cardiac myocyte α_1-adrenergic receptor by catecholamines induces hypertrophic growth of neonatal rat cardiac myocytes and initiates selective increases in contractile protein gene transcription. These effects on growth and gene expression do not depend on contractile activity. The cardiac myocytes contain at least two subtypes of α_1-adrenergic receptors and at least three isoforms of protein kinase C (PKC). A distinct α_1 receptor subtype may mediate hypertrophy and gene transcription. Different isoforms of PKC are translocated to different intracellular sites on activation, and there is evidence that the β-PKC isoform may be an element in the signal transduction pathway from an α_1 receptor at the surface to the cardiac myocyte nucleus. Growth regulation through a β-adrenergic receptor can also be demonstrated in the culture model. The growth response mediated through a β-adrenergic receptor differs in several respects from that transduced through an α_1-adrenergic receptor.

Introduction

An increase in the mass of the heart, myocardial hypertrophy, occurs in many common cardiovascular diseases, including hypertension, myocardial infarction, valve disease, and cardiomyopathy. Although the hypertrophic process in disease is generally a compensatory response to maintain the pump function of the organ in the face of changes in hemodynamic demands, it is eventually associated with disorders of excitation, relaxation, and contraction, and hence is 'pathological'. In patients with hypertension, for example, the development of ventricular hypertrophy is associated with a significant increase in morbidity and mortality. In striking contrast, the 'physiological' myocardial hypertrophy that accompanies normal development and exercise training appears not to be associated with functional abnormalities [1]. These observations have raised fundamental questions relevant to clinical treatment of this important growth process: can abnormal (or pathological) hypertrophy be prevented and/or switched to a more physiological form; and can physiological heart growth be stimulated when it is needed?

Answers to these questions require detailed information on the molecular mechanisms that regulate myocardial growth, an area that has attracted scientists for many years. At the organ level, growth of the heart can be manifest by an increase in cavity size and/or by a focal or generalized increase in wall thickness. At the cell level, most evidence suggests that the increase in mass of the post-natal heart is produced by hypertrophy of cardiac myocytes and hyperplasia of certain other types of cells in the heart [1]. Although changes in the numbers of fibroblasts and vascular cells and the products they produce (collagen, for example) are of obvious importance in the hypertrophic response, most work has focused on hypertrophic growth of cardiac myocytes.

Cardiac myocytes comprise only about 25% of the total cells in the heart, even though they make up the bulk of its mass [1]. Assays of proteins and mRNAs specific to cardiac myocytes during hypertrophy *in vivo* have suggested that physiological and pathological hypertrophy are associated with differences in myocyte gene expression, as outlined subsequently. These differences in gene expression have not been proved to be causative for normal and abnormal function. However, the differences in gene expression do indicate that hypertrophy is a heterogeneous process and that different molecular events may produce normal and abnormal myocyte hypertrophy. In our search for molecular mechanisms of cardiac myocyte growth, we have focused on adrenergic hormones.

Hemodynamic and non-hemodynamic signals for myocardial hypertrophy: adrenergic hormones

Because myocardial hypertrophy is usually seen in conditions associated with changes in hemodynamic loading conditions, it has been thought for many years that some physical stimulus, such as stretch, must initiate the cellular events producing myocyte growth. On the other hand, many studies *in vivo* have documented a dissociation between multiple hemodynamic loading parameters and heart mass [2]. Although the inconsistencies between hemodynamics and hypertrophy might be explained by incomplete study of hemodynamics, these studies have suggested that non-hemodynamic factors may be important signals in the initiation and/or maintenance of cardiac myocyte growth.

Adrenergic hormones, the catecholamines norepinephrine (NE) from sympathetic terminals and epinephrine (EPI) from the adrenal medulla, have received the most attention as potential non-hemodynamic stimuli for myocardial growth. The major evidence from studies *in vivo* falls into three categories [2]. First, there is increased sympathetic activity in several types of experimental hypertrophy produced by alterations in hemodynamic loading and in subsets of patients with hypertension. Second, hypertrophy is induced by catecholamine infusions that do not alter hemodynamics and by sympathetic innervation of the non-working, transplanted heart. Third, myocardial hypertrophy in hypertension is reduced by anti-hypertensive agents that decrease (or do not increase) sympathetic activity.

Thus, evidence exists for a direct effect of catecholamines on myocardial growth in some circumstances, independent of hemodynamic factors. On the other hand, other studies have failed to support this view; and subtle hemodynamic effects of catecholamine infusions might not be measured. In addition, it is very difficult to sort out catecholamine mechanisms *in vivo*. NE and EPI activate at least four types of adrenergic receptors in the myocardium: α_1- and β_1-adrenergic receptors on cardiac myocytes and α_2- and β_2-adrenergic receptors on sympathetic terminals [1]. The α_2- and β_2-adrenergic receptors on sympathetic terminals mediate inhibition and stimulation of NE release, respectively. NE release is also modulated by calcium and by angiotensin II. Therefore, it is very difficult to be sure about the extent to which cardiac myocyte α_1- and β_1-adrenergic receptors are being activated when experiments are done *in vivo* with various agonists and antagonists. This difficulty is magnified by the relatively long time course required to observe changes in heart mass. To test the role of adrenergic hormones in cardiac myocyte growth in the absence of hemodynamic stimuli and to define the adrenergic receptor mediating the effect, we developed a cardiac myocyte culture system.

A cell culture model for cardiac myocyte hypertrophy

Our laboratory was the first to develop a culture model that could be used to test under defined conditions potential stimuli for cardiac myocyte hypertrophy, such as catecholamines, and to study their mechanism of action [3, 4]. Cells are obtained from the neonatal rat ventricle and maintained at low density in serum-free medium. Over 90% of the cells are cardiac myocytes. Potential growth factors are added to the culture medium, and myocyte size and gene expression are assayed after varying times. Myocyte size can be quantified by measuring total cell protein content, surface area, or volume [5]. An assay of radiolabeled protein is simple and sensitive and is used most often. Gene expression is measured at the mRNA and/or protein levels. There is no alteration in DNA synthesis or cell number under the culture conditions used. The control myocytes have no change in size over time and do not exhibit spontaneous contractility or beating. Thus, changes in both size and beating induced by growth factors can be detected with high sensitivity.

There are important cautions in the use of this or any other culture system, as reviewed recently in detail [6, 7, 8]. In particular, cell culture is an artificial system; and the properties of the cells vary with culture conditions. Thus, it is not possible to extrapolate directly from the culture model to the heart *in vivo*. It can be asked later whether basic mechanisms discovered in culture are realized *in vivo* and under what circumstances.

Catecholamines induce cardiac myocyte hypertrophy without hyperplasia through an alpha$_1$-adrenergic receptor

Treatment of the cultured cardiac myocytes with NE (or EPI) increases myocyte size by ~1.5- to 2-fold times control, as quantified by measurements of cell protein content, surface area, and volume [4, 5, 9]. This hypertrophic response develops over about 24 hr. The EC_{50} for NE is ~200 nM, a concentration that could be found in an active sympathetic synapse *in vivo* [10]. As noted, NE and EPI activate all subtypes of α- and β-adrenergic receptors. However, studies with selective adrenergic agonists and antagonists indicate that it is the α_1 receptor specifically that transduces cardiac myocyte hypertrophy in response to NE [5]. Chien and colleagues have confirmed this observation [11].

Proliferating cells undergo a transient doubling in size (hypertrophy) prior to division [12], and cell proliferation can be considered to reflect at least four components: hypertrophy, DNA synthesis, mitosis, and cytokinesis [1]. Interestingly, there is no increase in DNA synthesis, mitosis, or cytokinesis in the cultured neonatal cardiac myocytes in response to NE under any conditions tested so far [5, 9, 11, 13, 14], although α stimulation has been reported to increase the number of myocytes with positive staining for proliferating cell nuclear antigen (PCNA) [14]. These findings in cultured neonatal cardiac myocytes could be interpreted as arrest at the hypertrophic stage of growth. In contrast, other types of cultured cells complete all four stages of growth and proliferate in response to α_1 stimulation [1]. Interesting recent work suggests that cultured fetal rat cardiac myocytes proliferate in response to NE [15]. The mechanisms underlying the different growth responses to α_1 stimulation in fetal and neonatal cultured myocytes are important for future study. In the intact neonatal rat heart, data suggest that myocytes switch from hyperplastic to hypertrophic growth [16].

Myocyte contractility in alpha$_1$-adrenergic stimulated hypertrophy

As discussed in the Introduction, it has been very difficult using *in vivo* models to test the possibility that cardiac myocyte growth might be regulated by soluble growth factors, and not just by mechanical stimuli resulting from alterations in hemodynamic loading or contractility. The fact that hypertrophy can be induced in culture indicates that cardiac myocyte growth does not require hemodynamic stimuli. Additional studies show that α_1-stimulated hypertrophy does not require myocyte contractile activity [1, 5]. The control myocytes in serum-free medium are quiescent and do not contract spontaneously. With α_1 stimulation alone (NE in the presence of the β-adrenergic antagonist propranolol), there is induction of hypertrophy, without a significant change in contractile activity [5]. This observation that NE-stimulated myocyte hypertrophy does not require contractile activity has been confirmed in another laboratory [17]. In addition, phorbol myristate acetate (PMA), an activator of protein kinase C, induces hypertrophy of the cultured myocytes [18], but does not stimulate contractile activity (unpublished data).

The fact that a soluble growth factor (a catecholamine) can stimulate cardiac myocyte growth in the absence of contractility does not imply that mechanical stimuli have no effect on myocyte growth. In fact, there is evidence that mechanical stimuli can also activate growth processes in cardiac myocytes in culture [17, 19], as discussed elsewhere in this volume. Thus, emerging data suggest that both mechanical stimuli and soluble growth factors are important in cardiac myocyte hypertrophy.

Gene expression in alpha$_1$-stimulated hypertrophy: selective up-regulation of early developmental isogenes

As mentioned earlier and discussed elsewhere in this volume, there is heterogeneity in cardiac myocyte gene expression during hypertrophy *in vivo*. In particular, in the rat heart, there are differences in the isoforms of contractile proteins that are expressed preferentially in different types of hypertrophy [1]. For example, exercise, a form of physiological hypertrophy, is associated with up-regulation of α-myosin heavy chain (MHC), whereas hypertrophy due to pressure overload is characterized by induction of β-MHC and skeletal α-actin, two isogenes expressed normally in the immature heart. In the culture system, quantitation of myocyte contents of specific mRNAs and proteins indicates that α_1-adrenergic stimulation is coupled to selective changes in gene expression during hypertrophy [20, 21, 22]. Specifically, α_1-adrenergic stimulation increases markedly the levels of mRNAs encoded by genes expressed characteristically during early developmental stages and pressure-load hypertrophy *in vivo:* c-*myc,* skeletal α-actin, and β-MHC. Thus, the pattern of gene expression in α_1-stimulated hypertrophy in culture and in pressure overload hypertrophy *in vivo* appear to be similar. This raises the possibility that similar mechanisms are operative *in vivo* and in culture.

Several findings indicate that the preferential up-regulation of skeletal α-actin and β-MHC is not just the nonspecific result of growth in culture but reflects selective coupling of the α_1 receptor to the expression of particular genes. First, there is selectivity in the α_1 effects on gene expression. α_1 stimulation does not change α-MHC mRNA or protein content [20], and the increase in cardiac α-actin mRNA is much less than that in skeletal α-actin [22]. Second, the fold-increases in specific mRNAs are much greater than the increases in total RNA. Third, the myocyte phenotype defined by the pattern of gene expression in culture is not the same for all hypertrophic stimuli. For example, in the culture model, both thyroid hormone and serum produce hypertrophy; but thyroid hormone increases α-MHC and decreases β-MHC, whereas serum up-regulates both α- and β-MHC. Both patterns are distinct from that produced by α_1 receptor activation [20]. Thus, in culture as *in vivo,* there is heterogeneity of cardiac myocyte gene expression during hypertrophy.

Are catecholamines, acting through the α_1-adrenergic receptor, the only growth factors coupled to early developmental isogene expression in cardiac myocytes? There is recent evidence that certain peptide growth factors can produce effects similar to those of catecholamines in cultured myocytes [23]. Are there other growth factors, besides thyroid hormone, coupled to preferential expression of adult isoforms of MHC and α-actin? How is the myocyte phenotype altered by growth factor interactions? Are differences in gene expression responsible for pathological and physiological hypertrophy? There are many questions to be addressed.

Gene transcription in alpha$_1$-stimulated hypertrophy

To investigate the mechanism of α_1-adrenergic induction of hypertrophy and regulation of gene expression, it was necessary to determine whether receptor activation was altering protein or RNA synthesis or degradation. In the case of growth factors that increase DNA synthesis, it is clear that the growth factor must transmit a signal to the nucleus. The existence of a nuclear signal is not clear for growth factors that produce hypertrophy only.

An increase in total protein content defines hypertrophy of myocytes and other cells [12]. This increase could be produced in theory by alterations in either protein synthesis or degradation. Chien and colleagues [11] and we [24] have found that α_1 stimulation increases total and myofibrillar protein synthesis in this system, with no change in protein degradation. Augmented capacity for protein synthesis is provided by an α_1-stimulated increase in the per cell contents of total RNA (which is largely rRNA) [22] and specific mRNAs [13, 20], as discussed above. Thus, α_1-stimulated hypertrophy in this model system is regulated at the pre-translational level, at least in part.

Increases in the level of a particular RNA can be produced by increases in its synthesis or decreases in its degradation. To distinguish these possibilities, we have used the nuclear run-on assay to quantify the initiation of transcription after α_1-adrenergic stimulation [13]. The major finding is a selective increase in transcription of the skeletal α-actin isogene, consistent with the selective increase in skeletal α-actin mRNA content. Other notable observations are as follows: (1) isogene transcription is activated rapidly, within 1–3 hr after receptor stimulation; (2) activation of transcription does not require protein synthesis, consistent with the postulate that receptor stimulation leads to modification of a pre-existing factor or factors that activates transcription; (3) there is a difference in the time course of increased transcription of skeletal α-actin and cardiac α-actin, with transcription of skeletal α-actin being much more prolonged; and (4) the α_1-stimulated increase in mRNA synthesis (α-amanitin-sensitive transcription) is transient, whereas the increase in rRNA and tRNA synthesis (α-amanitin-insensitive transcription) is sustained.

These studies establish that stimulation of the α_1 receptor on the cell surface transmits a signal or signals to specific genes in the myocyte nucleus. The disparate changes in transcription of skeletal α-actin and cardiac α-actin are consistent with the disparate changes in steady state mRNA levels, emphasizing further the selectivity of α_1 effects. On the other hand, the overall increases in RNA transcription suggest that activation of transcription can be a critical regulatory event in the overall increase in cell protein that defines hypertrophy, in agreement with work by others [25]. The complex time course of transcription of different RNA species after α_1 stimulation does raise the possibility that multiple mechanisms are involved.

The intracellular pathway from the alpha$_1$ receptor to gene transcription: a possible role for a specific isoform of protein kinase C

What is the intracellular pathway for activation of transcription by the α_1-adrenergic receptor? Evidence from many systems suggests a common theme in the regulation of gene expression by hormones and growth factors: post-translational modification of pre-existing protein transcription factors that interact with promoter elements of a particular gene [26]. In our system, the rapid initiation of transcription after α_1 stimulation and the independence from new protein synthesis are consistent with this theme [13]. Thus, we have examined early second messenger events after activation of the α_1 receptor.

Stimulation of the α_1 receptor in our system, as in other cardiac myocyte preparations, activates phosphoinositide-phospholipase C (PI-PLC), as assayed by production of mono- and poly-phosphorylated inositols [27]. Activation of PI-PLC is not sensitive to inhibition by pertussis toxin, although a pertussis toxin substrate is abundant in our cells. Similarly, α_1-induced hypertrophy is not inhibited by pertussis toxin [27]. Thus, if a G protein couples the myocyte α_1 receptor to PI-PLC and to the growth response, it appears not to be a pertussis toxin substrate.

Activation of PI-PLC implies production of diacylglycerol (DAG), the endogenous activator of protein kinase C (PKC). We have not assayed DAG in the system, and phospholipids other than PIs may be sources of DAG. However, we have documented that stimulation of the α_1 receptor activates myocyte PKC. Important findings in these studies concern differences in PKC activation by a phorbol ester and an α_1-adrenergic agonist and differences in translocation site of various PKC isoforms.

Both phorbol myristate acetate (PMA) and an α_1-adrenergic agonist (NE) stimulate acute translocation of PKC activity to the particulate fraction [18]. However, acute activation of PKC by PMA differs from that by NE in several respects. PMA-induced activation is more pronounced and persistent and is accompanied by loss of PKC activity and generation of calcium/lipid-independent kinase activity. PMA and NE also differ in their chronic effect on PKC. Chronic treatment with NE upregulates PKC, measured as total PKC activity per cell and ^{3}H-phorbol dibutyrate binding sites per cell, whereas chronic treatment with PMA down-regulates PKC. Thus, PKC is activated differently by PMA and NE, both acutely and chronically [18]. These results suggest caution in extrapolating from PMA effects to those of natural PKC-activating agonists, such as NE.

The PKC family comprises several isoforms, including α, β, γ, δ, ε, and ς. The cultured cardiac myocytes express at least three different PKC isoforms, α-PKC, β-PKC, and a PKC isoform recognized by the monoclonal antibody CK 1.4 [28]. The isoform recognized by CK 1.4 is not α, β, or γ, but its identity is unknown. We have used immunofluorescence with monoclonal antibody CK 1.4, and a different monoclonal specific for β-PKC, to determine the intracellular site of translocation of PKC isoforms after stimulation of the myocytes with PMA and NE [28]. By immunofluorescence, the PKC isoform recognized by CK 1.4 is translocated to myofibrils upon activation, whereas β-PKC is translocated to the surface membrane and the perinuclear area. These different intracellular locations of PKC isoforms upon activation raise the possibility that PKC isoforms might phosphorylate different substrates at these different sites, and thus have distinct functions in the cardiac myocytes.

Translocation of β-PKC to the perinuclear area is intriguing with respect to an involvement of this PKC isoform in α_1-stimulated transcription. Very recently we have obtained data that provide direct support for a role of this PKC isoform in activation of the β-MHC gene [29]. A co-transfection approach was used. Expression plasmids encoding normal α- or β-PKC, or constitutively-activated mutants of α- or β-PKC, were introduced into the cultured cardiac myocytes, along with a plasmid containing 3300 bp of the β-MHC promoter fused to a reporter gene. Thus, assay of the protein product of this reporter gene served as an index of activation of the β-MHC promoter by one of the expressed PKCs. Transfection efficiency was monitored by another reporter gene driven by a viral promoter. The activated mutant of the β-PKC isoform increased expression from the β-MHC promoter, and was more than twice as potent as activated α-PKC. These data provide the first direct evidence that the β-PKC isozyme is in a pathway regulating transcription of the β-MHC isogene and are the first to document functional differences among PKC isoforms in cardiac myocytes. One hypothesis is that β-PKC modifies a transcription factor that interacts with the β-MHC promoter. The specific DNA sequences in the β-MHC promoter that mediate the β-PKC response are under study. Identification of the sequence element(s) should provide the means to identify the putative transcription activator(s) or inhibitor(s) that are modulated by β-PKC.

The alpha$_1$ receptor in myocardial hypertrophy *in vivo:* current evidence and future directions

Thus, in a culture model, stimulation of the cardiac myocyte α_1-adrenergic receptor induces hypertrophy and initiates specific changes in gene expression that simulate those seen in pressure-overload hypertrophy *in vivo*. Is there evidence that an α_1-adrenergic mechanism is operative *in vivo*? A variety of studies summarized recently provide strong circumstantial evidence for a role of the α_1 receptor in hypertrophy *in vivo* [2]. These include measurements of α_1-adrenergic receptors and responses in hypertrophy and experiments with α_1-adrenergic antagonists.

However, the situation has taken on new complexity with the realization that there are multiple subtypes of α_1-adrenergic receptors. Pharmacologic studies and molecular cloning indicate that there are at least two α_1 receptor subtypes in many cells and tissues, including the heart and cardiac myocytes [30, 31, 32]. Much remains to be learned about α_1 receptor subtypes in cardiac myocytes, including the number of subtypes, their relative proportions, physiologic functions, coupling to second messenger systems, and regulation. For example, it is unknown whether different subtypes are responsible for the growth, electrophysiologic, and contractile effects of α_1 stimulation. We have preliminary evidence that a particular subtype mediates α_1-stimulated growth, promoter activation, and inositol phosphate production in the cultured myocytes (unpublished data and [33]). α_1-Adrenergic subtypes in cardiac myocytes are a particularly important area for further study, since it might be possible to modulate selectively the various effects of α_1 stimulation.

A beta-adrenergic receptor in cardiac myocyte hypertrophy

Our work has supported the notion that catecholamines can be important non-hemodynamic factors in myocardial hypertrophy and has focused attention on the α_1-adrenergic receptor as a mediator of these effects. Prior emphasis had been on the β-adrenergic receptor, since the function of the myocyte α_1 receptor has been unclear [34] and since a β agonist (isoproterenol) induces hypertrophy *in vivo* [2]. We have proposed that isoproterenol-stimulated hypertrophy may be mediated, at least in part, by released NE accting on the myocyte α_1-adrenergic receptor, given that β-adrenergic stimulation increases NE release and that sympathetic activity is increased in isoproterenol-stimulated hypertrophy [5].

Recent studies may require revision of this postulate, since we have been able to show a growth-promoting effect of β-adrenergic stimulation in the cultured cardiac myocytes. This β-adrenergic growth effect differs from that mediated through the α_1 receptor in several respects. Cardiac myocyte hypertrophy induced by β stimulation in the culture system (1) is dependent on the particular culture medium used; (2) has a 1000-fold lower EC_{50}; (3) desensitizes; (4) may depend on myocyte contractile activity; (5) and appears not to be accompanied by activation of transcription (unpublished data and [13, 35]). We have not yet determined the β-adrenergic receptor subtype responsible for this effect, and this determination is critical.

Other growth factors for cardiac myocytes

Evidence is accumulating that cardiac myocyte growth and gene expression can be regulated by multiple soluble growth factors, as well as mechanical stimuli, at least in culture models. In addition to adrenergic hormones, cardiac myocyte growth, or processes related to growth, has been observed in response to peptide growth factors, angiotensin II, stretch, and contractility [17, 19, 23, 36, 37]. This list is almost certain to expand and indicates that cardiac myocytes are similar to other types of cells that respond to multiple growth factors [26]. Important areas for future work include the patterns of myocyte gene expression elicited by different growth factors; the intracellular transduction mechanisms for each growth factor, redundancy and synergy of growth factor effects; and the roles of these different growth factors *in vivo*.

Acknowledgements

We thank Charles Ordahl, Daria Mochly-Rosen, and Curt Henrich for their contributions to the work reviewed here. These studies were supported by the Department of Veterans Affairs Research Service and the NIH. PCS is a Clinical Investigator of the Department of Veterans Affairs, and CSL is a Research Associate of the Department of Veterans Affairs. KK is the recipient of the Comroe Fellowship from the Cardiovascular Research Institute at the University of California, San Francisco.

References

1. Simpson PC: Molecular mechanisms in myocardial hypertrophy. Heart Failure 5: 113–129, 1989
2. Simpson PC, Kariya K, Karns LR, Long CS: The α_1-adrenergic receptor in left ventricular hypertrophy. J Vasc Med Biol 1991 (in press)
3. Simpson P, Savion S: Differentiation of rat myocytes in single cell cultures with and without proliferating nonmyocardial cells: cross-striations, ultrastructure, and chronotropic response to isoproterenol. Circ Res 50: 101–116, 1982
4. Simpson P, McGrath A, Savion S: Myocyte hypertrophy in neonatal rat heart cultures and its regulation by serum and by catecholamines. Circ Res 51: 787–801, 1982
5. Simpson P: Stimulation of hypertrophy of cultured neonatal rat heart cells through an α_1-adrenergic receptor and induction of beating through an α_1- and β_1-adrenergic receptor interaction: evidence for independent regulation of growth and beating. Circ Res 56: 884–894, 1985
6. Simpson PC: Measurement of pharmacological effects in isolated myocytes. In: WA Clark, RS Decker and TK Borg (eds) Biology of Isolated Adult Cardiac Myocytes. Elsevier Science Publishing Co., Inc., New York, 1988, pp 108–117
7. Simpson PC: α_1-Adrenergic stimulated hypertrophy in neonatal rat heart muscle cells. In: H Toshima and BJ Maron (eds) Hypertrophic Cardiomyopathy, Cardiomyopathy Update 2. University of Tokyo Press, Tokyo, 1988, pp 73–96
8. Simpson PC, Karns LR, Long CS: An approach to the molecular regulation of cardiac myocyte hypertrophy. In: S Chien (ed.) Molecular Biology of the Cardiovascular System. Lea & Febiger, Philadelphia, 1990, pp 53–81
9. Simpson P: Norepinephrine-stimulated hypertrophy of cultured rat myocardial cells is an alpha$_1$-adrenergic response. J Clin Invest 72: 732–738, 1983
10. Bevan JA: Norepinephrine and the presynaptic control of adrenergic transmitter release. Fed Proc 39: 187–190, 1978
11. Meidell RS, Sen A, Henderson SA, Slahetka MF, Chien KR: α_1-adrenergic stimulation of rat myocardial cells increases protein synthesis. Am J Physiol 251: H1076–H1084, 1986
12. Baserga R: Growth in size and cell DNA replication. Exp Cell Res 151: 1–5, 1984
13. Long CS, Ordahl CP, Simpson PC: α_1-Adrenergic receptor stimulation of sarcomeric actin isogene transcription in hypertrophy of cultured rat heart muscle cells. J Clin Invest 83: 1078–1082, 1989
14. Marino TA, Walter RA, Cobb E, Palasiuk M, Parsons T, Mercer WE: Effects of norepinephrine on neonatal rat cardiocyte growth and differentiation. In Vitro Cell Dev Biol 26: 229–236, 1990
15. Marino TA, Walter RA, D'Ambra K, Mercer WE: Effects of catecholamines on fetal rat cardiocytes *in vitro*. Am J Anat 186: 127–132, 1989
16. Clubb FJ Jr, Bishop SP: Formation of binucleated myocardial cells in the neonatal rat: an index for growth hypertrophy. Lab Invest 50: 571–577, 1984
17. McDermott PJ, Morgan HE: Contraction modulates the capacity for protein synthesis during growth of neonatal heart cells in culture. Circ Res 64: 542–553, 1989
18. Henrich CJ, Simpson PC: Differential acute and chronic response of protein kinase C in cultured neonatal rat heart myocytes to α_1-adrenergic and phorbol ester stimulation. J Mol Cell Cardiol 20: 1081–1085, 1988
19. Komuro I, Kaida T, Shibazaki Y, Kurabayashi M, Katoh Y, Hoh E, Takaku F, Yazaki Y: Stretching cardiac myocytes stimulates protooncogene expression. J Biol Chem 265: 3595–3598, 1990
20. Waspe LE, Ordahl CP, Simpson PC: The cardiac β-myosin heavy chain isogene is induced selectively in α_1-adrenergic receptor-stimulated hypertrophy of cultured rat heart myocytes. J Clin InNvest 85: 1206–1214, 1990
21. Starksen NF, Simpson PC, Bishopric N, Coughlin SR, Lee WMF, Escobedo JA, Williams LT: Cardiac myocyte hypertrophy is associated with c-*myc* proto-oncogene expression. Proc Natl Acad Sci USA 83: 8348–8350, 1986
22. Bishopric NH, Simpson PC, Ordahl CP: Induction of the skeletal α-actin gene in α_1-adrenoceptor-mediated hypertrophy of rat cardiac myocytes. J Clin Invest 80: 1194–1199, 1987
23. Parker TG, Schneider MD: Peptide growth factors can provoke 'fetal' contractile protein gene expression in rat cardiac myocytes. J Clin Invest 85: 507–514, 1990
24. Simpson P: An alpha$_1$-adrenergic receptor regulates RNA, protein and myosin synthesis in cultured neonatal rat heart cells (abstract). Clin Res 33: 90A, 1984
25. Lee HR, Henderson SA, Reynolds R, Dunnmon P, Yuan D, Chien KR: α_1-adrenergic stimulation of cardiac gene transcription in neonatal rat myocardial cells: effects on myosin light chain-2 gene expression. J Biol Chem 263: 7352–7358, 1988
26. Simpson PC: Proto-oncogenes and cardiac hypertrophy. Ann Rev Physiol 51: 189–202, 1989
27. Karliner JS, Kagiya T, Simpson PC: Effects of pertussis toxin on α_1-agonist-mediated phosphatidylinositide turnover and myocardial cell hypertrophy in neonatal rat ventricular myocytes. Experientia 46: 81–84, 1990
28. Mochly-Rosen D, Henrich CJ, Cheever L, Khaner H, Simpson PC: A protein kinase C isozyme is translocated to cytoskeletal elements on activation. Cell Regulation 1: 693–706, 1990
29. Kariya K, Karns LR, Simpson PC: Activation of the β-myosin heavy chain (MHC) promoter by the β isoform of protein kinase C (abstract). Circulation 82: III–687, 1990

30. Minneman KP: α_1-Adrenergic receptor subtypes, inositol phosphates, and sources of cell Ca^{2+}. Pharmacol Rev 40: 87–119, 1988
31. Schwinn DA, Lomasney JW, Lorenz W, Szklut PJ, Fremeau RT Jr, Yang-Feng TL, Caron MG, Lefkowitz RJ, Cotecchia S: Molecular cloning and expression of the cDNA for a novel α_1-adrenergic receptor subtype. J Biol Chem 265: 8183–8189, 1990
32. del Balzo U, Steinberg SF, Rosen MR: The α_1-adrenergic increase in neonatal ventricular automaticity: role of a distinct α_1-receptor subtype (abstract). Circulation 80: II-164, 1989
33. Simpson PC, Cuenco RG, Paningbatan MO, Murphy MD: An α_1-adrenergic receptor subtype sensitive to WB-4101 transduces cardiac myocyte growth (abstract). Circulation 82: III-561, 1990
34. Benfey BG: Minireview: function of myocardial α-adrenoceptors. Life Sciences 46: 743–757, 1990
35. Rocha-Singh KJ, Simpson PC, Karliner JS: Cyclic AMP induces hypertrophy of cultured rat cardiac myocytes (abstract). Circulation 80: II-93, 1989
36. Henrich CJ, Simpson PC: Cardiac nonmyocytes produce a heparin-binding growth factor for cardiac myocytes (abstract). Circulation 80: II-451, 1989
37. Aceto JF, Baker KM: [Sar[1]] angiotenson II receptor-mediated stimulation of protein synthesis in chick heart cells. Am J Physiol 258: H806–H813, 1990

Address for offprints: P.C. Simpson, VA Medical Center (111-C), San Francisco, CA 94121, USA

Molecular and Cellular Biochemistry **104**: 45–50, 1991.

Angiotensin converting enzyme inhibition in Dahl salt-sensitive rats

Thomas C. Peeler[1], Kenneth M. Baker[1], Carolina F. Esmurdoc[2] and Mitchell I. Chernin[2]
[1] *Weis Center for Research, Danville, PA 17822, USA;* [2] *Bucknell University, Lewisburg, PA 17889, USA*

Key words: hypertension, Dahl rats, angiotensin converting enzyme, cardiac hypertrophy, angiotensin II

Abstract

Angiotensin II has previously been reported to have *in vivo* and *in vitro* cardiac hypertrophic effects. We used the salt-sensitive Dahl rat genetic strain to separate mechanical (pressure overload) vs. hormonal (renin-angiotensin system) input in cardiac hypertrophy. Blood pressure was significantly increased and left ventricular hypertrophy, as indexed by LV/BW ratios, was present at 7 and 15 days in rats receiving 4% and 8% NaCl compared to the 1% controls. There was no effect of the angiotensin converting enzyme inhibitor, enalapril maleate, on lowering the blood pressure in 8% NaCl-treated animals, however, there was a significant reduction in LV/BW ratio in 8% NaCl-treated animals that received this drug. Left ventricular angiotensinogen mRNA activity was significantly reduced in rats receiving 4% and 8% NaCl. In this model of hypertension the cardiac hypertrophy which develops is largely dependent on mechanical forces though there remains a significant contribution to this process from either circulating or localized angiotensin II production. Regulation of angiotensinogen gene expression in the hypertrophied left ventricle suggests that volume and electrolyte control of angiotensinogen gene expression in the heart and/or hereditary factors are predominant in the control of regulation of this gene in the left ventricle of Dahl rats.

Introduction

The renin-angiotensin system (RAS) has an important role in the regulation of blood pressure and volume, and fluid homeostasis. The peptide hormone produced, angiotensin II (AII), has a number of specific cardiac functions. Receptor-mediated actions of AII in the heart include inositol phosphate release, mobilization of intracellular Ca^{2+}, and activation of voltage-sensitive Ca^{2+} channels [1–6]. In addition, AII is a direct stimulus of cardiac muscle growth, independent of the effects of this hormone on increasing vascular resistance and cardiac afterload [7–9]. Left ventricular hypertrophy resulting from cardiac afterload can be prevented by the angiotensin converting enzyme (ACE) inhibitor, enalapril maleate, demonstrating the key role that AII plays in the development of cardiac hypertrophy [7]. In this latter study, angiotensinogen mRNA expression in left ventricular tissue was up-regulated by aortic constriction, suggesting that a localized cardiac RAS may have a paracrine or autocrine function in pressure overload left ventricular hypertrophy.

We wanted to determine the effects of elevated mean arterial pressure on left ventricular hypertrophy and angiotensinogen gene expression in a genetic model in which circulating AII levels were very low, in assiciation with marked suppression of the circulating RAS. In the Dahl salt-sensitive rat, inbred lines have been developed which become hypertensive when raised on diets high in NaCl [10–12]. Previous work has shown that the circulating RAS is suppressed in these animals, and does not change in response to increases in blood pressure resulting from high salt diet [13]. Initially, we wanted to determine if salt-induced increases in mean arterial pressure resulted in left ventricular hypertrophy, in the absence of circulating AII. If left ventricular hypertrophy developed in response to higher blood pressure, what effect would treatment with enalapril maleate have on hypertrophy in this suppressed RAS model? We followed the expression of angiotensinogen mRNA in right and left ventricles, kidney, and liver in these animals as an indirect determinant of the effects of increased blood pressure and volume on localized AII processing systems.

Materials and methods

Animals

Adult Dahl salt-sensitive rats (Hsd : DS) were purchased from Harlan Sprague-Dawley, Inc., (Frederick, MD). They were divided into three groups and fed rat chow (modified Agway RMH 3200 pelleted, INAGRA, Inc.) ad lib, containing either 1%, 4%, or 8% NaCl. Rats on the 8% NaCl diet were further subdivided into two groups; one that received enalapril maleate (0.2 mg/ml) in their drinking water, and one with no addition to their drinking water. This dose of enalapril maleate has previously been shown to fully inhibit *in vivo* conversion of AI to AII in the rat [7]. Tail cuff blood pressures were recorded weekly by previously reported methods [14]. After four weeks on the various diets, animals were sacrificed and left ventricular weights (septum plus left ventricular free wall) obtained.

RNA Preparation

Rats were euthanized with nembutal (100 mg/kg) and the heart, liver, and right kidney removed and immediately frozen in liquid nitrogen. Total RNA was extracted by the method of Chomczynski and Sacchi [15]. Tissues were placed in 10 ml of acid guanidinium thiocyanate (4 M guanidinium-thiocyanate, 2 mM sodium citrate, pH 7.0, 0.5% sarcosyl, 0.1 M 2-mercaptoethanol; solution D), and were disrupted with a Brinkman Polytron at a setting of 6 for 30 sec (× 2). Following disruption, 1.0 ml 2 M sodium acetate, pH 4.0, 10 ml water-saturated phenol, and 2.0 ml chloroform : isoamyl alcohol (49 : 1) were added sequentially to 10 ml of the 15–20% tissue homogenate, the mixture vortexed and then centrifuged at 15,000 g for 20 min at 4° C. The resulting aqueous phase was transferred to a sterile Corex tube, precipitated with isopropanol (1: 1) at 20° C for 1 hr, and centrifuged at 15,000 g for 20 min at 4° C. The resulting pellet was resuspended in 0.3 ml solution D, transferred to an eppendorf tube, and precipitated in isopropanol as above. Total RNA was pelleted from the mixture at 4° C, washed with 75% ethanol, and resuspended in 200 μl sterile H_2O. RNA samples were quantified spectrophotometrically at 260 nm. Agarose-formaldehyde gel electrophoresis [16] was used to verify RNA integrity.

RNA hybridization analysis

The RNA dot-blot analysis [17] was used to quantify the amount of angiotensinogen mRNA in heart, liver, and kidney. The probe used for the dot-blot analysis was the cDNA insert to angiotensinogen [18], which was labeled by random priming (Pharmacia) to a specific activity of 1–4 × 10^8 cpm/μg. Twenty micrograms of RNA from right ventricle, left ventricle, kidney, or liver were serially diluted and applied to a Gene Screen Plus™ hybridization membrane, using a Schleicher and Schuell Minifold I.

Membranes were prehybridized for 24 hr at 42° C in 15 ml of 50% formamide, 10% dextran sulfate, 1% SDS and 1 M NaCl. Hybridization was performed in the above solution with the addition of denatured radiolabeled probe and 100 μg/ml denatured salmon sperm DNA at 42° C for 48 hr. Following hybridization, the membrane was washed twice in 2 X standard saline citrate (SSC) for 5 min, 25° C; twice in 2 X SSC and 1% SDS for 30 min at 60° C; and twice in 0.1 X SSC for 30 min at 25° C. Hybridization was visualized by exposing the blot to Kodak XAR-5 x-ray film at − 70° C. The resulting autoradiograms were scanned with a densitometer (Molecular Dynamics), and the relative levels of the mRNAs quantified by integrating the areas

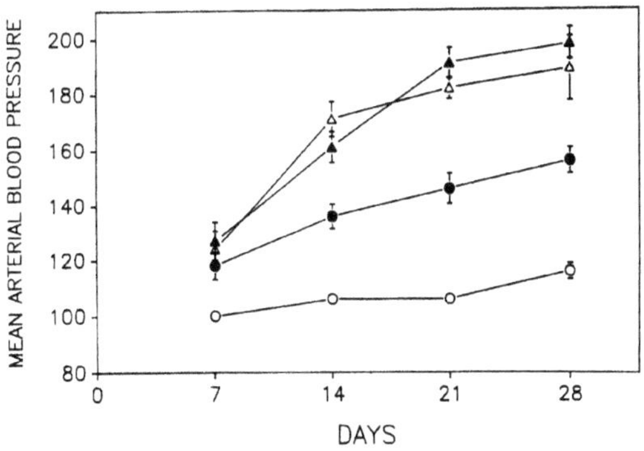

Fig. 1. Effect of NaCl diet and enalapril on mean tail artery blood pressure (mmHg). The blood pressures f rats receiving 1%, 4%, and 8% NaCl were significantly different at 14, 21, and 28 days from the beginning of the dietary treatments ($p < 0.01$). Blood pressures of rats on the 8% NaCl diet receiving enalapril (8% NaCl + enalapril) were not significantly different from 8% rats not receiving enalapril at any time point. Values are given as means ± SEM. 1% NaCl, ○; 4% NaCl, ●; 8% NaCl △; 8% NaCl + enalapril △.

under the peaks. RNA concentrations at or below 5 μg were scanned. After exposure of the autoradiograph for angiotensinogen, the blots were stripped in 0.01% SDS, 0.01 × SSC at 100°C, prehybridized, and then rehybridized with a β-actin cDNA probe. Angiotensinogen mRNA expression was normalized to the expression of β-actin mRNA.

Statistical analysis

An unpaired t-test was used to compare blood pressures and angiotensinogen mRNA expression from rats receiving the various concentrations of NaCl. Left ventricle/body weight ratios were compared using ANOVA. P values < 0.05 were considered significant.

Results

The effects of different NaCl diets on mean arterial blood pressures are shown in Fig. 1. Rats on the 1% NaCl diet exhibited only a slight increase in blood pressure over the four week treatment period. Rats on the 4% and 8% NaCl diets had significantly higher blood pressures than the 1% group throughout the study, with mean pressures approaching 200 mmHg in the 8% animals. Animals receiving 8% NaCl and treated with enalapril (0.2 mg/ml) had no significant change in mean arterial pressure.

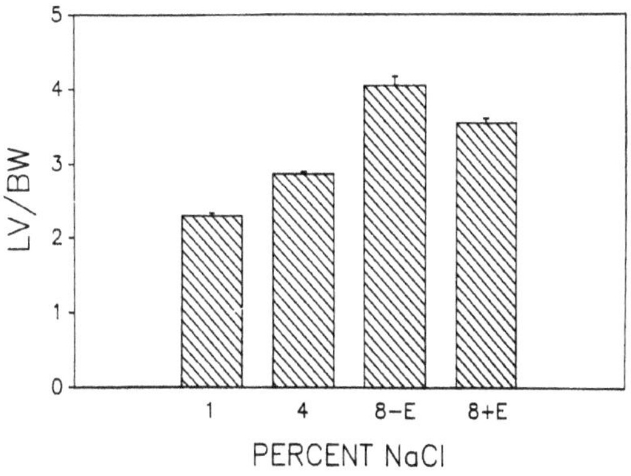

Fig. 2. Left ventricular (free wall plus septum)/body weight ratios (LV/BW; mg/g) for the 4 groups, 28 days following initiation of the dietary NaCl. The values are given as means ± SEM, and all differed significantly from each other by ANOVA ($p < 0.01$).

Hypertensive Dahl rats developed left ventricular hypertrophy (Fig. 2) as evidenced by increases in left ventricular to body weight ratios. The increase in left ventricular mass was in proportion to the increase in mean arterial blood pressure resulting from the 4% and 8% NaCl diets. Treatment of 8% NaCl diet rats with the ACE inhibitor, enalapril maleate, significantly reduced left ventricular hypertrophy, but not to the 1% NaCl (control) diet values.

Angiotensinogen mRNA was quantified by dot-blot analysis in Dahl rats on the different NaCl diets. The expression of angiotensinogen mRNA in the left ventricle (Fig. 3a) decreased significantly in response to higher percentages of NaCl in the diet. Angiotensinogen mRNA levels also decreased, but not significantly, in response to increasing dietary NaCl in right ventricle (Fig. 3b). There were no significant differences in angiotensinogen mRNA levels in kidney (Fig. 4a) or liver (Fig. 4b) with the various dietary intakes of NaCl. All blots exhibited linearity to 5 μg RNA concentration (data not shown).

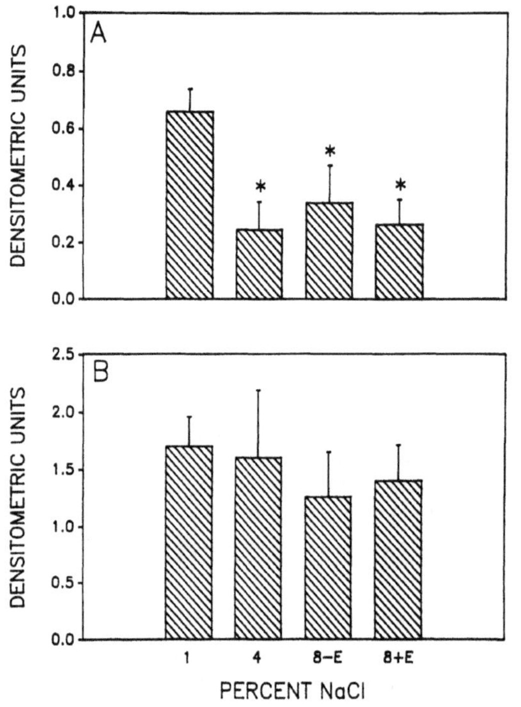

Fig. 3. Messenger RNA levels for angiotensinogen following treatment with 1%, 4%, or 8% NaCl with or without enalapril maleate (8 − E, 8 + E) in (A) left ventricular RNA and (B) right ventricular RNA. Left ventricular levels of angiotensinogen mRNA differed significantly from animals treated with 4%, 8% − E, and 8% + E NaCl from the 1% control, but not from each other (* $p < 0.05$; $n = 3$). There were no significant differences between any of the treatments in RNA isolated from right ventricles.

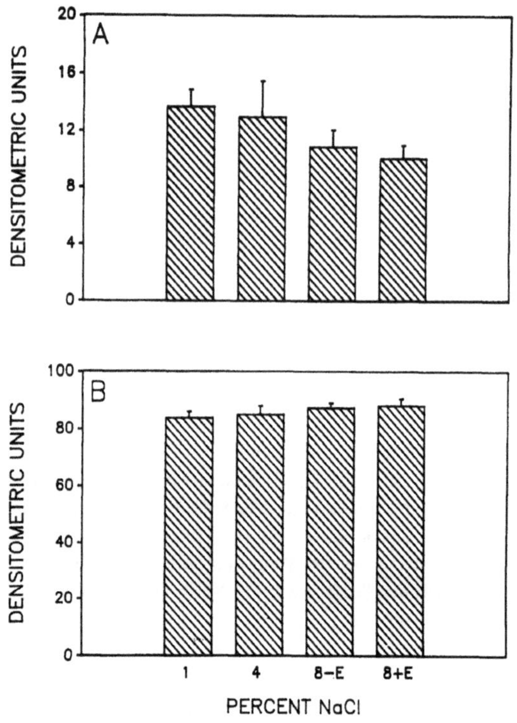

Fig. 4. Messenger RNA levels for angiotensinogen following treatment with 1%, 4%, or 8% NaCl with or without enalapril maleate (8 − E, 8 + E) in (A) kidney and (B) liver. There were no significant differences between any of the treatments in mRNA isolated from kidney or liver.

Discussion

Previous data suggest that AII may have a role in mediating left ventricular hypertrophy [7–9]. This direct effect of AII on stimulating cardiac muscle cell growth is receptor mediated [2, 8, 9]. There is *in vivo* evidence that the hypertrophic effect of AII could be mediated, at least in part, by increased activity of a localized intracardiac RAS with resultant increases in AII production [7]. The present study was performed to determine the role of AII in the development of cardiac hypertrophy in a salt-sensitive genetic model of volume-dependent hypertension. Since the circulating RAS is known to be suppressed in this model of hypertension, we postulated that any reductions in cardiac hypertrophy brought about by treatment with enalapril maleate would be due to ACE inhibition of a cardiac-localized system of AII production.

We utilized the outbred Dahl salt-sensitive model of hypertension to minimize genetic differences between animals, and to determine more directly the effect of varying concentrations of dietary Na^+ on the development and severity of hypertension. Animals that received 4% and 8% NaCl in their diet developed significant hypertension by 1–2 weeks, which became more severe over time (Fig. 1). These significant elevations in mean arterial blood pressure were accompanied with increases in LV/BW (Fig. 2) demonstrating that left ventricular hypertrophy is associated with hypertension in this genetic line. These data indicate that in Dahl salt-sensitive rats, left ventricular hypertrophy developes in the absence of increased levels of circulating AII, and is likely the result of mechanical (stretch) stimulation of protein synthesis in cardiac tissue, though there appears to be hormonal input from AII as well (see below).

Enalapril maleate is an ACE inhibitor which effectively decreases the concentration of AII in the circulation. Treatment of Dahl rats on the 8% diet with enalapril had no significant effect on mean arterial blood pressures (Fig. 1), a result that is consistent with the markedly suppressed circulating RAS in this genetic model. On the other hand, enalapril significantly inhibited the development of left ventricular hypertrophy in rats on the 8% NaCl diet (Fig. 2). Since this genetic line is known to have markedly suppressed circulating RAS activity, the reduction of cardiac hypertrophy by enalapril may be due to inhibition of AII production from a cardiac-localized RAS. In a recent study, Jackson, *et al.* [19] showed that a high salt diet in spontaneously hypertensive rats overcame the beneficial effects of an ACE inhibitor (ramapril) in reducing protein excretion in diabetic nephropathy. In contrast, our results show that enalapril still reduced cardiac hypertrophy, even in the presence of NaCl-induced elevation of mean arterial blood pressure. These results suggest that while in some systems the effects of ACE inhibitors are solely due to the blood pressure lowering effect, in the case of cardiac hypertrophy ACE inhibitors have a specific effect, presumably due to a reduction in the concentration of AII. However, we cannot completely exclude other effects of ACE inhibition (increased levels of bradykinin, etc.) that could conceivably influence cardiac muscle growth. In animals on the 8% NaCl diet, left ventricular hypertrophy was not completely inhibited by enalapril as evidenced by significantly higher LV/BW in these animals compared to 1% NaCl. These results are in contrast to the pressure overload model of cardiac hypertrophy, where enalapril treatment completely inhibited the development of left ventricular hypertrophy [7]. The effect that ACE inhibition had on left ventricular hypertrophy in 8% NaCl diet Dahl salt-sensitive rats may be due to the suppression of an intracardiac RAS in this genetic line, but at mean arterial pressures approaching 200 mmHg, pressure-induced stimulation of left ventricular hypertrophy may be overriding any other contribution (hormonal growth factor, etc.) to growth.

We have previously shown that angiotensinogen mRNA levels in the left ventricle were significantly increased in abdominal aortic constricted animals that developed cardiac hypertrophy [7]. In contrast to these findings, in the present study, angiotensinogen mRNA levels in the left ventricle were significantly reduced in animals that developed left ventricular hypertrophy (Fig. 3a). There were no significant changes in angiotensinogen mRNA expression in right ventricle, kidney, or liver, indicating a differential regulation of gene expression between these tissues and left ventricle. Treatment with enalapril had no effect on left ventricular angiotensinogen mRNA expression, in contrast to our earlier study [7]. In that case, enalapril caused an up-regulation of angiotensinogen mRNA in cardiac tissue, apparently in response to decreased levels of circulating AII. The differences in left ventricular angiotensinogen mRNA expression between the Dahl rat volume-dependent and the aortic constriction model of hypertension may be due to the overriding effects of the high mean arterial blood pressure in the 8% NaCl diet animals. In the previous study, angiotensinogen mRNA expression was determined for rats with mean arterial pressures of approximately 144 mmHg [7], whereas in the Dahl rat model, gene regulation was followed in animals with 200 mmHg mean arterial pressure. An alternative explanation is that the Dahl salt-sensitive line may be fundamentally different in the expression and regulation of the precursor genes of the renin angiotensin system as compared to other genetic strains of rats.

In summary, the ACE inhibitor enalapril maleate reduced left ventricular hypertrophy in hypertensive Dahl salt-sensitive rats, even though the enalapril treatment did not reduce mean arterial pressure. Expression of angiotensinogen mRNA was down-regulated in the left ventricle likely in response to elevated arterial pressure and systemic volume in this genetic model. These results suggest that AII may play a role in cardiac hypertrophy, even in a sodium/volume-dependent form of hypertension in which the circulating RAS is suppressed. Additionally, these results suggest that locally produced AII in the heart may directly contribute to cardiac hypertrophy, in part, independent of changes in cardiac afterload.

Acknowledgements

We wish to thank Kathleen Conrad and Gary Cooper for their expert technical assistance and Debbie McCaffery for typing the manuscript. This work was supported by grants from the National Institutes of Health, the American Heart Association, the Pennsylvania Affiliate of the American Heart Association, and by the Geisinger Clinic and Mars Foundations.

References

1. Allen IS, Cohen N, Dhallan R, Gaa ST, Lederer WJ, Rogers TB: Angiotensin II increases spontaneous contractile frequency and stimulates calcium current in cultured neonatal rat heart myocytes: Insights into the underlying biochemical mechanisms. Circ Res 62: 524–534, 1988
2. Baker KM, Aceto JF: Characterization of avian angiotensin II cardiac receptors: Coupling to mechanical activity and phosphoinositide metabolism. J Mol Cell Cardiol 21: 375–382, 1989
3. Baker KM, Campanile CP, Trachte GJ, Peach MJ: Identification and characterization of the rabbit angiotensin II myocardial receptor. Circ Res 54: 286–293, 1984
4. Baker KM, Singer HA: Identification and characterization of guinea pig angiotensin II ventricular and atrial receptors: coupling to inositol phosphate production. Circ Res 62: 896–904, 1988
5. Baker KM, Singer HA, Aceto JF: Angiotensin II receptor-mediated stimulation of inositol phosphates and cytosolic-free calcium in cultured chick myocytes. J Pharm Exp Ther 251: 578–585, 1989
6. Freer R, Pappano A, Peach M, Bing K, McLean M, Vogen S, Sperelakis N: Mechanism for the positive inotropic effect of angiotensin II on isolated cardiac muscle. Circ Res 39: 178–183, 1976
7. Baker KM, Chernin MI, Wixson SK, Aceto JF: Renin-angiotensin system involvement in pressure overload cardiac hypertrophy in rats. Am J Physiol 259: H324–H332, 1990
8. Baker KM, Aceto JF: Angiotensin II stimulation of protein synthesis and cell growth in chick heart cells. Am J Physiol 259: H610–H618, 1990
9. Aceto JF, Baker KM: [Sar^1] Angiotensin II receptor-mediated stimulation of protein synthesis in chick heart cells. Am J Physiol 258: H806–H813, 1990
10. Dahl LK, Heine M, Tassinari L: Role of genetic factors in susceptibility to experimental hypertension due to chronic excess salt ingestion. Nature 194: 480–482, 1962
11. Dahl LK, Heine M, Tassinari L: Effect of chronic salt ingestion. Evidence that genetic factors play an important role in susceptibility to experimental hypertension. J Exp Med 115: 1173–1190, 1962
12. Dahl LK, Knudsen KD, Heine MA, Leitl GJ: Effects of chronic excess salt ingestion. Modification of experimental hypertension in the rat by variations in the diet. Circ Res 22: 111–118, 1968
13. Iwai J, Dahl LK, Knudsen KD: Genetic influence on the renin-angiotensin system. Circ Res 32: 678–683, 1973
14. Jackson B, Johnston CI: The contribution of systemic hypertension to progression of chronic renal failure in the rat remnant kidney: Effect of treatment with an angiotensin converting enzyme inhibitor or a calcium inhibitor. J Hypertens 6: 495–501, 1988
15. Chomczynski P, Sacchi N: A single-step method of RNA isolation by acid guanidinium thiocyanate-phenol/chloroform extraction. Anal Biochem 162: 156–159, 1987
16. Rave NR, Crkvenjakov R, Boedtker H: Identification of procollagen mRNAs transferred to diazobenzyloxymethyl paper from formaldehyde agarose gels. Nucleic Acids Res 6: 3559–3567, 1979
17. Kafatos FC, Jones CW, Efstradiadis E: Determination of nucleic acid sequence homologies and relative concentrations by dot hybridization procedure. Nucleic Acids Res 7: 1541–1552, 1979
18. Lynch KR, Seminad VI, Ben-Ari ET, Garrison JC: Localization of preangiotensinogen messenger RNA sequences in the rat brain. Hypertension 8: 540–543, 1986
19. Jackson B, Fabris B, Paxton D, Franze L, Johnston CI: High salt diet ameliorates effects of angiotensin converting enzyme inhibition in spontaneously hypertensive streptozotocin diabetic rats. Clin Exp Pharmacol Physiol 17: 229–234, 1990

Address for offprints: K.M. Baker, Weis Center for Research, North Academy Avenue, Danville, PA 17822, USA

Molecular and Cellular Biochemistry **104**: 51–56, 1991.

Activation of adenylate cyclase during swelling of S49 cells in hypotonic medium is not involved in subsequent volume regulation

Peter A. Watson, Kathryn E. Giger, and Carole M. Frankenfield
Weis Center for Research, Geisinger Clinic, Danville PA 17822, USA

Key words: cyclic AMP, adenylate cyclase, cell volume; cellular swelling

Abstract

Experiments in S49 mouse lymphoma cells indicate that adenylate cyclase activity is increased following swelling in hypotonic medium through a mechanism independent of the G-proteins which are involved in hormonal regulation of the enzyme. An intact actin cytoskeleton is apparently required for stimulation of adenylate cyclase by mechanical forces. It was hypothesized that this increase in cAMP may be involved in triggering subsequent volume regulatory events. Manipulation of intracellular cAMP content and protein kinase A activity in S49 cells prior to swelling or during the regulatory volume decrease following swelling provided no evidence of a significant role for cAMP in regulating the extent of initial volume increase or the subsequent regulatory volume decrease. Treatment of S49 cells with 10–200 μM miconazole, previously shown to inhibit adenylate cyclase activity, attenuated the initial volume increase with medium dilution and accelerated the rate of regulatory decrease in a dose-dependent and time-dependent manner. However, incubation with 100 μM miconazole for 20 min, which completely inhibited swelling-induced increases in cAMP content, had no significant effect on either the initial volume expansion or the extent of regulatory volume decrease.

Introduction

Adenylate cyclase, the enzyme responsible for the generation of intracellular second messenger adenosine 3′, 5′-cyclic monophosphate (cAMP), is the target for regulation by a multitude of hormones (for review, see 1). Hormonal modulation of cAMP content and subsequent activation of cAMP-dependent protein kinase (protein kinase A) is involved in regulation of numerous cellular processes. Protein kinase A influences cellular biochemistry at many levels through phosphorylation of proteins such as enzymes involved in energy metabolism, receptors for hormones, ion channels, and trans-acting factors involved in regulation of specific gene expression.

Recently, the accumulation of cAMP following mechanical deformation of a number of cell types and tissues has implicated cAMP in two cell regulatory responses to mechanical stimuli. Increased tissue contents of cAMP have been observed in response to distention in the ventricle of rat heart (pressure overload *in vitro* [2–4]) and the lung (compensatory ventilation following partial pneumonectomy [5]). Both interventions have been shown to stimulate compensatory increases in protein synthesis in these tissues. In the case of the heart, deformation-induced increases in cAMP have been shown to be required for accelerated formation of ribosomes [3]. Increased cAMP content has also been shown to occur in S49 mouse lymphoma cells [6] and turkey erythrocytes [7] following swelling in hyposmolar medium.

We have utilized the S49 mouse lymphoma cell

lines to investigate which components of the hormonal regulatory pathway for cAMP metabolism may be required for mechanical stimulation of cAMP accumulation. Results of these studies indicate that cell deformation stimulates cAMP synthesis through a direct effect on adenylate cyclase, independent of the G-proteins involved in regulation of cAMP metabolism by hormones [8].

S49 mouse lymphoma cells show a regulatory volume decrease following an initial swelling in hypotonic medium. It was hypothesized that the swelling-induced accumulation of cAMP may be involved in modulating the ion channels involved in this volume regulatory response in S49 cells. Activation of certain ion conductances involved in regulatory volume decreases in cells has been shown to occur as a result of accumulation of cAMP and activation of protein kinase A [9]. S49 mouse lymphoma cells were preincubated with cell-permeant analogs of cAMP to determine if the ability of these cells to respond to medium osmolarity reduction was altered in response to activation of protein kinase A. These responses to a reduction in medium osmolarity include [1] an initial volume increase, and [2] a subsequent regulatory volume reduction. S49 cells were also exposed to forskolin, which rapidly increases cAMP content of cells, during the initial phases of the regulatory volume decrease to determine if complete activation of protein kinase A significantly accelerated the rate of volume regulation. Additionally, S49 cells pretreated with miconazole, a nitroimidazole derivative previously shown to inhibit adenylate cyclase [8, 10], were swollen to determine if inhibition of cAMP accumulation influenced either the initial extent of swelling or the rate or degree of regulatory volume decreases. No evidence was obtained which would indicate that cAMP accumulation of increased protein kinase A activity influences the regulatory volume decrease seen in S49 cells following swelling in hypotonic medium.

Materials and methods

Cell culture

S49 mouse lymphoma cells were grown in Dulbecco's modified Eagle's medium (DMEM) with 10% horse serum at 37°C in 10% CO_2 at densities between 2×10^5 and 2×10^6 cells per ml. Experiments were performed in defined medium mimicking DMEM with 0.75% bovine serum albumin [6] with modifications as indicated in the Tables and Figs.

Determination of cAMP and protein content

Following the experimental period, cells were extracted with trichloroacetic acid (TCA) for determination of cAMP content and cellular protein. S49 cells were isolated from suspension by centrifugation at 900 g, and the cell pellets extracted with cold 6% TCA. Proteins were pelleted from the extract, the supernatant washed with water-saturated ether to remove TCA, and lyophilized for analysis of cAMP content by the protein binding assay of Gilman [11]. Protein content was assessed by the method of Lowry [12].

Cell volume in S49 swelling experiments

Cells were incubated in either normosmolar or 33% hyposmolar defined medium for durations indicated in the Tables and Figs. Cells were diluted (3×10^5 cell into 20 ml total volume) in identical medium, and sized at the indicated times using a Coulter Counter equipped with a Channelyzer calibrated with latex beads [6].

Results

Increased cAMP content in response for forskolin does not modify the rate or extent of regulatory volume responses in S49 cells

The possible involvement of swelling-induced increases in cAMP content in the volume regulatory responses of S49 cells was examined by elevating cAMP content in cells prior to and following swelling in hypotonic medium. Results of experiments in which cells were pretreated with forskolin (2×10^{-5} M) prior to dilution of the medium osmolarity indicate that activation of protein kinase A prior to reduction of medium osmolarity did not alter the ability of cells to accumulate volume (results not shown). Treatment of swollen cells with forskolin

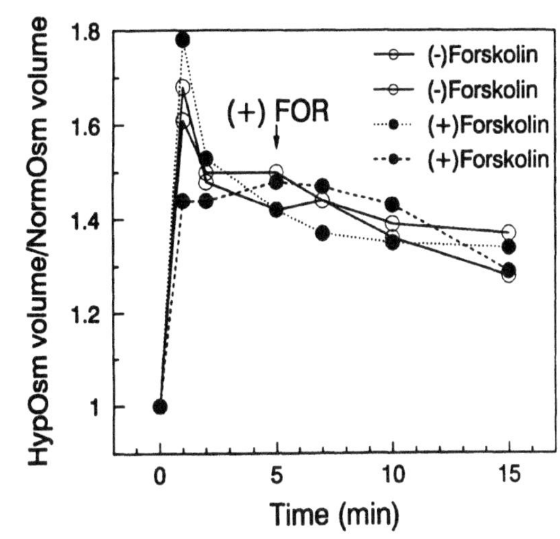

Fig. 1. Forskolin-induced increases in cAMP do not influence the rate of regulatory volume reduction in S49 cells. S49 wild type mouse lymphoma cells were incubated in defined medium containing BSA for 10 min at 37° C in 10% CO_2. Cells were diluted with defined medium or water to reduce osmolarity by 33%. Forskolin (2×10^{-5} M) was added to the indicated hyposmolar suspensions 5 min after dilution of cell suspension. Aliquots of the cell suspension were removed at the times indicated and cell volume determined as described in the 'Methods'. Data are presented as Hyposmolar cell volume divided by the volume of the paired cells of the Normosmolar control group, with each data point representing the mean of 4 observations. Lines connect data points generated by serial samples from each of 2 seperate experiments. Standard errors were less than 5% of their respective means.

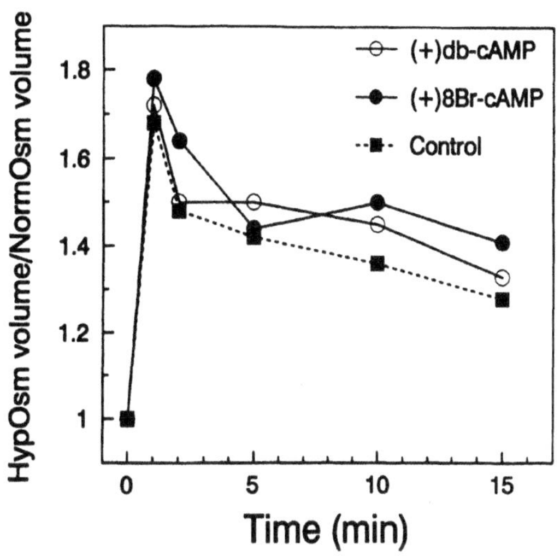

Fig. 2. Activation of protein kinase A by addition of dibutyryl-cAMP or 8-bromo-cAMP prior to swelling does not influence initial volume expansion or rate of regulatory volume reduction. S49 wild type cells were incubated for 10 min in defined medium containing BSA and, where indicated, either 2×10^{-4} M dibutyryl-cAMP or 8-bromo-cAMP was included. Cell suspensions were diluted with either medium or water to reduce osmolarity by 33%. Aliquots of the suspension were removed at the times indicated and cell volume determined as described in the 'Methods'. Data are expressed as described in Fig. 1.

5 min after exposure to medium of reduced osmolarity did not change cell volume at the time of addition or the rate of regulatory volume decrease (Fig. 1).

Treatment of S49 cells with cAMP analogs does not alter the extent of initial volume expansion or subsequent volume regulatory responses

Prior exposure of S49 wild type cells to cAMP analogs dibutyryl-cAMP or 8-bromo-cAMP at concentrations previously shown to completely activate protein kinase A [8] had no effect on the ability of cells to respond to hypotonic conditions with an initial increase in volume or in their ability to regulate volume subsequent to swelling (Fig. 2). These results are identical to those obtained when cells were pretreated with forskolin as described above.

Miconazole inhibits cAMP accumulation, initial volume expansion and modifies volume regulatory responses following hypotonic swelling

Treatment of S49 cells with miconazole in concentrations from 10–200 μm for durations of 20–60 min attenuated the initial degree of swelling in response to reduced medium tonicity in a dose-dependent and time-dependent manner (Fig. 3). Increases in either miconazole concentration or duration of exposure also resulted in an altered rate of regulatory volume decrease following swelling (Fig. 3). Further experiments were performed to address the possibility that altered swelling responsiveness in cells pretreated with miconazole might account for the previously noted ability of miconazole (200 μM for 2 h) to prevent swelling-induced cAMP accumulation in S49 cells. S49 cells treated for 20 min with 100 μM miconazole still demonstrated initial increases in volume and volume regulatory responses with reduction of medium osmolarity (Figs

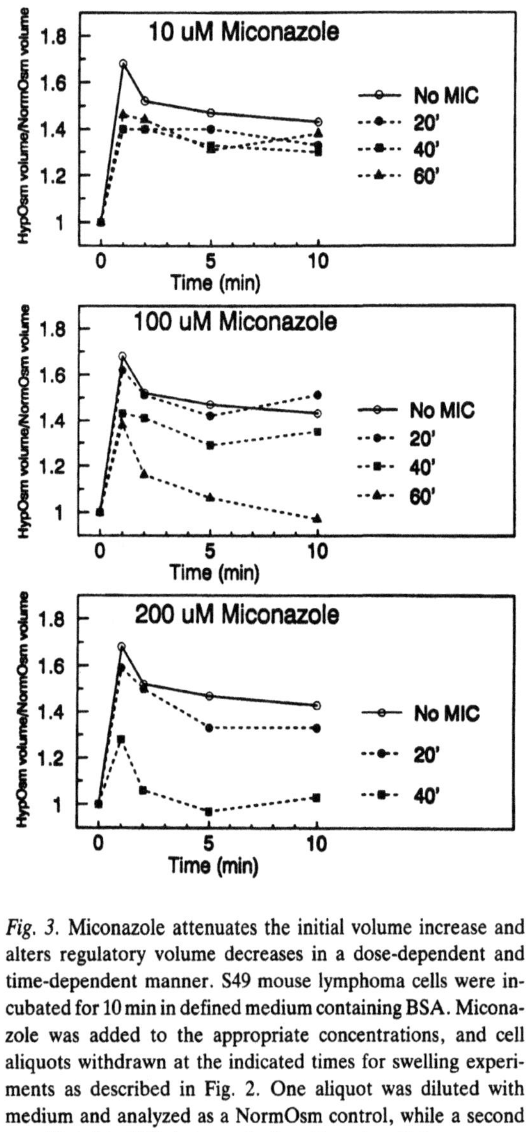

Fig. 3. Miconazole attenuates the initial volume increase and alters regulatory volume decreases in a dose-dependent and time-dependent manner. S49 mouse lymphoma cells were incubated for 10 min in defined medium containing BSA. Miconazole was added to the appropriate concentrations, and cell aliquots withdrawn at the indicated times for swelling experiments as described in Fig. 2. One aliquot was diluted with medium and analyzed as a NormOsm control, while a second was diluted to 33% hyposmolarity with water. Cells were analyzed for cell volume as indicated in the 'Methods'. Data are expressed as described in Fig. 1.

3 and 4), but showed no concomitant increase in cAMP content (Fig. 4).

Discussion

Changes in cellular shape and size resulting from either the imposition of external forces (such as load in muscle, [13]), volume expansion in response to osmolar imbalance [14], or as the result of tractile forces generated by cells against extracellular matrices [15–17], can result in profound alterations in the biochemistry of cells. Mechanical deformation of a number of cell types and tissues, including heart [2–4], lung [5], avian erythrocytes [7], and S49 mouse lymphoma cells [6, 8], has been shown to result in the accumulation of the intracellular second messenger cAMP. While the increases in cAMP resulting from stretch of the ventricular myocardium and the lung have been implicated in anabolic responses in these tissues, the role of cAMP accumulation following swelling of S49 cells and erythrocytes is not known. One possible role for this response may be to activate specific ion channels and ion conductances involved in volume regulation in these cells. Indeed, activation of a specific Cl-conductance in erythrocytes by protein kinase A is implicated in volume regulation [9]. Experiments were performed to determine if modulation of cAMP by pharmacologic interventions would alter the ability of cells to respond to hyposmolar conditions with a rapid increase in cell volume, or the ability to initiate volume regulatory events which lead to a decrease in cellular volume following swelling.

Pretreatment with 8-bromo cAMP (2×10^{-4} M) leads to maximal acute activation of protein kinase A activity in S49 cells such that protein kinase A activity cannot be further elevated by cAMP in the *in vitro* assay [8]. Such treatment of S49 cells prior to swelling was without effect on the ability of S49 cells to swell in response to medium osmolarity reduction or to volume regulate following swelling. Additionally, treatment of cells prior to or during the volume regulatory responsee with forskolin (which causes a 200-fold increase in cAMP content in these cells) did not alter the pattern of regulatory volume decrease in these cells.

Previously reported experiments demonstrate that treatment of S49 cells for 2 h with the nitroimidazole derivative miconazole prevents forskolin and swelling-induced increases in adenylate cyclase activity [8]. Results reported here demonstrate that treatment with as little as 100 μM miconazole for 20 min prevented swelling-induced accumulation of cAMP in S49 cells, while having little or no effect

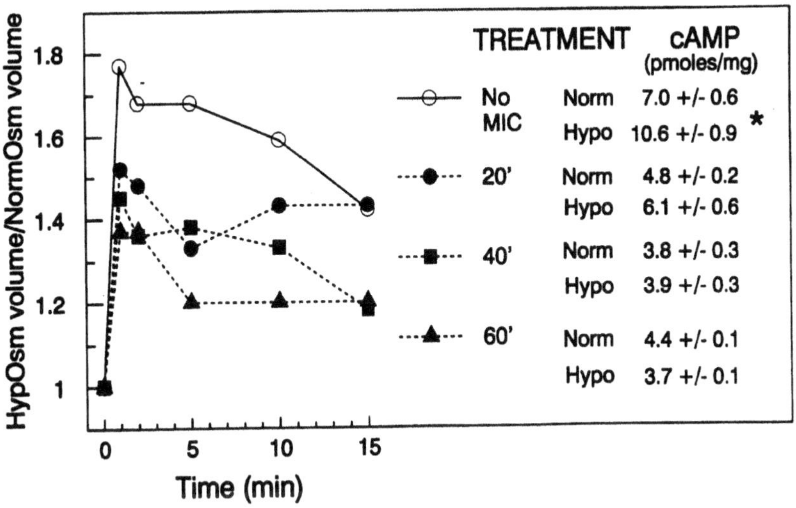

Fig. 4. Deformation-induced increases in adenylate cyclase activity, swelling, and regulatory volume reduction are differentially sensitive to inhibition by miconazole. Cells were treated as described in Fig. 3, with an additional aliquot of cells being withdrawn for analysis of cAMP and protein content as described in the 'Methods'. Volume data are expressed as in Fig. 1, and cAMP content data are expressed as pmoles/mg cell protein. All values and data points represent the mean ± SE of 4 observations. * represents $p < 0.05$ vs NormOsm control.

on the ability of S49 cells to respond to hyposmolar conditions with a rapid increase in cell volume, or in the ability of cells to initiate volume regulation. It was concluded from the results of the experiments described above that increases in cAMP in S49 cells with swelling do not play a significant role in initiating volume regulatory events. It was thought that the degree of volume increase and subsequent regulatory volume decrease might be an inverse function of cAMP content. This was dismissed following consideration of results which show that elevation of cAMP content prior to swelling or during the regulatory volume decrease had no effect on either of these responses to hypotonic swelling.

It was concluded in previous published work that deformation-induced accumulation of cAMP in S49 cells was the result of acceleration of cAMP synthesis by adenylate cyclase, occurring independent of the G-proteins responsible for hormonal regulation of cAMP synthesis [1]. Prevention of deformation-induced increases in cAMP following swelling by miconazole, which was shown to inhibit adenylate cyclase independent of effects on G-proteins [10] confirms this conclusion. Mechanical stimulation of adenylate cyclase is modified by cytoskeletal disruption in a manner reminiscent of the response of stretch-activated ion channels to cytoskeletal disruption [18]. While cytoskeletal disruption by either colchicine or cytochalasin B increased cAMP content in S49 cells in normosmotic medium, only cells exposed to colchicine maintained responsiveness to deformation-induced cAMP accumulation. Adenylate cyclase activity was no longer responsive to swelling in S49 cells following treatment with cytochalasin B. This implies that deformational forces require intact actin microfilaments for modulation of adenylate cyclase. Disruption of microfilaments has also been shown to influence the activity of stretch-activated ion channels in embryonic skeletal muscle, while disruption of microtubules was without affect [18]. A similar hypothesis has been presented with regards to observations regarding shape-dependent modulation of Na^+/H^+ exchanger activity [15]. We speculate that cAMP may initiate long term adaptive responses to cellular deformation, such as increased protein synthesis or expression of genes for structural proteins such as cytoskeletal proteins.

References

1. Gilman AG: G-proteins: transducers of receptor generated signals. Annu Rev Biochem 56: 615–649, 1987
2. Xenophontos XP, Watson PA, Chua BHL, Haneda T, Morgan HE: Increased cAMP content accelerates protein synthesis in rat heart. Circ Res 65: 647–656, 1989
3. Watson PA, Haneda T, Morgan HE: Effect of higher aortic pressure on ribosome formation and cAMP content in rat heart. Am J Physiol 256: C1257–C1261, 1989
4. Haneda T, Watson PA, Morgan HE: Elevated aortic pressure preferentially accelerates ribosome formation through a cAMP-dependent mechanism. In: Alan R. Liss: Molecular Biology of the Cardiovascular System. Inc 1990, pp 21–28
5. Russo LA, Rannels SR, Laslow KS, Rannels DE: Stretch-related changes in lung cAMP after partial pneumonectomy. Am J Physiol 257: E261–E268, 1989
6. Watson PA: Accumulation of cAMP and calcium in S49 mouse lymphoma cells following hyposmotic swelling. J Biol Chem 264: 14735–14740, 1989
7. Morgan HE, Xenophontos XP, Haneda T, McGlaughlin S, Watson PA: Stretch-anabolism transduction. J Appl Cardiol 4: 415–422, 1989
8. Watson PA: Direct stimulation of adenylate cyclase by mechanical forces in S49 mouse lymphoma cells during hyposmotic swelling. J Biol Chem 265: 6569–6575, 1990
9. London RD, Lipkowitz MS, Sinert RH, Abramson RG: Modulation of ionic permeability in a nonpolarized cell: effect of cAMP. Am J Physiol 257: F985–F993, 1989
10. Stalla GK, Stalla J, von Werder K, Muller OA, Gerzer R, Hollt V, Jakobs KH: Nitroimidazole derivatives inhibit anterior pituitary cell function apparently by a direct effect on the catalytic subunit of the adenylate cyclase holoenzyme. Endocrinology 125: 699–706, 1989
11. Gilman AG: A protein binding assay for adenosine 3′: 5′-cyclic monophosphate. Proc Nat Acad Sci USA 67: 305–312, 1970
12. Lowry OH, Rosebrough NJ, Farr AL, Randall RJ: Protein measurement with the folin phenol reagent. J Biol Chem 193: 265–275, 1951
13. Goldspink DF: The influence of passive stretch on the growth and protein turnover of the denervated extensor digitorum longus muscle. Biochem J 174: 595–602, 1978
14. Grinstein SA, Rothstein B, Sarkadi B, Gelfand EW: Responses of lymphocytes to anisotonic media: volume-regulating behavior. Am J Physiol 246: C204–215, 1984
15. Ingber DE, Folkman J: How does extracellular matrix control capillary morphogenesis? Cell 58: 803–805, 1989
16. Schwartz MA, Both G, Lechene CP: pH regulation in spread cells and round cells. J Biol Chem 265: 1327–1332, 1990
17. Ingber DE, Prusty D, Frangioni JV, Cragoe EJ, Lechene C, Schwartz MA: Control of intracellular pH and growth by fibronectin in capillary endothelial cells. J Cell Biol 110: 1803–1811, 1990
18. Sachs F: Baroreceptor mechanisms at the cellular level. Fed Proc 46: 12–16, 1987

Address for offprints: P.A. Watson, Weis Center for Research, Geisinger Clinic, Danville, PA 17822, USA

Molecular and Cellular Biochemistry **104**: 57–60, 1991.

Mechanical transduction by membrane ion channels: a mini review

F. Sachs
SUNY Biophysical Sciences, Buffalo, N.Y. 14214, USA

Key words: ion channels, mechanical, transduction, membrane, stress, volume

Abstract

There are ion channels in the cell membrane that are sensitive to stress in the membrane cytoskeleton. Some channels turn on with stress, others turn off. In specialized receptors such as those involved in hearing, touch, etc. the role of the channels is clear. However, virtually all cells have these channels, and we don't yet know the physiological role of the channels although it is reasonable to suppose that they are involved in the control of cell size, either acutely as in volume regulation, or trophically as in the control of cell division.

Review

Mechanical transduction is a widespread property of cells. The exterosenses of higher organisms (hearing, touch, etc.) are familiar examples. The enterosenses feedback from the skeletal musculature (muscle spindles, tendon organs, etc.) and the visceral musculature (detection of blood pressure, filling of the lungs, bladder, etc.) are even more vital to survival. These senses inform the central nervous system about the state of the external and internal environment and undoubtedly involve the activation of mechanically sensitive ion channels in the appropriate nerves. Additionally, mechanical transduction plays a role in the life of single cells. Paramecia are classical example. Much like people, they will speed up when prodded posteriorly and will backup when prodded anteriorly. This response is known to involve two sets of mechanically sensitive permeabilities, presumably channels, one selective for potassium and the other for calcium [22].

There are a great many cellular processes that are dependent upon mechanical transduction, but are not necessarily dependent upon ion channels. Cell volume regulation, for example, is a problem for all cells. Cells in the renal system can transport many cell volumes per minute. A slight imbalance between influx and efflux will lead to rapid changes in cell volume. Yet, how does a cell know its volume? Volume is an extensive variable and cannot be measured by the concentration of a soluble substance. Volume must be measured mechanically, perhaps by strain in the cytoskeleton [18]. This is a general problem in cell physiology that relates not only to acute volume regulation but to the general requirement for trophic regulation: how does a cell know how big it is? Cell growth can be driven by mechanical forces as seen in cardiac hypertrophy due to overload [4, 17] or skeletal muscle hypertrophy due to passive stretch [32]. Mechanical effects are seen in all cells, in plants as well as animals. The rust fungus, Uromyces, can precisely sense a 0.5 μm ridge signifying the entrance to a stomate [15]. In response to light touch, higher plants will transcribe massive amounts of a few genes [2].

It is difficult to work out the ways in which forces influence cell growth because cell growth is generally slow and involves the interplay of a host of second and higher order messengers. Effect cannot be readily distinguished from cause. If anabolic rates exceed catabolic rates by 1%, there will be major changes in cell growth, buf effects of this magnitude would be missed in an acute experiment. In S49 cells, for example, hypotonic stress leads to a three fold increase in cAMP [33]. Is that a significant change? Is hypotonic stress mechanically modulating cyclase or is there another messenger. Since we cannot readily stress membrane fragments, and whole cells have so many interacting systems, deciphering cause and effect can be difficult.

In the case of stretch sensitive ion channels, we have a primary transducer. The state of the channel is changed directly by the applied forces – no second messengers are required [26]. If we understand what makes a channel mechanically sensitive, then we may be able to understand what makes other enzymes mechanically sensitive. We have good evidence that SACs (Stretch Activated Channels) are linked in series with some component of the cytoskeleton. This may be a general rule for those enzymes that are highly sensitive to stress. In order for external forces to affect the gating of a channel, they must do work on the channel, and that work is dominated by the distance the force moves: work = force × distance. Howard and Hudspeth [16] estimated from elegant studies on the compliance of saccular hair cells, that the SA channel changes dimensions by 4 nm between the closed state and the open state! If conformational changes of 4 nm are indeed present, the channel protein should have some remarkable, and distinctive features.

Stretch sensitive ion channels are a distinct class of channels; that is, they do not represent a heretofore unknown property of a sodium channel (R. Horn, personal communication), an acetylcholine-activated channel [12], a calcium-activated potassium channel [13], or other channel that we know of. Stretch-sensitive channels are remarkably similar in their properties, independent of the source. They have similar number densities ($\approx 1/\mu^2$), are activated by similar membrane tensions ($\approx$ 1 dyn/cm) and display similar voltage sensitivities. The primary differences are those of ion selectivity. It seems as though a class of very similar channels is responsible for such diverse phenomena as hearing and osmosensing.

The obvious place to look for mechanosensitive channels is among the specialized mechanoreceptors such as the vestibulo-cochlear organs, muscle spindles, Pacinian corpuscles, etc. Unfortunately, it has been difficult to record single channel currents from these preparations. Ohmori has published records of single channel currents in vestibular hair cells [24], and non-selective cation SACs have been demonstrated in the cray-fish stretch receptor neuron [6]. But the most extensive and detailed data comes from non-specialized cells where the physiological role for these channels is not yet clear. There are no specific activators or blockers for mechanosensitive channels and we are likely looking for trophic effects.

General features of gating

There are two classes of mechanosensory ion channels: stretch-activated channels (SACs) that turn on when the membrane is stretched, and stretch-inactivated channels (SICs) that are tonically active and turn off when the membrane is stretched. Both classes may coexist, providing a mechanically adjustable setpoint for membrane potential [21].

Ionic selectivity

The selectivity of these mechanosensory channels varies from cell to cell and channel to channel. In animal cells, the SACs are either non-selective cation channels which pass the alkali cations and calcium, or are potassium selective. Anion selective SACs have been demonstrated in plant cells [11]. Because the non-selective channels also pass calcium, it is possible to produce a mechanosensitive chloride conductance through the activation of calcium-activated chloride channels [1, 3, 31]. This dual anion/cation pathway can produce salt transport which is essential for cell volume regulation [3].

The stress required to activate channels

Although the patch is stimulated by applying hydrostatic pressure, the actual stimulus is the membrane tension created by the pressure. SACs will open with either positive or negative pressure and SICS will close with either pressure [21]. We have made a direct demonstration of the tension dependence using quantitative imaging [29, 30].

There are no second messengers
Experiments with excised patches indicate that the transduction response is due to a direct action on the channel and not due to release of previously stored energy, such as might exist with a second messenger. Calcium is not involved since it can be removed from both sides of the membrane without eliminating the response [3, 5, 13].

The cytoskeletal meshwork
Guharay and Sachs [13] proposed that the high sensitivity of SA channels to membrane tension is best explained if forces are focussed on the channels by a cytoskeletal lattice probably of the intermediate filament category. Although this conclusion was originally based on a theory which we now know was incomplete (Sachs and Lecar, in preparation), the conclusion seems to be correct. We have been able to measure SA channel activation in patches in which the lipids were essentially unstressed and only the cytoskeleton had enough long range order to bear the tension [30]. The cytoskeletal meshwork linked to the channels is not f-actin or tubulin since reagents for these components do not block SAC activity [13], and there are no reagents for the other components of the cytoskeleton. We think the relevant cytoskeleton may be spectrin [28], but have not been able to prove it. If we could find SA channels in mammalian red cells we would have a good opportunity to solve the issue of cytoskeletal involvement. Unfortunately, we have had no success in forming tight seals on mammalian red cells or ghosts. Since other membrane proteins have been shown to bind to ankyrin [23], we might expect the SA channel to show a similar binding site.

Membrane structure
The mechanical structure of membranes is complicated, and in fact the membrane cannot be said to end at any particular place since the lipid component is linked by cytoskeleton to the cell interor, and by the extracellular matrix (ECM) to external structures. Mechanical studies of membrane properties have generally used pipette aspiration techniques and have been done almost entirely on pure lipid membranes [10] or mammalian erythrocytes [9]. In both cases, the lipid properties dominate the area elastic modulus. The lipids have a comparable or higher elastic modulus than does the red cell [7, 8]. Although the cytoskeleton is present, the lipid is less compliant and bears almost all the stress. There is almost no data on the mechanical properties of the intact cytoskeleton in more typical cells. The patch clamp provides a new tool to study cytoskeletal and ECM properties since the lipids are unconstrained, especially in excised patches.

In a recent experiment using a variety of stimuli, Morris and Horn [20] were unable to record whole cell SAC and SIC currents from isolated growth cones even though they could record SIC and SAC single channel currents from patches. This disparity suggests that formation of the patch may significantly alter important structures. It also suggests that in some tissues, mechanosensitive channels are protected from activation and perhaps local alteration in the cytoskeleton is necessary for activation. The Morris and Horn result is not general. We have been able to record single SA channels and to measure large inward currents in *Xenopus* myocytes in response to hypoosmotic stress. We have measured calcium uptake in chick heart cells stimulated with mechanical probes and hypoosmotic stress. Gustin and coworkers [14] have shown whole cell activation in yeast, and Okada and coworkers have shown large, Gd^{+3} blockable, inward currents and Ca^{+2} uptake in I407 cells in response to hypoosmotic stress [25].

For those interested in a more information on mechanosensitive ion channels, several recent reviews are available [19, 26, 27].

Acknowledgements

This work was supported by grants from the USARO 26099-LS and NIH DK-37792 and the Muscular Dystrophy Association.

References

1. Bader CR, Bertrand D, Schlichter R: Calcium-activated chloride current in cultured sensory and parasympathetic quail neurones. J Physiol (Lond) 394: 125–148, 1987
2. Braam J, Davis RW: Rain-, Wind-, and Touch-Induced Expression of Calmodulin and Calmodulin-Related Genes in Arabidopsis. Cell 60: 357–364, 1990
3. Christensen O: Mediation of cell volume regulation by Ca^{+2} influx through stretch-activated channels. Nature 330: 66–68, 1987
4. Cooper G: Cardiocyte adaptation to chronically altered load. Ann Rev Physiol 49: 501–518, 1987
5. Cooper KE, Tang JM, Rae JL, Eisenberg RS: A cation channel in frog lens epithelia responsive to pressure and calcium. J Membrane Biol 93: 259–269, 1986
6. Erxleben C: Stretch-activated current through single ion channels in the abdominal stretch receptor neuron of the crayfish. J Gen Physiology 94: 1071–1083, 1989
7. Evans E, Needham D: Giant Vesicle Bilayers composed of Mixtures of Lipids, Cholesterol and Polypeptides. Faraday Discuss Chem Soc 81: 267–280, 1986
8. Evans E, Needham D: Physical Properties of Surfactant Bilayer Membranes: Thermal Transitions, Elasticity, Rigidity, Cohesion, and Colloidal Interactions. J Physical Chemistry 91: 4219–4228, 1987
9. Evans E, Waugh R, Melnik L: Elastic area compressibility modulus of red cell membrane. Biophysical J 16: 585, 1976
10. Evans EA, Waugh R: Mechano-Chemistry of Closed, Vesicular Membrane Systems. J Colloid and Interface Sci 60, No 2: 286–298, 1988
11. Falke LC, Edwards KL, Pickard BG, Misler S: A stretch-activated anion channel in tobacco protoplasts. FEBS 237: 141–144, 1988
12. Guharay F, Sachs F: Mechanoreceptor ion channels are not nicotinic. Biophys J 47: 203a, 1984
13. Guharay F, Sachs F: Stretch-activated single ion channel currents in tissue-cultured embryonic chick skeletal muscle. J Physiol (Lond) 352: 685–701, 1984
14. Gustin MC, Zhou X-L, Martinac B, Kung C: A Mechanosensitive Ion Channel in the Yeast Plasma Membrane. Science 242: 762–766, 1988
15. Hoch H, Staples RC, Whitehead B, Comeau J, Wolf ED: Signaling for growth orientation and cell differentiation by surface topography in Uromyces. Science 235: 1659–1662, 1987
16. Howard J, Hudspeth AJ: Compliance of the Hair Bundle Associated with Gating of Mechanoelectrical Transduction Channels in the Bullfrog's Saccular Hair Cell. Neuron 1: 189–199, 1988
17. Kent RL, Hoober JK, Cooper GIV: Load Responsiveness of Protein Synthesis in Adult Mammalian Myocardium: Role of Cardiac Deformation Linked to Sodium Influx. Circ Res 64, no 1: 74–85, 1989
18. Mills JW, Lubin M: Effect of adenosine 3′, 5′-cyclic monophosphate on volume and cytoskeleton of MDCK cells. Am J Physiol 250: C319–C324, 1986
19. Morris CE: Mechanosensitive Ion Channels. J Membrane Biol 113: 93–107, 1990
20. Morris CE, Horn R: Voltage clamp of isolated growth cones. Biophysical J 57: 318a, 1990 (Abstract)
21. Morris CE, Sigurdson WJ: Stretch-Inactivated Ion Channels Coexist with Stretch-Activated Ion Channels. Science 243: 807–809, 1989
22. Naitoh Y: Mechanosensory transduction in protozoa. In: G. Colombetti, F. Lenci (ed.) Membranes and Sensory Transduction. New York: Plenum Press, 1984, p 113–134
23. Nelson WJ, Hammerton RW: A Membrane-Cytoskeletal Complex Containing Na^+, K^+-ATPase, Ankyrin, and Fodrin in Madin-Darby Canine Kidney (MDCK) Cells: Implications for the Biogenesis of Epithelial Cell Polarity. J Cell Biol 108: 893–902, 1989
24. Ohmori H: Mechanoelectrical transducer has discrete conductances in the chick vestibular hair cell. Proc Natl Acad Sci USA 81: 1888–1891, 1984
25. Okada Y, Hazama A, Yuan WL: Stretch-induced activation of Ca^{2+} permeable ion channels is involved in the volume regulation of hypotonically swollen epithelial cells. Neurosci Res in press: 1990
26. Sachs F: Mechanical transduction in biological systems. Crit Rev Biomed Eng 16: 141–169, 1988
27. Sachs F: Ion Channels as Mechanical Transducers. In: Stein WD, Bronner F (ed.) Cell Shape: Determinants, Regulation and Regulatory Role, San Diego, NY, Berkeley, Boston: Academic Press, 1989, p 63–92.
28. Shen BW, Josephs R, Steck TL: Ultrastructure of the Intact Skeleton of the Human Erythrocyte Membrane. J of Cell Biology 102: 997–1006, 1986
29. Sokabe M, Sachs F: Stress and strain in patch clamped membranes. Biophys J 57: 265a, 1990
30. Sokabe M, Sachs F, Jing Z: Quantitative video microscopy of patch clamped membranes – stress, strain, capacitance and stretch channel activation. Biophys J 59: 722–728, 1991
31. Taleb O, Feltz P, Bossu J-L, Feltz A: Small-conductance chloride channels activated by calcium on cultured endocrine cells from mammalian pars intermedia. Pflugers Arch 412: 641–646, 1988
32. Vandenburg H, Kaufman S: *In vitro* model for stretch-induced hypertrophy of skeletal muscle. Science 203: 4377–265, 1979
33. Watson PA: Accumulation of cAMP and calcium in S49 mouse lymphoma cells following hypoosmotic swelling. J Biol Chem 246: 14735–14740, 1989

Address for offprints: F. Sachs, SUNY Biophysical Sciences, Buffalo, NY 14214, USA

SIGNAL TRANSDUCTION

Molecular and Cellular Biochemistry **104**: 63–72, 1991.

The effect of culture and membrane potential on $G_{o\alpha}$ expression in neonatal rat cardiac myocytes

K.A. Foster, P.J. McDermott[1] and J.D. Robishaw
Weis Center for Research, Geisinger Clinic, Danville, Pennsylvania 17822, USA;
[1] *Present address: Dept. of Cardiology, Medical University of South Carolina, Veterans Administration Medical Center, Charleston, SC 29403, USA*

Key words: immunoblotting, inositol phosphate, muscarinic receptors

Abstract

The effects of culture and membrane potential on $G_{o\alpha39}$ expression were examined in neonatal rat cardiac myocytes. During six days of culture, the amount of $G_{o\alpha39}$ in myocytes increased six-fold. The increase in $G_{o\alpha39}$ appeared to be programmed, since $G_{o\alpha39}$ of rat hearts also increased *in vivo* within three days after birth before declining by six days after birth. Furthermore, the age of the rat from which cardiac myocytes were isolated determined the amount of $G_{o\alpha39}$ that accumulated in cultured cells with myocytes from two day-old rats producing more $G_{o\alpha39}$ than myocytes from six day-old rats. In addition, agents which alter membrane potential (KCl and bupivacaine) inhibited the accumulation of $G_{o\alpha39}$ in cultured myocytes. In an attempt to identify the signaling pathway in which cardiac $G_{o\alpha39}$ is involved, muscarinic receptor-stimulated inositol phosphate production was examined, but was found to be comparable in myocytes that had six-fold differences in $G_{o\alpha39}$ content. Thus $G_{o\alpha39}$ does not appear to couple muscarinic receptors to phospholipase C in rat cardiac myocytes.

Introduction

Guanine nucleotide binding (G) proteins are transducers of extracellular signals to intracellular effectors. In the heart, the heterotrimeric G proteins, consisting of α, β, and γ subunits, are known to couple several neurotransmitters and hormones to effectors such as adenylyl cyclase [1], phospholipases A_2 and C [2, 3], and Ca^{2+} and K^+ channels [4, 5]. Five different G proteins, including $G_{s\alpha52}$, $G_{s\alpha45}$, $G_{i\alpha40}$ ($G_{i\alpha2}$), $G_{i\alpha41}$ ($G_{i\alpha1}$, $G_{i\alpha3}$), and $G_{o\alpha39}$ have been identified in the hearts of various species, including rat [6–8], cow [9, 10], and chick [11], by immunoblotting analysis. However, the functions of most of the cardiac G proteins have not yet been definitively determined. To date, the only G protein function that is known with certainty is that of G_s. Both $G_{s\alpha52}$ and $G_{s\alpha45}$ are known to activate adenylyl cyclase [12] and cardiac Ca^{2+} channels [4]. In contrast, specific functions have not been assigned to the three forms of G_i or G_o. G_i, G_o, or other as yet unidentified G proteins are likely to be involved with the regulation of myocardial phospholipase C activity and the activation of atrial K^+ channels [3, 13].

One way to discern the function of a particular G protein is to develop a model in which the expression of the G protein can be altered. We have shown previously that $G_{o\alpha39}$ increases dramatically when neonatal rat cardiac myocytes are placed in culture [14]. In addition, this increase in $G_{o\alpha39}$ can be inhibited by culturing myocytes in the presence of 50 mM KCl. In the present study, we have further characterized the effect of culture and KCl on

$G_{o\alpha39}$ expression in myocytes and have begun to examine a possible role for $G_{o\alpha39}$ in muscarinic receptor-stimulated inositol phosphate production.

Materials and methods

Cell culture

Primary cell cultures were prepared from ventricles of 2- to 4-day-old neonatal rats as described previously [15]. Briefly, the cells were dissociated in a Celstir apparatus (Wheaton Scientific) with a mixture of 2.4 U/ml trypsin (Worthington Biochemical), 2.7 U/ml chymotrypsin, and 0.94 U/ml elastase (Sigma) dissolved in Ca^{2+}-Mg^{2+}-free Hank's salt solution buffered with 30 mM N-2-hydroxyethylpiperazine N'-2-ethanesulfonic acid (HEPES), pH 7.4. The dissociated cells from five 20 min incubations were centrifuged at 500 g, resuspended in Eagle's minimum essential medium (MEM, Gibco) containing 10% newborn calf serum, and pooled. The cells were enriched for cardiac myocytes by differential plating for 1.5 h, after which nonadherent cells were plated at a concentration of 4.5×10^6 cells per 60 mm dish in MEM containing 10% newborn calf serum and 0.1 mM 5-bromo-2'-deoxyuridine (BrdU, Sigma). After an overnight incubation, the adherent cells were rinsed and maintained in a serum-free medium [15]. High $[K^+]$ medium was prepared by adding a stock solution of KCl in MEM to final concentrations of 10, 20, 30, 40, or 50 mM. Bupivacaine was added as a stock solution in MEM to final concentrations of 0.1 or 0.3 mM. The media were changed routinely every 48 h.

Membrane preparation

Membranes were isolated from cultured myocytes and from hearts of rat pups at days 17 to 21 *in utero* and postnatal days 0 through 10 by lysing cells in a hypotonic buffer [14]. The cell suspension was homogenized with a Potter-Elvehjem tissue grinder in the presence of protease inhibitors (leupeptin and phenylmethylsulfonyl fluoride). The lysate was layered on 40% sucrose and centrifuged at 80,000 g for 1 h. The interfacial material was collected and pelleted at 120,000 g for 30 min. The resulting pellet was resuspended at approximately 1 mg/ml in 20 mM HEPES, 2 mM $MgCl_2$, 1 mM EDTA and was used for immunoblotting and protein determination [16].

Immunoblotting

For Western blotting, G proteins from myocyte membranes were resolved on 11% (33 : 1) sodium dodecyl sulfate (SDS)-polyacrylamide gels and were transferred to nitrocellulose and immunoblotted as previously described [14]. The primary antipeptide antibody used in these studies, A-10, is specific for $G_{o\alpha39}$ [14]. A-10 does not recognize other G protein α subunits. However, the $G_{o\alpha39}$ antibody does recognize an unidentified 60 kDa protein, which probably contains the same or similar peptide sequence as $G_{o\alpha39}$. Antibody binding was detected with goat antibody to rabbit immunoglobulin G labeled with ^{125}I (New England Nuclear) at 5×10^5 cpm/ml. Autoradiographic images of the blots were obtained with Kodak XAR-5 film after exposure with an intensifying screen. Blots were quantified using AMBIS Radioanalytic Imaging System with pure recombinant $G_{o\alpha}$ subunit (a generous gift from Drs. Maureen Linder and Alfred Gilman) as standard.

Muscarinic receptor binding

For receptor binding studies, myocytes were homogenized in the hypotonic buffer described above, centrifuged at 80,000 g for 1 h, and resuspended at approximately 1 mg/ml in incubation buffer containing 50 mM Tris-HCl (pH 7.4), 1 mM $CaCl_2$, 1 mM $MgCl_2$, 140 mM NaCl, and 10 mM NaH_2PO_4. Myocyte homogenates (100 μg) were preincubated in the presence or absence of 10 μM atropine sulfate (Sigma) for 10 min at 37° C. Assays were initiated by the addition of 0.01 to 1.5 nM $[^3H]$-quinuclidinylbenzilate (QNB, New England Nuclear) in incubation buffer to a final volume of

400 μl. At the end of the 1 h incubation at 37° C, the samples were diluted 10-fold with ice-cold incubation buffer and immediately filtered over Whatman GF/C filters. The tubes and filters were rinsed three times with 4 ml of ice-cold buffer and the radioactivity retained on the filters was determined by liquid scintillation spectrometry. Specific binding of [^{3}H]-QNB was defined as the difference between total binding (in the absence of atropine) and nonspecific binding (in the presence of atropine). Nonspecific binding accounted for less than 2% of total binding. The dissociation constants (K_d) and maximum number of binding sites (B_{max}) for [^{3}H]-QNB were determined by Scatchard analysis [17].

Inositol phosphate production

Muscarinic receptor-stimulated inositol phosphate production was examined in day 6 KCl-depolarized and contracting myocytes in culture. Myocytes were incubated with 1 μCi/ml [^{3}H]-myoinositol (New England Nuclear) in MEM for 2 h, rinsed three times with MEM, and incubated with 10 mM LiCl for 10 min. Inositol phosphate production was stimulated by addition of the muscarinic agonist carbachol to a final concentration of 2.5 mM during a 1 h incubation. Myocytes were scraped from the dishes in methanol and two phases were formed by adding chloroform and H_2O to a final ratio of 10 : 10 : 5 ($CHCl_3$: MeOH : H_2O). The aqueous phase was put over a column containing 180 mg of AGl-x8 (formate form, BioRad) ion exchange resin. The column was washed with 10 ml H_2O to remove free inositol and the inositol phosphate was eluted with 8 ml 200 mM ammonium formate/100 mM formic acid. The eluted inositol phosphates were measured by liquid scintillation spectrometry.

To determine the specific activity of the phosphatidylinositol pool, the phospholipids in the organic phase were resolved on silica gel 60 thin layer chromatography plates using the solvent system chloroform : methanol : acetic acid : H_2O (81 : 10 : 45 : 1). The phosphatidylinositol spot was either counted for radioactivity or assayed for phosphorus content [18].

Results

The $G_{o\alpha 39}$ content of cardiac myocytes isolated from 2- to 4-day old neonatal rats was determined by immunoblotting analysis. Isolated, spontaneously contracting myocytes were cultured in a chemically defined medium at a density of 4.5 × 10^6 cells per 60 mm dish. On days 1 through 6 of culture, membranes were prepared from myocytes, and the proteins were resolved on 11% polyacrylamide gels and immunoblotted (Fig. 1, inset). The $G_{o\alpha 39}$ bands were quantified using pure recombinant $G_{o\alpha}$ standard by the AMBIS Radioanalytic Imaging System, which directly counts the amount of radioactivity on the blot. Interestingly, the amount of $G_{o\alpha 39}$ in myocytes increased 6-fold during the first 6 days of culture from 282 ng/mg on day 1 to 1691 ng/mg on day 6 (Fig. 1). The greatest increase in $G_{o\alpha 39}$ occurred between days 3 and 5 of culture. Although not shown, $G_{o\alpha 39}$ continues to accumulate at a slower rate until at least day 16 of culture. The amounts of the other G protein α subunits did not change significantly during 6 days of culture [14].

To determine whether the increase in myocyte $G_{o\alpha 39}$ observed during 6 days in culture also occurred *in vivo*, cardiac membranes from prenatal and neonatal rats at − 5 to + 10 days of age were immunoblotted with the $G_{o\alpha 39}$ antibody (Fig. 2). $G_{o\alpha 39}$ increased slightly (1.5-fold) between − 5 and − 1 days before birth and then increased rapidly (2.5-fold) from birth to a maximum at 2–3 days after birth. By 5 days of age, the cardiac $G_{o\alpha 39}$ had rapidly declined to prenatal levels.

Since the amount of $G_{o\alpha 39}$ changes dramatically in the neonatal period, we tested whether the amount of $G_{o\alpha 39}$ expressed in cultured myocytes depends on the age of the rat pup from which the myocytes were isolated. Figure 3 shows that myocytes isolated from 6 day-old rats (in which the $G_{o\alpha 39}$ expression had declined) accumulate less $G_{o\alpha 39}$ during 6 days of culture than cells isolated from 2 day-old rats (in which $G_{o\alpha 39}$ is maximally expressed). These findings suggest that at least part of the increase in $G_{o\alpha 39}$ is programmed and is not merely an effect of culture. Moreover, culturing myocytes *in vitro* appears to remove the signal for

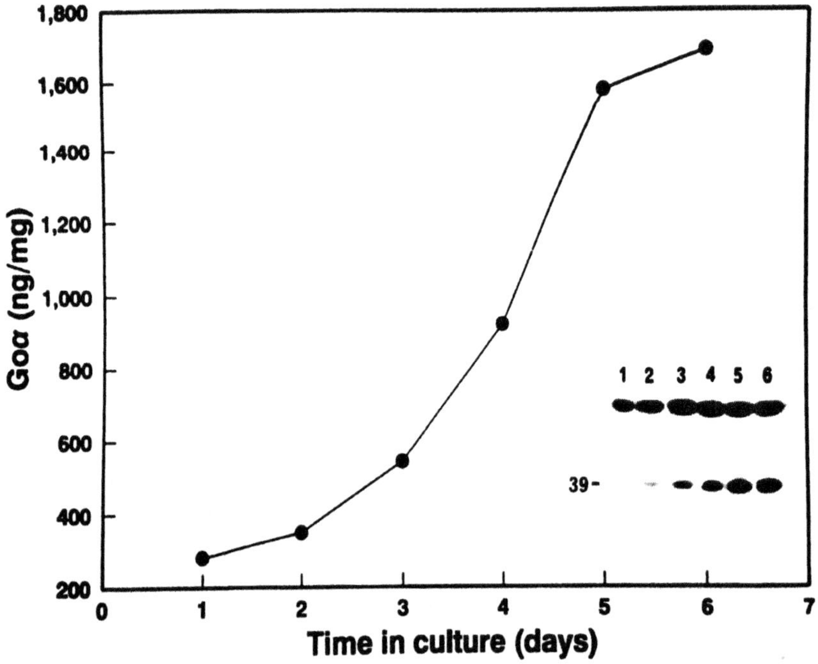

Fig. 1. Effect of time in culture on $G_{o\alpha 39}$ expression in cardiac myocytes. Membrane proteins (15 μg) from myocytes in culture for 1 through 6 days were separated by SDS-PAGE, transferred to nitrocellulose, and immunoblotted with the $G_{o\alpha 39}$-specific antiserum, A-10, at 1 : 500 dilution. The amount of ^{125}I-labeled secondary antibody bound to the blot was quantified on AMBIS Radioanalytic Imaging System using 5 to 25 ng of pure recombinant $G_{o\alpha}$ subunit as standard. *Inset*, film was exposed to blot with an intensifying screen overnight at − 70° C. Number to left indicates apparent molecular weight (× 10^3).

inhibition of $G_{o\alpha 39}$ expression that occurs *in vivo* at 4–5 days of age.

In a previous study, we showed that depolarization of cultured cardiac myocytes by 50 mM KCl inhibits $G_{o\alpha 39}$ accumulation [14]. In the present study, we examined the dependence of KCl concentration on the inhibition of $G_{o\alpha 39}$ accumulation by culturing myocytes in normal (4 mM) KCl medium or in the presence of increasing concentrations of KCl (10 to 50 mM) for 6 days. Spontaneous contractions were observed in myocytes cultured in 4, 10, or 20 mM KCl, but no contractions were observed in cells cultured in 30, 40, or 50 mM KCl. KCl treatment of myocytes for 6 days inhibited $G_{o\alpha 39}$ accumulation in a dose-dependent manner (Fig. 4). The effect of KCl on $G_{o\alpha 39}$ expression was half maximal at approximately in 16 mM KCl. KCl did not have a comparable effect on any other G protein in cardiac myocytes [14].

To further demonstrate that the inhibition of $G_{o\alpha 39}$ accumulation by KCl was due to altered membrane potential, myocytes were cultured in the presence of the Na^+ channel blocker, bupivacaine, which prevents membrane depolarization. The addition of bupivacaine to the culture medium resulted in a slower rate of myocyte contraction at 0.1 mM bupivacaine and an inhibition of contraction at 0.3 mM bupivacaine. Bupivacaine at 300 μM was nearly as effective as KCl at 50 mM in inhibiting $G_{o\alpha 39}$ accumulation (Fig. 5). Although not shown, culturing myocytes in the presence of 10 μM verapamil, a Ca^{2+} channel blocker, was only partially effective as KCl in inhibiting $G_{o\alpha 39}$ accumulation. These findings provide further evidence that altered membrane potential is associated with the inhibition of $G_{o\alpha 39}$ accumulation in cardiac myocytes.

The ability to manipulate the abundance of $G_{o\alpha 39}$ by time in culture and membrane depolarization should allow us to study the function of $G_{o\alpha 39}$ in

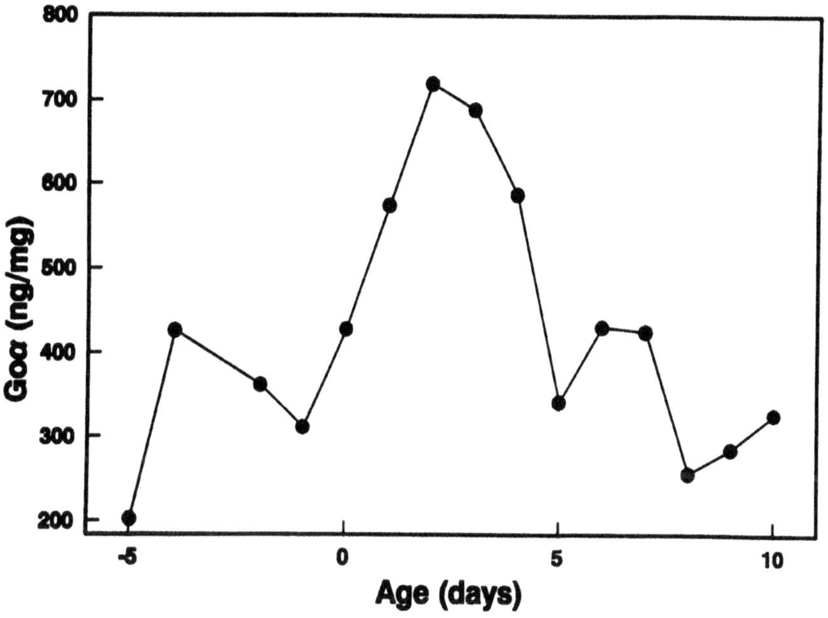

Fig. 2. Expression of $G_{o\alpha39}$ in pre-natal and neonatal rat hearts. Membranes were isolated from rat hearts at days 17 through 21 of gestation and each of the first 10 days after birth. Proteins were resolved by SDS-PAGE, transferred to nitrocellulose, and immunoblotted with the $G_{o\alpha39}$-specific antiserum, A-10, at 1 : 500 dilution. The amount of ^{125}I-labeled secondary antibody bound to the blot was quantified on AMBIS Radioanalytic Imaging System using 5 to 25 ng of pure recombinant $G_{o\alpha}$ subunit as standard.

cardiac myocytes. As a first step in determining the signaling pathway in which $G_{o\alpha39}$ is involved, muscarinic receptor-stimulated inositol phosphate production was examined in contracting and KCl-depolarized myocytes on day 6 of culture. Muscarinic receptor number was determined by [^{3}H]-quinuclidinyl benzilate (QNB) binding in the presence or absence of atropine. Interestingly, total [^{3}H]-QNB binding was significantly higher in KCl-depolarized myocytes, ($p < 0.05$ by Student's t test for paired observations, $n = 5$) while nonspecific binding was comparable in KCl-depolarized and contracting myocytes. Scatchard analysis of [^{3}H]-QNB specific binding revealed a 20% higher density of binding sites on KCl-depolarized myocytes (319 ± 67 fmol/mg protein) when compared to contracting myocytes (267 ± 49 fmol/mg). This was associated with no change in affinity (K_d) for [^{3}H]-QNB between the two groups (0.11 ± 0.02 nM for contracting and 0.12 ± 0.02 nM for KCl-depolarized). Thus, muscarinic receptor expression does not appear to down-regulate coordinately with $G_{o\alpha39}$ expression.

Although the muscarinic receptor is not coordinately expressed with $G_{o\alpha39}$, it does not rule out the possibility that $G_{o\alpha39}$ could be coupled to muscarinic receptors. Therefore, muscarinic receptor-

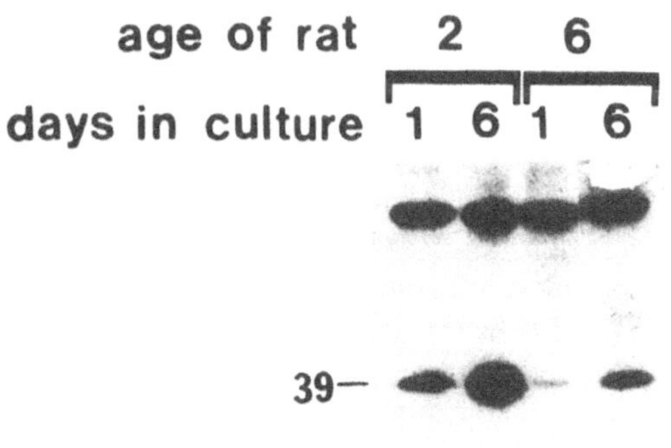

Fig. 3. Immunoblot showing the effect of age of rat at time of myocyte isolation on extent of $G_{o\alpha39}$ accumulation in culture. Myocytes were isolated from hearts of rats at 2 or 6 days of age and placed in culture for 1 or 6 days. Membrane proteins (12 μg) were resolved by SDS-PAGE, transferred to nitrocellulose, and immunoblotted with the $G_{o\alpha39}$-specific antiserum, A-10, at 1 : 500 dilution. Number to left indicates apparent molecular weight ($\times 10^3$). Film was exposed to blot with an intensifying screen for 3 days at $-70°$C.

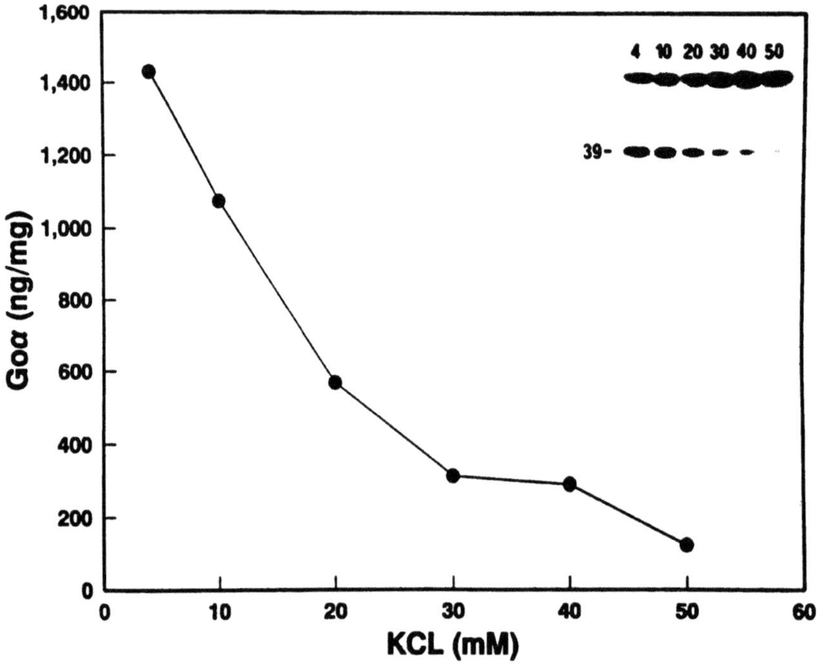

Fig. 4. Effect of KCl concentration on $G_{o\alpha39}$ accumulation in cultured cardiac myocytes. Myocytes were cultured for 6 days in medium containing 4, 10, 20, 30, 40, or 50 mM KCl. Membrane proteins (15 μg per lane) were separated by SDS-PAGE, transferred to nitrocellulose, and immunoblotted with the $G_{o\alpha39}$-specific antiserum, A-10, at 1 : 500 dilution. The amount of ^{125}I-labeled secondary antibody bound to the blot was quantified on AMBIS Radioanalytic Imaging System using 5 to 25 ng of pure recombinant $G_{o\alpha}$ subunit as standard. *Inset*, film was exposed to blot with an intensifying screen overnight at − 70° C. Number to left indicates apparent molecular weight ($\times 10^3$).

stimulated inositol phosphate production was examined in day 6 KCl-depolarized and contracting myocytes. Myocytes were incubated with 1 μCi/ml [^{3}H]-myoinositol for 2 h followed by a 10 min incubation with 10 mM LiCl prior to addition of the muscarinic agonist carbachol to a final concentration of 2.5 mM. Total inositol phosphates were measured by separating inositol phosphates from free inositol by ion exchange chromatography and counting the radioactivity. Specific activity of the phosphatidylinositol (PI) pool was determined by resolving PI from the other phospholipids by thin layer chromatography, counting the PI spot for radioactivity, and assaying a duplicate PI spot for phosphorus. No significant differences were observed between KCl-depolarized and contracting myocytes in the amount of inositol phosphate released per mg protein in the presence or absence of 2.5 mM carbachol (Fig. 6). Thus $G_{o\alpha39}$ does not appear to couple muscarinic receptors to phospholipase C in cardiac myocytes.

Discussion

The effect of culture on $G_{o\alpha39}$ expression was studied in neonatal rat cardiac myocytes. Immunoblotting analysis with $G_{o\alpha39}$-specific antiserum showed that $G_{o\alpha39}$ increases 6-fold between days 1 and 6 of culture. Preliminary evidence suggests that $G_{o\alpha39}$ continues to accumulate through day 16 of culture. $G_{o\alpha39}$ increases at a greater rate than total cell protein during 6 days in culture [15], suggesting that the accumulation of $G_{o\alpha39}$ during 6 days of culture is not in proportion to cell growth. In contrast, other G proteins, e.g. G_s and G_i, did not increase with time in culture [14], indicating that the accumulation is specific to $G_{o\alpha39}$. The increase in cardiac

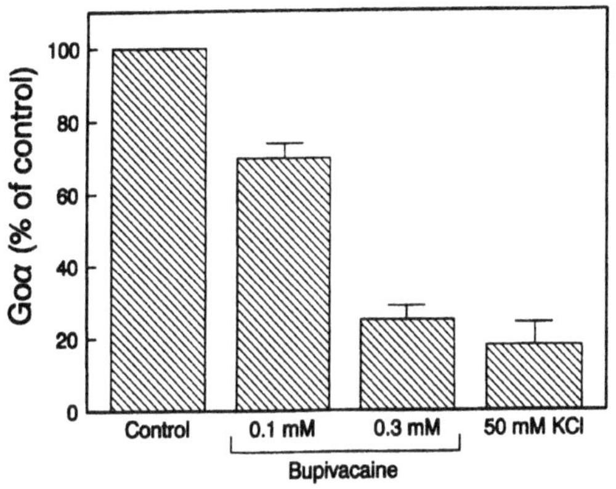

Fig. 5. Effect of bupivacaine on $G_{o\alpha39}$ expression in cultured cardiac myocytes. Myocytes were cultured for 6 days in medium containing 4 mM KCl, 0.1 mM bupivacaine, 0.3 mM bupivacaine, or 50 mM KCl. Membrane proteins (15 μg) were separated by SDS-PAGE, transferred to nitrocellulose, and immunoblotted with the $G_{o\alpha39}$-specific antiserum, A-10, at 1 : 500 dilution. $G_{o\alpha39}$ values were obtained by densitometric scanning of autoradiographs from 3 different cell preparations and are expressed as percent of control ± SE with the 4 mM KCl group being 100%.

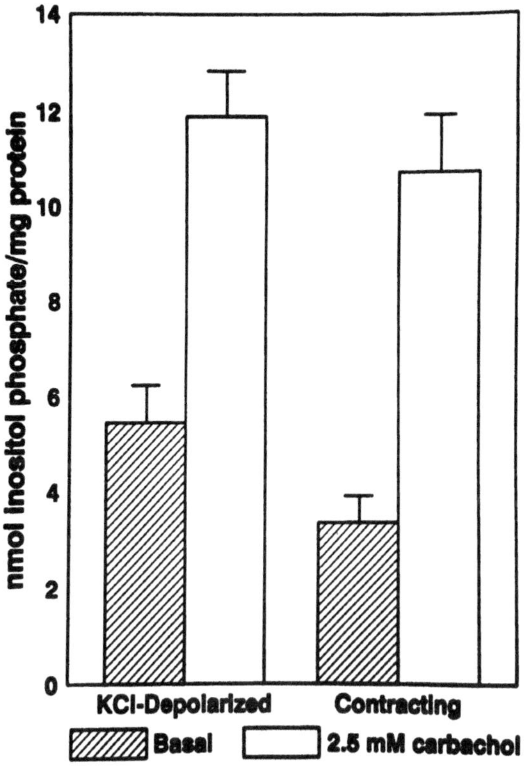

Fig. 6. Muscarinic receptor-stimulated inositol phosphate production in KCl-depolarized and contracting myocytes. Day 6 KCl-depolarized and contracting myocytes were incubated with [^{3}H]-myoinositol prior to addition of 2.5 mM carbachol or vehicle (MEM) (basal). Values are means ± SE for n = 5.

$G_{o\alpha39}$ also appears to be specific to the myocytes, since no $G_{o\alpha39}$ was detected at any time in non-muscle heart cells [14].

The accumulation of $G_{o\alpha39}$ in cardiac myocytes may reflect what happens normally *in vivo*. Cardiac $G_{o\alpha39}$ content was determined in prenatal and neonatal rats from −5 to +10 days of age. Examination of cardiac $G_{o\alpha39}$ *in vivo* revealed that $G_{o\alpha39}$ content increased to a maximum at 2–3 days after birth and then declined rapidly by 5–6 days of age. Thus $G_{o\alpha39}$ does appear to be programmed to increase in the heart within the first three days after birth. Similar findings were reported by Asano *et al.* [6] who found an increase in $G_{o\alpha39}$ just after birth. However, in Asano's study, the $G_{o\alpha39}$ remained elevated for 10 days before gradually declining to prenatal levels by 30 days of age [6]. The difference between the results of the present study and those of Asano may be that our data is expressed per mg protein whereas Asano's data in expressed per g tissue.

The signal which slows down cardiac $G_{o\alpha39}$ expression by day 5 *in vivo* is evidently absent in cultured myocytes. The signal which regulates $G_{o\alpha39}$ expression in unknown, but may have a neural origin. Both parasympathectomy and co-culture of rat cardiac myocytes with sympathetic nerves were found to increase the amount of pertussis toxin substrate ($G_{i\alpha}$ and/or $G_{o\alpha}$) [19, 20]. Unfortunately, these studies did not employ specific probes for $G_{i\alpha}$ or $G_{o\alpha}$, so the identity of the increased pertussis toxin substrate(s) is unknown.

Interestingly, the increased expression of $G_{o\alpha39}$ observed in the present study could be prevented by culturing myocytes in the presence of agents that alter the membrane potential, such as KCl, bupivacaine, and to a lesser extent, verapamil. The potential or electrical activity of the membrane has been found to alter the expression of several other membrane-associated proteins. Thus, culturing neuroblastoma cells in the presence of the Na^+ channel activator veratridine resulted in increases in both $G_{o\alpha39}$ [21] and muscarinic acetylcholine receptors [22]. Neuroblastoma cells [22] and muscle cells [23] depolarized by KCl exhibited increases in muscarinic receptors and Na^+ channels, respectively. Nicotinic acetylcholine receptors of cultured rat skeletal muscle [24] and voltage-sensitive Na^+ channels of muscle cells [23] were increased by the Na^+ channel blockers, tetrodotoxin and bupivacaine, and were decreased by the calcium ionophore, A23187. Thus several membrane proteins appear to be regulated by electrical activity. Although the mechanism by which these agents affect the expression of membrane proteins is unknown, a change in the membrane potential has been associated with a change in the cytosolic Ca^{2+} level.

At least two studies report that the expression of the muscarinic acetylcholine receptor changes concomitantly with that of $G_{o\alpha39}$ under conditions where membrane potential is altered. Chronic membrane depolarization with veratridine resulted in increases in muscarinic receptor number and $G_{o\alpha39}$ content of neuroblastoma cells [21, 22]. Likewise, parasympathectomy of canine heart caused an increase in muscarinic receptor number accompanied by an increase in pertussis toxin sensitive G protein(s) [19]. Therefore, studies were undertaken to determine whether chronic depolarization of rat cardiac myocytes with KCl alters muscarinic receptor number concomitantly with the decrease in $G_{o\alpha39}$. Analysis of muscarinic receptor binding, however, revealed a 20% increase in muscarinic receptor density in KCl-depolarized myocytes, similar to that seen in KCl- and veratridine-depolarized neuroblastoma cells [22]. Thus the expression of muscarinic receptors in cardiac myocytes does not down-regulate coordinately with $G_{o\alpha39}$ during KCl-depolarization in cardiac cells.

The lack of coordinate expression of muscarinic receptors and $G_{o\alpha39}$ does not preclude the possibility that $G_{o\alpha39}$ may couple muscarinic receptors to intracellular effectors. Recent evidence suggests that muscarinic receptors can stimulate phospholipase C through $G_{o\alpha39}$. Injection of $G_{o\alpha39}$ into Xenopus oocytes evoked a Cl^- current by mobilizing the release of inositol triphosphate-dependent stores of intracellular calcium [25]. However, in the present study, no difference in muscarinic receptor-stimulated inositol phosphate production was observed between KCl-depolarized myocytes that contain a small amount of $G_{o\alpha39}$ and contracting myocytes that contain a 6-fold greater amount of $G_{o\alpha39}$. These findings suggest that $G_{o\alpha39}$ does not couple muscarinic receptors to phospholipase C in cardiac myocytes.

Thus the specific function of cardiac $G_{o\alpha39}$ is still unknown. Although $G_{o\alpha39}$ does not appear to couple muscarinic receptors to phospholipase C in the heart, $G_{o\alpha39}$ may couple another receptor, such as α_1-adrenergic [26] or angiotensin II [27], to the regulation of an ion channel(s). $G_{o\alpha39}$ has been shown to inhibit neuronal calcium currents through neuropeptide Y receptors [28] and activate brain potassium channels through an unidentified receptor(s) [29]. The correlation of $G_{o\alpha39}$ expression with altered membrane ion fluxes suggests that $G_{o\alpha39}$ may be coupled to an ion channel in cardiac myocytes. Further studies will be needed to determine which ion channel is acted on by $G_{o\alpha39}$ and through which receptor $G_{o\alpha39}$ is activated.

The elucidation of the signaling pathway in which $G_{o\alpha39}$ is involved has been further complicated by the discovery of a second messenger RNA encoding a $G_{o\alpha39}$-like protein. $G_{o\alpha1}$ and $G_{o\alpha2}$ are both 354 amino acids in length [30] and thus are likely to have the same electrophoretic mobility on SDS-polyacrylamide gels. The two $G_{o\alpha}$'s differ in 26 amino acids in the carboxyl terminal portion of the protein [30]. Since the antipeptide antibody used in the present study was made to a sequence near the amino terminus of $G_{o\alpha39}$, the antibody probably recognizes both $G_{o\alpha1}$ and $G_{o\alpha2}$. Therefore, it is unknown whether one or both $G_{o\alpha}$'s are found in cardiac myocytes and which one(s) are increasing with time in culture. Further studies of the $G_{o\alpha}$ signaling pathway will require the development of antibodies specific for $G_{o\alpha1}$ and $G_{o\alpha2}$.

Acknowledgements

We thank Eric Balcueva, Lois Carl, and Mike

Stashko for expert technical assistance. This work was supported by an American Heart Association Established Investigatorship award and a Pennsylvania Affiliate American Heart Association Grant to J.D. Robishaw.

References

1. Robishaw JD, Foster KA: Role of G proteins in the regulation of the cardiovascular system. Annu Rev Physiol 51: 229–244, 1989
2. Kim D, Lewis DL, Graziadei L, Neer EJ, Bar-Sagi D, Clapham DE: G-protein $\beta\gamma$-subunits activate the cardiac muscarinic K^+-channel via phospholipase A_2. Nature 337: 557–560, 1989
3. Jones LG, Goldstein D, Brown JH: Guanine nucleotide-dependent inositol trisphosphate formation in chick heart cells. Circ Res 62: 299–305, 1988
4. Mattera R, Graziano MP, Yatani A, Zhou Z, Graf R, Codina J, Birnbaumer L, Gilman AG, Brown AM: Splice variants of the α subunit of the G protein G_s activate both adenylyl cyclase and calcium channels. Science 243: 804–807, 1989
5. Codina J, Yatani A, Grenet D, Brown AM, Birnbaumer L: The α subunit of the GTP-binding protein G_k opens atrial potassium channels. Science 236: 442–445, 1987
6. Asano T, Kamiya N, Semba R, Kato K: Ontogeny of the GTP-binding protein G_o in rat brain and heart. J Neurochem 51: 1711–1716, 1988
7. Luetje CW, Tietje KM, Christian JL, Nathanson NM: Differential tissue expression and developmental regulation of guanine nucleotide binding regulatory proteins and their messenger RNAs in rat heart. J Biol Chem 263: 13357–13365, 1988
8. Mumby S, Pang IH, Gilman AG, Sternweis PC: Chromatographic resolution and immunologic identification of the α_{40} and α_{41} subunits of guanine nucleotide-binding regulatory proteins from bovine brain. J Biol Chem 263: 2020–2026, 1988
9. Mumby SM, Kahn RA, Manning DR, Gilman AG: Antisera of designed specificity for subunits of guanine nucleotide-binding regulatory proteins. Proc Natl Acad Sci USA 83: 265–269, 1986
10. Price SR, Tsai SC, Adamik R, Angus CW, Serventi IM, Tsuchiya M, Moss J, Vaughan M: Expression of $G_{o\alpha}$ mRNA and protein in bovine tissues. Biochemistry 28: 3803–3807, 1989
11. Luetje CW, Gierschik P, Milligan G, Unson C, Spiegel A, Nathanson NM: Tissue-specific regulation of GTP-binding protein and muscarinic acetylcholine receptor levels during cardiac development. Biochemistry 26: 4876–4884, 1987
12. Graziano MP, Freissmuth M, Gilman AG: Expression of $G_{s\alpha}$ in *Escherichia coli*: Purification and properties of two forms of the protein. J Biol Chem 264: 409–418, 1989
13. Yatani A, Mattera R, Codina J, Graf R, Okabe K, Padrell E, Iyengar R, Brown AM, Birnbaumer L: The G protein-gated atrial K^+ channel is stimulated by three distinct $G_{i\alpha}$-subunits. Nature 336: 680–682, 1988
14. Foster KA, McDermott PJ, Robishaw JD: Expression of G proteins in rat cardiac myocytes: effect of KCl depolarization. Am J Physiol 259: H432–H441, 1990
15. McDermott PJ, Morgan HE: Contraction modulates the capacity for protein synthesis during growth of neonatal heart cells in culture. Circ Res 64: 542–553, 1989
16. Schaffner W, Weissmann C: A rapid, sensitive, and specific method for the determination of protein in dilute solution. Anal Biochem 56: 502–504, 1973
17. Scatchard G: The attraction of protein for small molecules and ions. Ann NY Acad Sci 51: 660–672, 1949
18. Rouser G, Fleischer S, Yamamoto A: Two dimensional thin layer chromatographic separation of polar lipids and determination of phospholipids by phosphorus analysis of spots. Lipids 5: 494–496, 1970
19. Hodges TD, Bailey JC, Fleming JW, Kovacs RJ: Selective parasympathectomy increases the quantity of inhibitory guanine nucleotide-binding proteins in canine cardiac ventricle. Mol Pharmacol 36: 72–77, 1989
20. Steinberg SF, Drugge ED, Bilezikian JP, Robinson RB: Acquisition by innervated cardiac myocytes of a pertussis toxin-specific regulatory protein linked to the α_1-receptor. Science 230: 186–188, 1985
21. Luetje CW, Nathanson NM: Chronic membrane depolarization regulates the level of the guanine nucleotide binding protein $G_{o\alpha}$ in cultured neuronal cells. J Neurochem 50: 1775–1782, 1988
22. Liles WC, Nathanson NM: Regulation of muscarinic acetylcholine receptor number in cultured neuronal cells by chronic membrane depolarization. J Neurosci 7: 2558–2563, 1987
23. Sherman SJ, Catterall WA: Electrical activity and cytosolic calcium regulate levels of tetrodotoxin-sensitive sodium channels in cultured rat muscle cells. Proc Natl Acad Sci USA 81: 262–266, 1984
24. Rubin LL: Increases in muscle Ca^{2+} mediate changes in acetylcholinesterase and acetylcholine receptors caused by muscle contraction. Proc Natl Acad Sci USA 82: 7121–7125, 1985
25. Moriarty TM, Padrell E, Carty DJ, Omri G, Landau EM, Iyengar R: G_o protein as signal transducer in the pertussis toxin-sensitive phosphatidylinositol pathway. Nature 343: 79–82, 1990
26. Brown JH, Buxton IL, Brunton LL: α_1-adrenergic and muscarinic cholinergic stimulation of phosphoinositide hydrolysis in adult rat cardiomyocytes. Circ Res 57: 532–537, 1985
27. Baker KM, Singer HA: Identification and characterization of guinea pig angiotensin II ventricular and atrial receptors:

coupling to inositol phosphate production. Circ Res 62: 896–904, 1988
28. Ewald DA, Sternweis PC, Miller RJ: Guanine nucleotide-binding protein G_o-induced coupling of neuropeptide Y receptors to Ca^{2+} channels in sensory neurons. Proc Natl Acad Sci USA 85: 3633–3637, 1988
29. VanDongen AMJ, Codina J, Olate J, Mattera R, Joho R, Birnbaumer L, Brown AM: Newly identified brain potassium channels gated by the guanine nucleotide binding protein G_o. Science 242: 1433–1437, 1988
30. Hsu WH, Rudolph U, Sanford J, Bertrand P, Olate J, Nelson C, Moss LG, Boyd III AE, Codina J, Birnbaumer L: Molecular cloning of a novel splice variant of the α subunit of the mammalian G_o protein. J Biol Chem 265: 11220–11226, 1990

Address for offprints: J.D. Robishaw, Weis Center for Research, Geisinger Clinic, Danville, PA 17822, USA

Molecular and Cellular Biochemistry **104**: 73–79, 1991.

The adenylyl cyclase family

J. Krupinski
Weis Center for Research, Geisinger Clinic, Danville, PA 17822-2610, USA

Key words: signal transduction, membrane proteins, cAMP, G proteins

Abstract

Hormone-sensitive adenylyl cyclase is a model system for the study of receptor-mediated signal transduction. It is comprised of three types of components: 1) receptors for hormones that regulate cyclic AMP (cAMP) synthesis, 2) regulatory GTP binding proteins (G proteins), and 3) the family of enzymes, the adenylyl cyclases. Concentrations of cAMP are altered by at least 35 different stimulatory or inhibitory hormones and neurotransmitters. Other signalling pathways may also influence cAMP production through regulation of particular adenylyl cyclase subtypes. The second messenger, cAMP propagates the hormone signal through the effects of cAMP-dependent protein kinase.

While structural information on the adenylyl cyclases is limited, a cDNA clone for a calmodulin-sensitive form of bovine brain adenylyl cyclase has been isolated. The amino acid sequence encoded by the Type I cDNA is approximately 40% identical to those specified by three other adenylyl cyclase cDNAs that have been cloned subsequently. This degree of structural variation implies that there must be functional differences between the adenylyl cyclases.

It has been estimated that 'about 80% of all known hormones and neurotransmitters, as well as neuromodulators and autocrine and paracrine factors, elicit their cellular responses by combining with receptors which are coupled to effector functions through G proteins [1]'. These three component systems (receptor, G protein, effector) were originally defined by investigations into the hormonal regulation of adenylyl cyclase [2]. While the list of relevant effector enzymes is still growing, the paradigm for the G-protein-linked-effector system remains adenylyl cyclase. About three dozen different hormones modulate cyclic AMP concentrations through regulation of this enzyme's activity [1]. The β-receptor, G_s, adenylyl cyclase system is one of only two examples in which highly-purified protein preparations have been combined in phospholipid vesicles to reconstitute the complete three-component signal transduction chain [3, 4]. This serves as the most rigorous proof that no additional protein components are required for the generation of a signal in this system.

Adenylyl cyclase activity is both stimulated and inhibited in a hormone-dependent manner. These effects are mediated by distinct, heterotrimeric G proteins: G_s (for *s*timulatory) and G_i (for *i*nhibitory). A number of observations are summarized in the following description of the mechanism by which adenylyl cyclase is regulated [5]. The binding of hormone to its receptor enhances the exchange of bound GDP for GTP at the nucleotide binding site on the G protein. This promotes subunit dissociation of the heterotrimeric G protein into separate $G_\alpha \cdot$ GTP and βγ complexes. The $G_\alpha \cdot$ GTP complexes are capable of binding to adenylyl cyclase and exerting direct effects on its catalytic activity, although for $G_{i\alpha} \cdot$ GTP inhibition is weak. Inhibition is predominantly mediated by the βγ pool that is generated upon activation of G_i by nucleotide binding. The activity of cyclase remains altered as long as GTP is intact. Hydrolysis to GDP at the G_α nucleotide binding site returns cyclase to its basal state and results in the release of $G_\alpha \cdot$ GDP. Free βγ may then serve as a sink to complex the free $G_\alpha \cdot$ GDP and regenerate the initial state. While this scheme has not been proven rigorously, it is a testable working model consistent with the available data.

The ubiquitous occurrence of adenylyl cyclase activity is consistent with its involvement in a variety of important physiological processes. Adenylyl cyclase activity was originally discovered by investigations of the mechanism by which sympathomimetic amines and glucagon regulate glycogen metabolism in liver [6]. Cyclic AMP synthesized by adenylyl cyclase activates cAMP-dependent protein kinase. The latter enzyme then initiates a cascade of phosphorylation reactions that results in glycogen degradation. This is only one example of the general mechanism in which the activation of adenylyl cyclase results in changes in cellular biochemistry mediated through the phosphorylation of key proteins by cAMP-dependent protein kinase [7]. Multiple proteins may be phosphorylated by this kinase in a given cell type and the particular substrates for phosphorylation can differ depending on the tissue. In adipocytes the hormone-dependent stimulation of adenylyl cyclase eventually results in the release of free fatty acids by the pathway involving cAMP-dependent protein kinase [8]. Activation of adenylyl cyclase by β-adrenergic agents in heart leads to the phosphorylation of phospholamban [9] and voltage-dependent Ca^{2+} channels [10]. The subsequent effects on Ca^{2+} influx have been proposed, at least in part, to account for correlations between β-adrenergic-stimulated cAMP accumulation, increased rate of relaxation, and increased force of contraction in heart. Even adenylyl cyclase itself may be regulated by the kinase. Direct phosphorylation of adenylyl cyclase by cAMP-dependent protein kinase may be partially responsible for the heterologous desensitization of glucagon-stimulated adenylyl cyclase activity in hepatocytes [11]. This type of feedback mechanism is appealing as a means of regulating adenylyl cyclase activity in the presence of a persistent stimulus. Numerous other examples could be given illustrating the physiological effects of the activation of cAMP-dependent protein kinase through the stimulation of adenylyl cyclase.

However, not all hormones capable of stimulating adenylyl cyclase in a given cell type will have the same effects on cAMP-dependent protein kinase. A compartmentalization of cAMP within cardiac myocytes has been postulated from the differential ability of certain hormones to activate cAMP-dependent kinase. Brunton and coworkers have found that both isoproterenol and prostaglandin E_1 enhance cAMP accumulation in cardiac myocytes, but only the former agonist stimulates the breakdown of glycogen [12]. Only the activation of the particulate cAMP-dependent protein kinase can mediate activation of phosphorylase [13]. This effect has been correlated with the selective phosphorylation of proteins that occurs in response to isoproterenol, but not the prostaglandin [14].

Cross-talk with other signalling pathways can also modulate the effects of adenylyl cyclase. A 130 kDa species that co-purified with adenylyl cyclase activity was phosphorylated when frog erythrocytes were treated with the protein kinase C activator, 12-O-tetradecanoyl phorbol-13-acetate [15]. A sensitization of the enzyme activity accompanies the phosphorylation [15]. *In vivo* this effect could be mediated by hormones capable of activating protein kinase C through the generation of diacyl-glycerol from the hydrolysis of phosphatidylinositol [4, 5]-bisphosphate. However, the latter reaction will also generate inositol trisphosphate which will cause the release of calcium from internal stores. Calcium may also influence adenylyl cyclase activity by complexing with calmodulin and activating other pathways which can modulate adenylyl cyclase activity [16], or adenylyl cyclase may be stimulated directly by Ca^{2+}/calmodulin in some tissues [17]. The activity of adenylyl cyclase purified from liver is insensitive to Ca^{2+}/calmodulin [11], while that from brain is stimulated 8–10 fold by Ca^{2+}/calmodulin relative to the activity observed in the presence of Mg^{2+} alone [18]. The precise role of calmodulin-sensitive adenylyl cyclase in mammalian brain has not been defined. However, in *Drosophila*, a series of learning mutants have been discovered that exhibit defects in responses involving cyclic nucleotides [19]. In particular the mutant, *rutabaga*, fails a negatively reinforced test of associative learning [20]. Enzyme activity assays performed on body parts from this mutant fly indicate a loss of Ca^{2+}/calmodulin responsive adenylyl cyclase [20]. The site of the *rutabaga* mutation has been mapped to a specific locus on the X chromosome in the *Drosophila* genome [21]. A cDNA for a calmodulin-sensitive form of bovine brain adenylyl cyclase cross-hybridizes to the *rutabaga* locus in preparations of *Drosophila* polytene chromosomes (Randall R. Reed, unpublished observation). The mechanism by which adenylyl cyclase activity may be translated into associative learning remains obscure.

Although essential for the complete characterization of the physiological role of adenylyl cyclase, information on the detailed structure of the enzyme is limited. This reflects the relatively low abundance (0.001–0.01% of *membrane* protein) as well as the instability of the enzyme. Monoclonal antibodies raised against purified adenylyl cyclase from bovine brain indicate that more than one member of this enzyme family may be expressed in a given tissue [22]. Immunoblots of solubilized membranes from bovine cerebral cortex indicate that three distinct species of 115, 150, and 160 kDa react with the antibody [22]. In immunoblots of either heart or lung only the 150 kDa form is detected [22]. Olfactory cilia possess a 180 kDa form of adenylyl cyclase that cross-reacts weakly with the antibody raised to the brain preparation [23]. At least a portion of the molecular weight heterogeneity may be explained by differential post-translational modifications of particular adenylyl cyclase subtypes. A cDNA clone encoding a calmodulin-sensitive form of bovine brain adenylyl cyclase specifies a protein of 1134 amino acids with one potential N-linked glycosylation site that is exposed to the extracellular milieu [24]. Based on protein biochemistry this corresponds to the form with an apparent molecular weight of 115,000 that is detected with the monoclonal antibody [22, 24]. A cDNA clone encoding the prominent olfactory adenylyl cyclase specifies a protein of 1144 amino acids with three potential N-linked glycosylation sites [25]. Treatment of olfactory cilia with a glycosidase reduces the apparent molecular weight of this adenylyl cyclase from approximately 200,000 to 129,000, in agreement with the molecular weight determined from the cDNA [25].

Biochemical studies have revealed that adenylyl cyclase activities can be divided into different classes depending on subcellular localization (membrane-bound vs. soluble activity), and their responses to a variety of regulators. For example, spermatozoa possess a soluble, calmodulin-sensitive activity that is not stimulated by activated G_s [26]. They also possess a membrane-bound activity insensitive to G_s [27]. Brain has both a calmodulin-sensitive and a calmodulin-insensitive adenylyl cyclase activity, each of which can be stimulated by manganese, activated G_s, and forskolin [22, 28]. Some forms of adenylyl cyclase are sensitive to mechanical forces [29]. The complete range of sensitivities that may be exhibited by distinct adenylyl cyclases is not yet clear. One way to examine this would be to purify different adenylyl cyclases and characterize their activities in isolation.

The diterpene, forskolin, directly activates adenylyl cyclase in the absence of G_s [30]. Pfeuffer's laboratory demonstrated that myocardial adenylyl cyclase activity could be purified over 2000-fold in a single step by chromatography on an affinity resin synthesized by coupling forskolin to Sepharose [31]. Forskolin-Sepharose chromatography has been combined with chromatography on ion exchange resins [11, 31], wheat germ agglutinin agarose [32, 28, 11], or calmodulin-Sepharose [28], to obtain highly-purified preparations of adenylyl cyclase.

The characteristics of the final forskolin-eluate differ depending on the source of the starting material. This is illustrated by comparing forskolin eluate obtained from heart and brain, the sources of the two best characterized preparations of highly purified enzyme. 1) Myocardial adenylyl cyclase can only be obtained in significant yield by forskolin-Sepharose chromatography if the enzyme is first preactivated with G_s prior to adsorption to the resin [33]. A preparation of adenylyl cyclase not preactivated by G_s, is readily purified from brain. This has been interpreted as a reflection of increased stability of the brain enzyme [34] and a synergistic activation of myocardial cyclase by the combination of forskolin and G_s [31]. This synergism has also been observed and analyzed in platelet membranes [35]. 2) The myocardial enzyme has an apparent molecular weight of 150,000 while that from brain appears to be about 115,000 [31, 34]. Thus one of the species observed in immunoblots of membranes from bovine cerebral cortex is selectively enriched by forskolin-Sepharose chromatography of brain membrane extracts [31, 34, 22]. The 150 kDa form from brain may be purified by chromatography on forskolin-Sepharose if the membranes are pre-activated with guanine nucleotide prior to detergent extraction of the protein [22]. 3) Forskolin eluate from brain is calmodulin-sensitive while that purified from heart is not [33], despite evidence for a calmodulin-sensitive adenylyl cyclase activity in myocardial membranes [36]. However, more than one calmodulin-sensitive adenylyl cyclase exists. The olfactory adenylyl cyclase can be stimulated by Ca^{2+}/calmodulin albeit with a much lower affinity than the calmodulin-sensitive form from brain [37]. It is possible that a distinct calmodulin-sensitive subtype may be found in heart.

A cDNA for a calmodulin-sensitive form of adenylyl cyclase has been isolated by starting with an oligonucleotide probe based on a peptide sequence obtained from the purified protein [24]. This form of the enzyme was named Type I adenylyl cyclase when it became clear that a cDNA for another form was present in the same brain library [24]. The structure predicted for the Type I enzyme based on the amino acid sequence is surprising considering that the only known function of adenylyl cyclase is to synthesize cAMP from ATPat the intracellular face of the membrane. The protein is predicted to have a short amino-terminal cytoplasmic tail followed by two large alternating sets of hydrophobic and hydrophilic domains. Each hydrophobic domain is divisible into six transmembrane spans. The two large hydrophilic regions are 30% identical (55% similar) to one another over a 200 amino acid region, and approximately 55% similar to the catalytic domains of the family of guanylyl cyclases [24, 38]. Based on this similarity, each of the two large hydrophilic regions within adenylyl cyclase has been proposed to encode a nucleotide binding domain. This general structure is strikingly reminiscent of that proposed for the product of the multidrug resistance gene [39], the product of the cystic fibrosis gene [40], and the product of the yeast *ste* 6 gene [41]. Interestingly, each of these proteins has been implicated in a transport process. The voltage-sensitive channels, another class of G-protein-regulated entities, are also organized in sets of six transmembrane spans [42], though they lack putative nucleotide binding domains. How this channel- or transporter-like structure contributes to adenylyl cyclase function is not clear.

The unanticipated protein structure of Type I adenylyl cyclase is not unique to this particular form of the enzyme. Thus far four adenylyl cyclases have been cloned from higher eukaryotes, and all are predicted to have the same general topography [24; 25; Randall R. Reed *et al.*; Boning Gao and Alfred G. Gilman, unpublished observations). However, their primary sequences vary significantly. This is best illustrated by a more detailed comparison of the amino acid sequences of two adenylyl cyclases, Type I and the prominent olfactory form, Type III.

The Type III cDNA was isolated from a rat olfactory library utilizing probes derived from both Type I and II cDNAs [25]. Figure 1 shows the percentage of amino acid identity calculated for the different domains within the common overall structure. The observed sequence diversity most likely is not the result of species differences since clones for both bovine and rat Type II have been isolated, and they show over 90% amino acid identity (Randall R. Reed *et al.*, unpublished observations). No significant amino acid similarity is found in a comparison of the short amino-terminal tails. The regions of highest overall identity between the cyclases are contained within the large cytoplasmic domains (Fig. 1). These correspond to the regions of over 200 amino acids which show a significant similarity to each other and to the catalytic domain of the guanylyl cyclases [24, 38]. However, the similarity between the same structural domain in two different cyclases is much greater than the similarity between the distinct cytoplasmic domains of a single cyclase. For example the *central* domain of

Type I more closely resembles the *central* domain of Type III, (53% identical) while the *central* domain of Type I is more distantly related to its own *carboxy* terminal domain (about 30% identical). This suggests that each of these domains may have its own unique nucleotide binding properties. Studies on the kinetics of inhibition of adenylyl cyclase by adenosine also are consistent with the existence of distinct nucleotide binding sites with only one catalytic domain [43]. Thus one large cytoplasmic domain may be the catalytic site while the other may be a regulatory domain. Alternatively each nucleotide binding site might be shared between the two large cytoplasmic domains.

If the putative nucleotide binding regions are excluded, the remainder of the molecules show only a limited degree of amino acid identity. A region extending over three hundred amino acids is only approximately 19% identical in the two forms. This includes over one hundred amino acids of hydrophilic sequence beginning from the end of the first putative nucleotide binding domain and continuing through more than two hundred amino acids of the second large hydrophobic domain (Fig. 1). Following the second putative nucleotide binding domain there is another region exhibiting minimal amino acid identity. When the complete amino acid sequences of Types I and III are aligned, they are found to be only 38% identical overall. Type I exhibits approximately 40% amino acid identity to any of the three other cloned subtypes. Only twelve stretches of five or more identical amino acids are observed in the overall sequence alignment of Types I and III. The longest of these is only 14 amino acids. To put this into perspective consider the sequence variation among other components of G-protein linked systems. M1 muscarinic receptors are linked to the release of calcium through the generation of inositol phosphates. M2 receptors are coupled to the hormone-dependent inhibition of adenylyl cyclase activity. M1 receptors are 43% identical to M2 receptors [44]. $G_{s\alpha}$, which mediates the hormone-dependent stimulation of adenylyl cyclase is about 40% identical to the members of the $G_{i\alpha}$ family, which mediate inhibition of adenylyl cyclase [45]. The three $G_{i\alpha}$ proteins are at least 85% identical to each other [45]. The olfactory specific G_α protein, G_{olf}, which is believed to mediate stimulation of the olfactory adenylyl cyclase, is 88% identical to the more widely distributed $G_{s\alpha}$ [46]. Considering that adenylyl cyclases are all assumed to have the same function, it was not anticipated that their sequences would vary as widely as proteins with distinct, or even antagonistic functions. The complete extent of the amino acid variation among the adenylyl cyclases requires an expla-

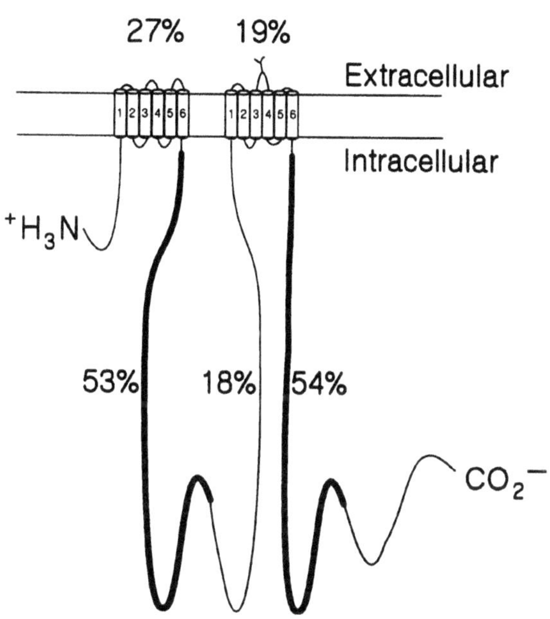

Fig. 1. A general comparison of the amino acid sequences of Types I and III adenylyl cyclase. Analysis of the amino acid sequences encoded by the cDNAs of both Types I and III adenylyl cyclase indicates that they share the indicated common topography [24, 25]. The location of the putative nucleotide binding domains has been highlighted with bold lines. Putative transmembrane spans are indicated as cylinders within the membrane. Only the one glycosylation site that is located in the same extracellular loop of both Types I and III adenylyl cyclase has been indicated. Type III has two additional glycosylation sites [25]. Individual domains within the common overall structure were aligned at the amino acid level with the program Gap using the default parameters [47]. The percentage of amino acid identity calculated from the alignments is indicated adjacent to the corresponding domain.

nation in terms of distinct molecular structures and their associated functions.

Acknowledgements

I would like to thank Alfred G. Gilman, Randall R. Reed and all of the members of their laboratories for communicating results prior to publication.

References

1. Iyengar R, Birnbaumer L: G Proteins. Academic Press, New York, 1990, pp 1–14
2. Gilman AG: G proteins and regulation of adenylyl cyclase. JAMA 262: 1819–1825, 1989
3. May DC, Ross EM, Gilman AG, Smigel MD: Reconstitution of catecholamine-stimulated adenylate cyclase activity using three purified proteins. J Biol Chem 260: 15829–15833, 1985
4. Feder D, Im M-J, Klein HW, Hekman M, Holzhofer A, Dees C, Levitzki A, Helmreich EJM, Pfeuffer T: Reconstitution of β_1-adrenoceptor-dependent adenylate cyclase from purified components. EMBO J 5: 1509–1514, 1986
5. Gilman AG: G proteins: transducers of receptor-generated signals. Ann Rev Biochem 56: 615–649, 1987
6. Rall TW, Sutherland EW, Berthet J: The relationship of epinephrine and glucagon to liver phosphorylase. IV. Effect of epinephrine and glucagon on the reactivation of phosphorylase in liver homogenates. J Biol Chem 224: 463–475, 1957
7. Krebs EG: Role of cyclic AMP-dependent protein kinase in signal transduction. JAMA 262: 1815–1818, 1989
8. Corbin JD, Reimann EM, Walsh DA, Krebs EG: Activation of adipose tissue lipase by skeletal muscle adenosine 3′, 5′-monophosphate-stimulated protein kinase. J Biol Chem 245: 4849–4851, 1970
9. Tada M, Katz AM: Phosphorylation of the sarcoplasmic reticulum and sarcolemma. Annu Rev Physiol 44: 401–423, 1982
10. Azuma J, Sawamura A, Harada H, Tanimoto T, Ishiyama T, Morita Y, Yamamura Y, Sperelakis N: Cyclic adenosine monophosphate modulation of contractility via slow Ca^{2+} channels in chick heart. J Mol Cell Cardiol 13: 577–587, 1981
11. Premont RT, Iyengar R: Adenylyl cyclase and its regulation by G_s. In: R Iyengar, L Birnbaumer (eds) G Proteins. Academic Press, New York, 1990, pp 147–179
12. Bode DC, Brunton LL: Post-receptor modulation of the effects of cyclic AMP in isolated cardiac myocytes. Mol and Cell Biochem 82: 13–18, 1988
13. Buxton ILO, Brunton LL: Compartments of cyclic AMP and protein kinase in mammalian cardiomyocytes. J Biol Chem 258: 10233–10239, 1983
14. Hayes JS, Bowling N, King KL, Boder GB: Evidence for selective regulation of the phosphorylation of myocyte proteins by isoproterenol and prostaglandin E_1. Biochim et Biophys Acta 714: 136–142, 1982
15. Yoshimasa T, Sibley DR, Bouvier M, Lefkowitz RJ, Caron MG: Cross-talk between cellular signalling pathways suggesting by phorbol-ester-induced adenylate cyclase phosphorylation. Nature 327: 67–70, 1987
16. Felder CC, Kanterman RY, Ma AL, Axelrod J: A transfected ml muscarinic acetylcholine receptor stimulates adenylate cyclase via phosphatidylinositol hydrolysis. J Biol Chem 264: 20356–20362, 1989
17. Minocherhomjee AM, Shattuck RL, Storm DR: Calmodulin-stimulated adenylate cyclases. In: P Cohen and CB Klee (eds) Calmodulin. Elsevier Science Publishers, New York, 1988, pp 249–263
18. Yeager RE, Heideman W, Rosenberg GR, Storm DR: Purification of the calmodulin-sensitive adenylate cyclase from bovine cerebral cortex. Biochem 24: 3776–3783, 1985
19. Kiger JA, Salz HK: Cyclic nucleotide metabolism and physiology of the fruit fly Drosophila melanogaster. Adv Insect Physiol 18: 141–179, 1985
20. Livingstone MS, Sziber PP, Quinn WG: Loss of calcium/calmodulin responsiveness in adenylate cyclase of rutabaga, a Drosophila learning mutant. Cell 37: 205–215, 1984
21. Livingstone MS: Genetic dissection of Drosophila adenylate cyclase. Proc Natl Acad Sci USA 82: 5992–5996, 1985
22. Mollner S, Pfeuffer T: Two different adenylyl cyclases in brain distinguished by monoclonal antibodies. Eur J Biochem 171: 265–271, 1988
23. Pfeuffer E, Mollner S, Lancet D, Pfeuffer T: Olfactory adenylyl cyclase: Identification and purification of a novel enzyme form. J Biol Chem 264: 18803–18807, 1989
24. Krupinski J, Coussen F, Bakalyar HA, Tang W-J, Feinstein PG, Orth K, Slaughter C, Reed RR, Gilman AG: Adenylyl cyclase amino acid sequence: possible channel- or transporter-like structure. Science 244: 1558–1564, 1989
25. Bakalyar HB, Reed RR: Identification of specialized adenylyl cyclase that may mediate odorant detection. Science, 250: 1403–1406, 1990
26. Gross MK, Toscano DG, Toscano WA: Calmodulin-mediated adenylate cyclase from mammalian sperm. J Biol Chem 262: 8672–8676, 1987
27. Hildebrandt JD, Codina J, Tash JS, Kirchick HJ, Lipschultz L, Sekura RD, Birnbaumer L: The membrane-bound spermatozoal adenylyl cyclase system does not share coupling characteristics with somatic cell adenylyl cyclases. Endocrinology 116: 1357–1366, 1985
28. Rosenberg GB, Storm DR: Immunological distinction between calmodulin-sensitive and calmodulin-insensitive adenylate cyclase. J Biol Chem 262: 7623–7628, 1987
29. Watson PA: Direct stimulation of adenylate cyclase by mechanical forces in S49 mouse lymphoma cells during hyposmotic swelling. J Biol Chem 265: 6569–6575, 1990

30. Seamon KB, Daly JW: Activation of adenylate cyclase by the diterpene forskolin does not require the guanine nucleotide regulatory protein. J Biol Chem 256: 9799–9801, 1981
31. Pfeuffer E, Drehev R-M, Metzger H, Pfeuffer T: Catalytic unit of adenylate cyclase: purification and identification by affinity crosslinking. Proc Natl Acad Sci USA 82: 3086–3090, 1985
32. Smigel MD: Purification of the catalyst of adenylate cyclase. J Biol Chem 261: 1976–1982, 1986
33. Pfeuffer T, Metzger H: 7-O-Hemisuccinyl-deacetyl forskolin-Sepharose: a novel affinity support for purification of adenylate cyclase. 1982. FEBS Lett 146: 369–375, 1982
34. Pfeuffer E, Mollner S, Pfeuffer T: Adenylate cyclase from bovine brain cortex: purification and characterization of the catalytic unit. EMBO J 4: 3675–3679, 1985
35. Nelson CA, Seamon KB: Binding of [^{3}H]forskolin to human platelet membranes. J Biol Chem 261: 13469–13473, 1986
36. Panchenko MP, Tkachuk VA: Calmodulin activates adenylate cyclase from rabbit heart plasma membranes. FEBS Lett 174: 50–53, 1984
37. Anholt RRH, Rivers AM: Olfactory Transduction: Cross-talk between second-messenger systems. Biochem 29: 4049–4054, 1990
38. Chinkers M, Garbers DL: The protein kinase domain of the ANP receptor is required for signaling. Science 245: 1392–1394, 1989
39. Gottesman MM, Pastan I: The multidrug transporter, a double-edged sword. J Biol Chem 263: 12163–12166, 1988
40. Riordan JR, Rommens JM, Kerem B-S, Alon N, Rozmahel R, Grzelczak Z, Zielenski J, Lok S, Plavsic N, Chou J-L, Drumm ML, Iannuzzi MC, Collins FS, Tsui L-C. Identification of the cystic fibrosis gene: cloning and characterization of complementary DNA. Science 245: 1066–1073, 1989
41. McGrath JP, Varshavsky A: The yeast STE6 gene encodes a homologue of the mammalian multidrug resistance P-glycoprotein. Nature 340: 400–404, 1989
42. Catterall WA: Structure and function of voltage-sensitive ion channels. Science 242: 50–61, 1988
43. Johnson RA, Shoshani I: Kinetics of 'P'-site-mediated inhibition of adenylyl cyclase and the requirements for substrate. J Biol Chem 265: 11595–11600, 1990
44. Dohlman HG, Caron MG, Lefkowitz RJ: A family of receptors coupled to guanine nucleotide regulatory proteins. Biochemistry 26: 2657–2664, 1987
45. Jones DT, Reed RR: Molecular cloning of five GTP-binding protein cDNA species from rat olfactory neuroepithelium. J Biol Chem 262: 14241–14249, 1987
46. Jones DJ, Reed RR: G_{olf}, an olfactory neuron specific G protein involved in odorant signal transduction. Science 244: 790–795, 1989
47. Devereux J, Haeberli P, Smithies O: A comprehensive set of sequence analysis programs for the VAX. Nucleic Acids Res 12: 387–395, 1984

Address for offprints: J. Krupinski, Weis Center for Research, Geisinger Clinic, Danville, PA 17822-2610, USA

Molecular and Cellular Biochemistry **104**: 81–86, 1991.

Cell signalling through phospholipid breakdown

J.H. Exton, S.J. Taylor, G. Augert and S.B. Bocckino
Howard Hughes Medical Institute, Department of Molecular Physiology and Biophysics and Department of Pharmacology, Vanderbilt University School of Medicine, Nashville, TN 37232, USA

Key words: phosphoinositides, phosphatidylcholine, phospholipases, diacylglycerol, phosphatidate, calcium

Abstract

There is much evidence that G-proteins transduce the signal from receptors for Ca^{2+}-mobilizing agonists to the phospholipase C that catalyzes the hydrolysis of phosphoinositides. However, the specific G-proteins involved have not been identified. We have recently purified a 42 kDa protein from liver that activates phosphoinositide phospholipase C and cross-reacts with antisera to a peptide common to G-protein α-subunits. It is proposed that this protein is the α-subunit of the G-protein that regulates the phospholipase in this tissue.

Ca^{2+}-mobilizing agonists and certain growth factors also promote the hydrolysis of phosphatidylcholine through the activation of phospholipases C and D in many cell types. This yields a larger amount of diacylglycerol for a longer time than does the hydrolysis of inositol phospholipids. Consequently phosphatidylcholine breakdown is probably a major factor in long-term regulation of protein kinase C. The functions of phosphatidic acid produced by phospholipase D are speculative, but there is evidence that it is a major source of diacylglycerol in many cell types. The regulation of phosphatidylcholine phospholipases is multiple and involves direct activation by G-proteins, and regulation by Ca^{2+}, protein kinase C and perhaps growth factor receptor tyrosine kinases.

Introduction

It is widely accepted that many hormones, neurotransmitters and related agonists exert some of their biological effects by stimulating the hydrolysis of phosphatidylinositol 4,5-bisphosphate (PIP_2) a minor phospholipid in the plasma membrane of their target cells. This produces two signalling molecules: inositol trisphosphate (IP_3) which releases intracellular Ca^{2+}, and diacylglycerol (DAG) which activates protein kinase C [1]. There is much indirect evidence that the phospholipase C that catalyzes PIP_2 breakdown is regulated by a guanine nucleotide-binding regulatory protein (G-protein) that is activated by agonist occupation of the relevant plasma membrane receptors [2]. In addition to elevating cytosolic Ca^{2+} through the release of Ca^{2+} from intracellular stores, the agonists also promote Ca^{2+} influx across the plasma membrane. This is thought to occur through the opening of Ca^{2+} channels, perhaps through the direct or indirect mediation of a G-protein, although other mechanisms are possible [2].

Recent studies have indicated that many of the agonists that stimulate PIP_2 hydrolysis also promote phosphatidylcholine (PC) breakdown through the activation of phospholipases of the C and D type to yield DAG, phosphatidic acid (PA), choline and phosphocholine [3]. Although the physiological significance of PC breakdown is far from clear, it is probably a major factor in the elevation of PA and DAG and the activation of protein kinase C induced by many agonists in various cell types [3]. This article will review our investigations into the molecular mechanisms involved in agonist-stimulated PIP_2 and PC breakdown.

Phosphatidylinositol 4,5-bisphosphate breakdown

The Ca^{2+}-mobilizing agonist vasopressin induces a large decrease in PIP_2 within a few seconds in isolated rat hepatocytes, whereas phosphatidylinositol 4-phosphate (PIP) does not change significantly and phosphatidylinositol (PI) decreases much more slowly [4]. The rapid breakdown of PIP_2 is associated with a parallel increase in IP_3 [3], indicating the activation of a phospholipase C. There are slower changes in other inositol phosphates, representing the further metabolism of IP_3 to inositol 1,4-bisphosphate and inositol 1, 3, 4, 5-tetrakisphosphate. Other Ca^{2+}-mobilizing agonists such as α_1-adrenergic agonists, angiotensin II, and P_{2y} purinergic agonists induce similar changes.

Experiments utilizing isolated rat liver plasma membranes have shown that direct addition of GTP and its stable analogues activates a phospholipase C that acts on endogenous and exogenous PIP_2 [5, 6]. This activation is produced by submicromolar concentrations of the GTP analogues, but not other nucleotides, requires millimolar Mg^{2+} and is competitively inhibited by a stable analogue of GDP. Addition of Ca^{2+}-mobilizing agonists in the presence, but not the absence, of low concentrations of GTP analogues also activates the phospholipase. Further evidence that Ca^{2+}-mobilizing receptors are linked to the phospholipase by a G-protein(s) is provided by the observations that GTP and its analogues reduce the binding of agonists to these receptors [7], that Ca^{2+}-mobilizing agonists stimulate a low K_m GTPase activity in liver plasma membranes [8] and that AlF_4-an activator of G-proteins stimulates IP_3 formation and Ca^{2+} mobilization in hepatocytes [9].

It is now evident that most tissues contain several phosphoinositide phospholipase C activities [10]. These all hydrolyze PI, PIP and PIP_2 in a Ca^{2+}-dependent manner, but PIP_2 is the preferred substrate at low Ca^{2+} concentrations. There are five major forms (α–ε) which have different molecular weights and can be distinguished immunologically. The phospholipase C isozyme responsible for G-protein-stimulated PIP_2 hydrolysis in liver has been identified as the β_1 isozyme.

Recent work in our laboratory has identified the G-protein that regulates PIP_2 phospholipase C in rat and bovine liver [11]. This was possible because treatment of plasma membranes from these tissues resulted in the formation of an activator of the phospholipase which did not lose activity during purification through a series of chromatographic steps. The final purification step involved HPLC on Mono Q ion exchange resin after ADP-ribosylation with pertussis toxin. This procedure resulted in chromatographic separation of the activator from contaminating G_i α-subunits. Silver stained gels of the final preparation revealed proteins with molecular weights of 42K and 35K. The 42 kDa protein was recognized by an antiserum raised to a peptide common to many G-protein α-subunits, and the 35 kDa protein was recognized by an antiserum to a G-protein β-subunit peptide. The concentration of the 42 kDa protein in the final fractions correlated with the magnitude of the phospholipase activation.

Phosphatidylcholine breakdown

One of the first pieces of evidence indicating that Ca^{2+}-mobilizing agonists stimulate the breakdown of another phospholipid besides PIP_2 came from chemical measurements of the DAG accumulating in response to vasopressin in rat hepatocytes [12]. These showed a gross discrepancy between the time course of DAG accumulation and that of IP_3 formation. Furthermore, fatty acid analysis of the DAG indicated that it came from another source and that this was probably PC [12, 13]. More refined HPLC analyses of the molecular species of DAG [14] and chemical measurements of the changes in inositol phospholipids, DAG and PA [4, 12, 13] have confirmed this conclusion. More direct proof of PC breakdown to DAG in hepatocytes was provided by experiments in which PC was selectively labelled utilizing [^{3}H]alkyl-lyso-glycerophosphocholine. In these cells, vasopressin stimulated the production of [^{3}H]alkyl-acylglycerol [14]. Similar data were obtained when PC was labelled with [^{14}C]lyso-PC and [^{14}C]DAG formation was measured [14]. There was also a stimulation of the release of [^{3}H]phosphocholine when PC was previously labelled by incubating the cells with [^{3}H]choline for 90 min. In hepatocytes and other cell types, there is now evidence that PC is broken down to PA by a phospholipase D activity as well as to DAG by a phospholipase C activity [3]. The activation of phospholipase D accounts for the rapid formation of PA observed in many cells stimulated by agonists [3]. The ability of agonists to activate phospholipase D in many cell types is also demonstrated by the formation of phosphatidylethanol when the cells are incubated with agonists in the presence of ethanol [3]. The production of phosphatidylethanol is due to the transphosphatidylation activity of phospholipase D.

Mechanisms of hormonal stimulation of phosphatidylcholine breakdown

Evidence for three mechanisms of hormonal stimulation of PC breakdown has been presented [3]. The first is G-protein-mediated stimulation of PC phospholipase C and D activities. The second is protein kinase C-mediated activation of these phospholipases. The third is Ca^{2+} stimulation of the phospholipases. In addition, certain growth factors cause PC hydrolysis presumably as a direct or indirect result of the activation of receptor tyrosine kinases.

Evidence for control of PC breakdown through a G-protein(s) comes from studies with isolated rat liver plasma membranes. These show that addition of GTP analogues stimulates the production of DAG, PA, choline and phosphocholine [13, 15]. Since there is an associated decrease in PC, the data are consistent with the activation of PC breakdown by phospholipase C and D. The effect is observed with micromolar or submicromolar concentrations of GTP analogues and is inhibited by a stable GDP analogue [15]. P_2-purinergic agonists (ATP and ADP) also stimulate PC breakdown in liver plasma membranes, but only in the presence of a GTP analogue [13, 15]. Neither cholera nor pertussis toxin alters the hydrolysis of PC [15] indicating that the putative G-protein is not a substrate for these toxins.

GTP analogues stimulate PC phospholipase D activity in permeabilized pulmonary artery endothelial cells [16] and fibroblasts [17] and the stimulating effect of the nucleotides can be enhanced by P_2-purinergic and cholinergic agonists in the respective systems. These data suggest that the phospholipase can be controlled by G-proteins in these tissues also.

Evidence that protein kinase C controls PC breakdown comes from studies in which tumor promoting phorbol esters and synthetic DAGs promote the release of choline or phosphocholine from cells or stimulate the breakdown of [^{3}H]-alkyl-PC or alkyl-[^{32}P]PC to labeled DAG or PA [for references, see 3, 16, 17, 19–28]. In some systems, inhibitors of protein kinase C have been shown to partly or fully block the effects of agonists and phorbol esters [22–25, 29] and down-regulation of the enzyme has been demonstrated to attenuate the response [19, 22, 25, 27–29].

Since almost all the agonists that elicit PC breakdown also promote rapid PIP_2 hydrolysis, it is possible that the initial transient increase in DAG arising from PIP_2 acts as a trigger for PC breakdown through activation of protein kinase C. The resultant PC hydrolysis would generate more DAG providing a positive feedback to maintain elevated levels of DAG and continued protein kinase C activation for prolonged physiological effects. Unlike the phosphoinositide signalling system, agonist-induced PC hydrolysis does not seem to undergo negative feedback.

Studies in several cell types have indicated that an elevation in cytosolic Ca^{2+} can induce PC breakdown. For example, the Ca^{2+} ionophore A23187 elicits the conversion of [^{3}H]alkyl-PC to [^{3}H]alkyl-acyl-glycerol or [^{3}H]alkyl-PA in hepatocytes and neutrophils [14, 30] and Ca^{2+} ionophores promote PC breakdown in other cells [20, 22, 24]. Furthermore, the effects of agonists on DAG formation or PC breakdown in several systems are partly dependent upon the cellular content of Ca^{2+} [12, 14,

19, 31, 32]. These observations raise the possibility that part of the stimulatory action of Ca^{2+}-mobilizing agonists on PC breakdown in some cells is due to the stimulation of phospholipases by the increase in cytosolic Ca^{2+}. It remains unknown whether or not Ca^{2+} acts directly on the enzymes or by a mechanism involving a Ca^{2+}-binding protein e.g. calmodulin and/or a protein kinase.

Several growth factors and cytokines (epidermal growth factor, platelet-derived growth factor, interleukins 1 and 3, γ-interferon, α-thrombin) stimulate PC hydrolysis in fibroblasts and other cell types [for references, see 3, 28, 33–35]. This can result in a very sustained elevation of DAG. For example, in IIC9 fibroblasts stimulated with epidermal growth factor DAG is increased for at least 4 h [34]. The prolonged nature of the response and the fact that its concentration-dependence is similar to that for DNA synthesis [34], suggests that PC hydrolysis may be important for the mitogenic effect of the growth factor.

There have been no studies of the mechanism(s) by which growth factors stimulate PC hydrolysis. In those cells in which the growth factors induce a transient breakdown of phosphoinositides, the response could be secondary to the activation of protein kinase C, as discussed above. Alternatively, stimulation of growth factor receptor tyrosine kinases could activate PC phospholipases more directly, as found for the γ-isozyme of PIP_2 phospholipase C [10].

Discussion

Although we have purified the α-subunit of the G-protein that mediates the action of Ca^{2+}-mobilizing agonists on PIP_2 phospholipase C in liver and shown that it is a novel 42 kDa protein, it is clear that at least one other G-protein is involved in other tissues. This is because in some cell types (neutrophils, mast cells, mesangial cells) the G-protein is sensitive to pertussis toxin whereas in others (hepatocytes, adrenal, thyroid and pituitary cells) it is not [3]. Immunological studies (unpublished) show that the 42 kDa protein is distinct from any presently purified or expressed G-protein, but it is possible that it is identical to one for which a DNA sequence is available e.g. α_q.

It is becoming clear that many hormones, neurotransmitters and growth factors stimulate PC breakdown in addition to PIP_2 hydrolysis in a wide variety of cells. An intriguing discovery is that it involves both phospholipases C and D, with the production of DAG and PA. The mechanisms by which various agonists activate these phospholipases are relatively undefined and are probably multiple. Evidence has been obtained for mechanisms involving G-proteins, protein kinase C, Ca^{2+} and receptor tyrosine kinases.

Another major point that requires further work is the definition of the physiological significance of agonist-stimulated PC breakdown. One concept that has much support is that PC contributes more DAG for activation of protein kinase C than does PIP_2, and this appears to be particularly true for late times of agonist action. A surprising finding is that large amounts of PA are produced from PC breakdown at early times of agonist action when DAG accumulation is not detectable. This indicates that phospholipase D is activated before phospholipase C and has led to the view that some of the DAG that accumulates is derived from PA through the action of phosphatidate phosphohydrolase. Although the major function of DAG is undoubtedly activation of protein kinase C, the physiological roles of agonist-generated PA remain unknown. Many functions have been proposed for this lipid, including the stimulation of DNA synthesis, but it is possible that its major role has yet to be defined.

References

1. Berridge MJ: Inositol trisphosphate and diacylglycerol: two interacting second messengers. Ann Rev Biochem 56: 159–193, 1987
2. Exton JH: The roles of calcium and phosphoinositides in the mechanisms of α_1-adrenergic and other agonists. Rev Physiol Biochem Pharmacol 111: 118–224, 1988
3. Exton JH: Signaling through phosphatidylcholine breakdown. J Biol Chem 265: 1–4, 1990
4. Augert G, Blackmore PF, Exton JH: Changes in the concentration and fatty acid omposition of phosphoinositides

induced by hormones in hepatocytes. J Biol Chem 264: 2574–2580, 1989
5. Uhing RJ, Prpic V, Jiang H, Exton JH: Hormone-stimulated polyphosphoinositide breakdown in rat liver plasma membranes. J Biol Chem 261: 2140–2146, 1986
6. Taylor SJ, Exton JH: Guanine-nucleotide and hormone regulation of polyphosphoinositide phospholipase C activity of rat liver plasma membranes. Biochem J 248: 791–799, 1987
7. Lynch CJ, Prpic V, Blackmore PF, Exton JH; Effect of islet-activating pertussis toxin on the binding characteristics of Ca^{2+}-mobilizing hormones and on agonist activation of phosphorylase in hepatocytes. Mol Pharmacol 29: 196–203, 1986
8. Fitgerald TJ, Uhing RJ, Exton JH: Solubilization of the vasopressin receptor from rat liver membranes. J Biol Chem 261: 16871–16877, 1986
9. Blackmore FF, Bocckino SB, Waynick LE, Exton JH: Role of guanine nucleotide-binding regulating protein in the hydrolysis of hepatocyte phosphatidylinositol 4,5-bisphosphate by calcium-mobilizing hormones and the control of cell calcium. Studies utilizing aluminum fluoride. J Biol Chem 260: 14477–14483, 1985
10. Rhee SG, Suh P-G, Ryu S-M, Lee SY: Studies of inositol phospholipid-specific phospholipase C. Science 244: 546–550, 1989
11. Taylor SJ, Smith JA, Exton JH: Purification from bovine liver membranes of a guanine nucleotide-dependent activator of phosphoinositide specific phospholipase C. Immunologic identification as a novel G-protein α-subunit. J Biol Chem. In press, 1990
12. Bocckino SB, Blackmore PF, Exton JH: Stimulation of 1,2-diacylglycerol accumulation in hepatocytes by vasopressin, epinephrine and angiotensin II. J Biol Chem 260: 14201–14207, 1985
13. Bocckino SB, Blackmore PF, Wilson PB, Exton JH: Phosphatidate accumulation in hormone-treated hepatocytes via a phospholipase D mechanism. J Biol Chem 262: 15309–15315, 1987
14. Augert G, Bocckino SB, Blackmore PF, Exton JH: Hormonal stimulation of diacylglycerol formation in hepatocytes. Evidence for phosphatidylcholine breakdown. J Biol Chem 264: 21689–21698, 1989
15. Irving HR, Exton JH: Phosphatidylcholine breakdown in rat liver plasma membrane. J Biol Chem 262: 3440–3443, 1987
16. Martin TW, Michaelis K: P_2-Purinergic Agonists Stimulate Phosphodiesteratic Cleavage of Phosphatidylcholine in Endothelial Cells. J Biol Chem 264: 8847–8856, 1989
17. Diaz-Meco MT, Larrodera P, Lopez-Barahona M, Cornet ME, Barreno PG, Moscat J: Phospholipase C-mediated hydrolysis of phosphatidylcholine is activated by muscarinic agonists. Biochem J 263: 115–120, 1989
18. Qian Z, Drewes LR: Muscarinic acetylcholine receptor regulates phosphatidylcholine phospholipase D in canine brain. J Biol Chem 264: 21720–21724, 1989
19. Martinson EA, Goldstein D, Brown JH: Muscarinic receptor activation of phosphatidylcholine hydrolysis. J Biol Chem 264: 14748–14754, 1989
20. Matozaki T, Williams JA: Multiple sources of 1,2-diacylglycerol in isolated rat pancreatic acini stimulated by cholecystokinin. J Biol Chem 264: 14729–14734, 1989
21. Billah MM, Eckel S, Mullmann TJ, Egan RW, Siegl MI: Phosphatidylcholine hydrolysis by phospholipase D determines phosphatidate and diglyceride levels in chemotactic peptide-stimulated human neutrophils. J Biol Chem 264: 17069–17077, 1989
22. Uhing RJ, Prpic V, Hollenbach PW, Adams DO: Involvement of protein kinase C in platelet-activating factor-stimulated diacylglycerol accumulation in murine peritoneal macrophages. J Biol Chem 264: 9224–9230, 1989
23. Slivka SR, Meier KE, Insel PA: α_1-Adrenergic receptors promote phosphatidylcholine hydrolysis in MDCK-D1 cells: A mechanism for rapid activation of protein kinase C. J Biol Chem 263: 12242–12246, 1988
24. Billah MM, Jin-Keon P, Mullmann TJ, Egan RW, Siegel MI: Regulation of phospholipase D in HL-60 granulocytes. J Biol Chem 264: 9069–9076, 1989
25. Cabot MC, Welch CJ, Zhang Z-C, Cao H-T: Evidence for a protein kinase C-directed mechanism in the phorbol diester-induced phospholipase D pathway of diacylglycerol generation from phosphatidylcholine. FEBS Lett 245: 85–90, 1989
26. Guy GR, Murray AW: Tumor promoter stimulation of phosphatidylcholine turnover in HeLa cells. Canc Res 42: 1980–1985, 1982
27. Muir JG, Murray AW: Bombesin and phorbol ester stimulate phosphatidylcholine hydrolysis by phospholipase C: Evidence for a role of protein kinase C. J Cell Physiol 130: 382–391, 1987
28. Besterman JM, Duronio V, Cuatrecasas P: Rapid formation of diacylglycerol from phosphatidylcholine: A pathway for generation of a second messenger. Proc Natl Acad Sci USA 83: 6785–6789, 1986
29. Liscovitch M, Amsterdam A: Gonadotropin-releasing hormone activates phospholipase D in ovarian granulosa cells. J Biol Chem 264: 11762–11767, 1989
30. Agwa DE, McPhail LC, Chabot MC, Daniel LW, Wykle RL, McCall CE: Choline-linked phosphoglycerides: A source of phosphatidic acid and diglycerides in stimulated neutrophils. J Biol Chem 264: 1405–1413, 1989
31. Pai JK, Siegel MI, Egan RW, Billah MM: Phospholipase D catalyzes phospholipid metabolism in chemotactic peptide-stimulated HL-60 granulocytes. J Biol Chem 263: 12472–12477, 1988
32. Polverino AJ, Barritt GJ: On the source of the vasopressin-induced increases in diacylglycerol in hepatocytes. Biochim Biophys Acta 970: 75–82, 1988
33. Pessin MS, Raben DM: Molecular species analysis of 1,2-diglycerides stimulated by α-thrombin in cultured fibroblasts. J Biol Chem 264: 8729–8738, 1989
34. Wright TM, Shin HS, Raben DM: Sustained increase in

1,2-diacylglycerol precedes DNA synthesis in epidermal-growth-factor-stimulated fibroblasts: Evidence for stimulated phosphatidylcholine hydrolysis. Biochem J 267: 501–507, 1990
35. Pessin MS, Baldassare JJ, Raben DM: Molecular species analysis of mitogen-stimulated 1,2-diglycerides in fibroblasts: Comparison of α-thrombin, epidermal growth factor and platelet-derived growth factor. J Biol Chem 265: 7959–7966, 1990

Address for offprints: J.H. Exton, Howard Hughes Medical Institute, Research Laboratories, Vanderbilt University School of Medicine, Rudolph Light Hall, Room 802, Nashville, TN 37232-0295, USA

MODULATION OF GENE EXPRESSION BY INTRACELLULAR SIGNALS

Molecular and Cellular Biochemistry **104**: 89–100, 1991.

Regulation of phosphoenolpyruvate carboxykinase (GTP) gene transcription

J. Liu and R.W. Hanson
Department of Biochemistry, Case Western Reserve University School of Medicine, Cleveland, OH, USA

Key words: phosphoenolpyruvate carboxykinase, cAMP, insulin, transcription, promoter

Abstract

Transcription of the gene for phophoenolpyruvate carboxykinase is regulated by several hormones which control the level of glucose synthesis in vertebrate animals. A 490 bp segment located at the 5′ end of the structural gene contains the necessary regulatory elements to account for the pattern of transcriptional regulation characteristic of the phosphoenolpyruvate carboxykinase gene. Multiple cis binding sites within the promotoer and nuclear binding proteins have been identified and shown to play a role in the regulation of gene transcription. The interaction of these transcription factors with each other and with the phosphoenolpyruvate carboxykinase promoter is central to the regulated expression of this gene. The key role of cAMP and insulin in controlling the level of gene transcription will be discussed and related to the function of transcription factors currently known to regulate the tissue specific expression of the phosphoenolpyruvate carboxykinase gene.

Introduction

The control of eucaryotic gene expression has been the subject of intensive study over the past 10 years [1, 2]. In most cases this regulation occurs at the level of transcription, although there are some exceptions, such as alternative splicing, regulation of mRNA stability, and the translational control of protein synthesis which are important in controlling the expression of some genes. Gene transcription involves the binding of transcription factors (proteins) to specific sites on the promoter and the activity of these transcription factors can be regulated by hormones and tissue and developmental stage-specific factors [3, 4].

Our research has focused on the expression of the gene for phosphoenolpyruvate carboxykinase (GTP) (EC 4.1.1.32) (PEPCK), an important enzyme in maintaining glucose homeostasis in vertebrate animals [5]. Transcription of the PEPCK gene can be both positively and negatively regulated by multiple hormones in a tissue-specific manner. Because of the complexity of this regulation, we will limit this review to a discussion of the major factors which acutly control transcription of the PEPCK gene, cAMP and insulin.

Physiological basis for regulation of PEPCK gene expression

PEPCK catalyzes the conversion of oxalacetate, generated in the citric acid cycle, to P-enolpyruvate, which is then converted to glucose. There are two isozymes of PEPCK present in vertebrates; a cytosolic form, the synthesis of which is acutely regulated by diet and hormones [6], and a mitochondrial form which is usually constitutively expressed [7]. The unique physiological role played by each of the isozyme forms of PEPCK has been reviewed in detail previously [8]. Here we will focus on the regulated transcription of the gene for the cytosolic form of this enzyme.

The cytosolic form of PEPCK is present in the

liver, kidney cortex and adipose tissue and the jejunum of the small intestine [9]. It can also be found in low levels in a diverse number of tissues including skeletal muscle, brain, lung, and mammary gland [9]. In the liver and kidney cortex, PEPCK is involved in gluconeogenesis, whereas in adipose tissue and the jejunum it is involved in a pathway termed glyceroneogenesis, which generates 3-phosphoglycerol from lactate, pyruvate and amino acids during starvation [10]. PEPCK is at a branch point in several metabolic pathways and is considered to be the rate determining step in hepatic gluconeogenesis [11], a role which underlies its significance in the control of glucose homeostasis in vertebrates.

Gluconeogenesis does not occur in mammals before birth so that glucose must be supplied to the fetus via the maternal circulation. PEPCK, the last of the gluconeogenic enzymes to appear during development, is absent in the liver during most of fetal life [12]. There is a rapid increase in the transcription of the PEPCK gene at birth, resulting in an accumulation of mRNA for the enzyme, beginning at parturition and continuing during the first day after birth [13]. In contrast, PEPCK is present in the fetal kidney and additional amounts of the renal enzyme develop only gradually after birth [14].

Regulation of PEPCK gene transcription

The major factor controlling the synthesis of hepatic PEPCK is the concentration of glucose in the blood. During starvation there is a decrease in the level of blood glucose, resulting in a rise in glucagon and subsequently in the concentration of hepatic cAMP. At the same time the concentration of serum insulin is decreased in response to the decline in the levels of blood glucose. Both cAMP and insulin act on this system, by rapidly and acutely increasing (cAMP) or decreasing (insulin) transcription of the PEPCK gene [15, 16]. The molecular mechanism of cAMP action on PEPCK gene transcription, while complex, is better understood than that for insulin, so we will focus primarily on our current understanding of the action of cAMP, and will include a discussion of several possible mechanisms to explain the effect of insulin. Finally, while we will discuss the regulation of PEPCK gene transcription, it must be emphasized that hormones also control the expression of this gene at the posttranscriptional level. For example, cAMP [17] and glucocorticoids [18] have been shown to stabilize PEPCK mRNA.

Several approaches have been used to study the mechanism by which hormones and other effector molecules regulate PEPCK gene transcription. These include the introduction of chimeric genes containing the PEPCK promoter linked to a reporter gene into cells in culture to determine the sequences responsible for control of transcription [19–21] or the measurement of the tissue and developmental specific regulation of PEPCK gene expression by introducing chimeric PEPCK-bGH (bovine Growth Hormone) genes into transgenic mice [22, 23]. In addition, a number of transcription factors which control the functioning of the PEPCK promoter have been identified using both DNase I footprinting and gel retardation analysis [24, 25]. The results from these studies indicate the complexity of the transcriptional regulation of the PEPCK gene.

Multiple regulatory elements are contained within a 490 bp region of the PEPCK promoter

A relatively short region of the PEPCK promoter, from − 490 to + 73, contains the necessary information to account for the major transcriptional properties of the endogenous PEPCK gene. Figure 1 diagrams this segment of the PEPCK promoter, together with putative regulatory elements, linked to the structural gene for bacterial chloramphenicol acetyltransferase (CAT), in a plasmid used for transfection into hepatoma cells. Transcription from this segment of the PEPCK promoter is responsive to cAMP (see Fig. 2), glucocorticoids [26], thyroid hormone[1], insulin [27], phorbol ester[2] and vanadate [28]. The effect of cAMP on tran-

[1] Giralt, M., unpublished observations.
[2] Park, E.A., unpublished observations.

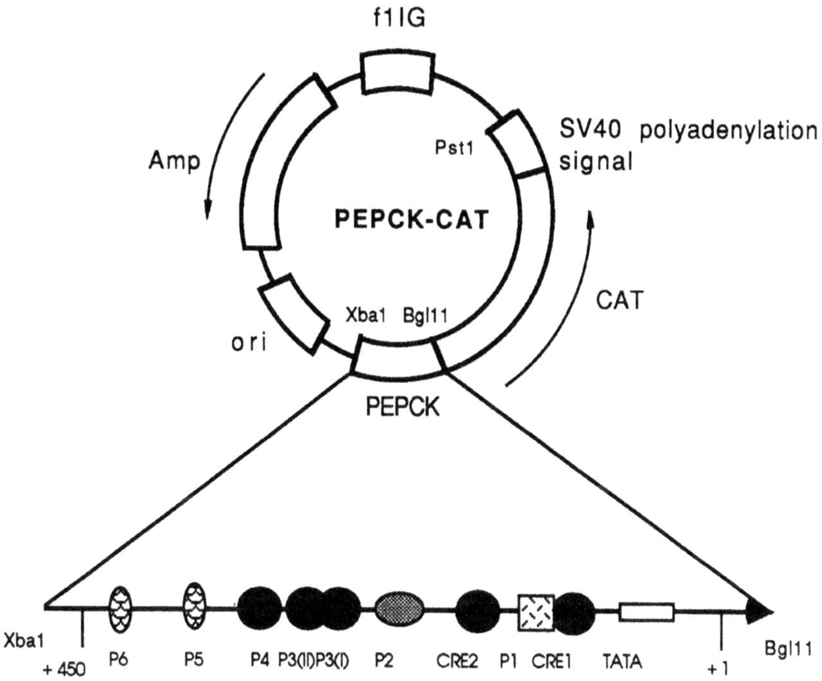

Fig. 1. PEPCK-CAT plasmid and protein binding domains in the PEPCK promoter. The XbaI-BglII fragment of PEPCK promoter (−490 to +73) was ligated into XbaI and BglII sites created in front of the CAT gene in the SV0CAT vector. The 2.5 kb XbaI-PstI fragment containing PEPCK promoter, CAT gene and SV40 polyadenylation signal were ligated into PstI and XbaI sites of high copy phagemid pTZ18R. Amp, ampicillin resistance gene; f1IG, the origin of replication for f1 phage; Ori, the origin of replication for plasmid; CAT, bacterial chloramphenicol acetyl transferase. The protein binding sites in the PEPCK promoter are outlined beneath the plasmid.

scription of a chimeric PEPCK-CAT gene, when introduced into HepG2 hepatoma cells, is shown in Fig. 2. The sequence from −460 to +73 of the PEPCK promoter can also direct the appropriate pattern of developmental, tissue specific and dietary regulation when it is introduced into the germ line of transgenic mice [22, 23]. Recently, Chalkley and his colleagues have described an enhancer element in the PEPCK promoter at −4000, which increased the expression of the PEPCK-CAT gene when transfected transiently into hepatoma cells [29]. The functional significance of this element on the expression of PEPCK in the animal remains to be determined.

A number of protein binding domains in the PEPCK promoter have been identified by DNase I footprinting analysis [24] and gel-shift assay [25]. There are at least 10 protein binding sites in the PEPCK promoter mapping between −490 and +73 (see Fig. 3 and Table 1) and several of these domains have the ability to bind multiple transcription factors. There is also a high degree of tissue specificity apparent in this binding. For example, protein(s) present in the nuclei from rat liver interact with the region of the PEPCK promoter between −260 to −230 (termed P3), while protein (s) from other tissues, such as kidney, do not bind to this region. Table 1 lists the proteins currently known to bind to the PEPCK promoter *in vitro* or to alter the rate of transcription from the PEPCK promoter in hepatoma cells in culture.

Characterization of cAMP responsive elements in the PEPCK promoter

The cAMP regulatory element (CRE) in the PEPCK gene was identified using a series of deletions in which the PEPCK promoter was linked to the ameno-3′-glycosyl phosphotransferase (*neo*)

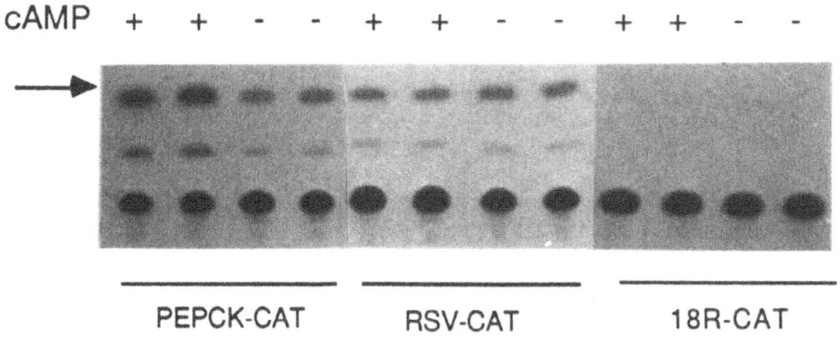

Fig. 2. Basal and cAMP induced transcription of PEPCK-CAT gene after transfection into HepG2 cells. Plasmid (10 μg) was transfected into HepG2 cells and 36 hours later the transfected cells were changed to 5% calf serum, either with or without 1 mM cAMP, and the incubation continued for another 12 hours. The cells were harvested and freeze-thawed three times to extract the enzyme. CAT activity was measured by the method described in reference 31.

structural gene and analyzed after stable transfection into hepatoma cells [21]. A CRE consensus sequence of T(G/T)ACGTCAs was ultimately proposed based on the results of these studies and the work of a number of investigators using other cAMP regulated genes [see reference 30 for a review]. An 18 base pair oligonucleotide synthesized from the sequence from − 98 to − 80 in the PEPCK promoter (designated CRE-1 in Fig. 1) can confer cAMP responsiveness when linked to a chimeric gene containing an enhancerless SV40 promoter and the CAT structural gene [24]. The activity of CAT was enhanced several fold by cAMP, with the overall response being dependent on the number of copies of the CRE (up to 3) present in the chimeric gene [24]. Thus, the information in the CRE from the PEPCK promoter is sufficient to confer cAMP regulation on a neutral promoter. Individual mutations in the CRE resulted in marked loss of binding affinity of nuclear proteins as well as induction of transcription by cAMP. Footprinting analysis showed that CRE-1 in the PEPCK promoter (− 91 to − 84) was protected from DNase I digestion by nuclear proteins from rat liver [23]. A more detailed analysis of the work establishing the CRE and its functional role in the cAMP regulation of gene transcription is contained in a review by Roesler *et al.* [30].

The transcription factors which interact with CRE-1 and regulate PEPCK gene transcription have been the subject of detailed investigation. The 18 bp oligonucleotide from the PEPCK promoter (− 98 to − 80) corresponding to CRE-1 produced two distinct bands when incubated with nuclear extracts from rat liver and analyzed by gel retardation assay [25]. Several proteins were isolated using CRE-1 oligonucleotide affinity chromatography, the major one having a molecular weight of approximately 43 kd. Further analysis of the binding of purified transcription factors to the PEPCK promoter has yielded surprising results. The CRE-1 from the PEPCK promoter can interact with several different binding proteins, including CREB (cAMP Regulatory Element Binding Protein), Jun/Fos[3] and C/EBP (CCAAT/Enhancer Binding Protein) [31, 32], to name only those for which we have direct evidence of binding by DNase I footprinting analysis (see Table 1). CREB belongs to family of transcription factors of which Jun (also known as AP-1) is the prototype [33]. Jun forms heterodimers with another transcription factor, Fos to regulate gene transcription [34, 35]. C/EBP not only binds to CRE-1, but also to P3(I) and the 5′ segment of P4 [31], a pattern identical to the binding of protein(s) isolated from rat liver nuclei [24]. C/EBP has a molecular weight of 43 kd, which is the size of the major protein band isolated from rat liver nuclei by affinity chromatography. CREB has the same molecular weight, complicating a simple analysis of the proteins present in liver nuclei which bind to CRE-1 of the PEPCK promoter. It is possible that both proteins compete at this

[3] Gurney, A., unpublished observations.

critical site in the PEPCK promoter to regulate gene transcription.

cAMP induction is mediated through multiple cis *sequences in the PEPCK promoter*

In order to determine whether multiple protein binding sites in the PEPCK promoter are required for the cAMP regulation of transcription, we introduced specific block mutations into all of the protein binding sites in the promoter (Fig. 1). These mutations prevented the binding of both proteins from liver nuclei and isolated transcription factors. A block mutation in CRE-1, did not completely inhibit cAMP responsiveness of a chimeric PEPCK-CAT gene in hepatoma cells. This suggested that other elements in the PEPCK promoter are also involved in cAMP regulation of transcription. Complete elimination of cAMP responsiveness was noted only when mutations were introduced into both CRE-1 and P3 (I) (Fig. 4). Oligonucleotide competition in DNase1 footprinting assays suggested that those two sites bind a similar protein (s) present in liver nuclei. CRE-2 and P4 also bound to nuclear proteins but with a lower affinity. Both CRE-1 and P3(I) bind purified C/EBP with a high affinity. Comparison of sequence of these four binding sites show that P4 and CRE-2 have two and one nucleotide differences respectively, as compared to CRE-1. In order to examine the structural and functional significance of these small differences at these two sites, the two nucleotides at P4 and one nucleotide at CRE-2 were changed to correspond to the sequence at CRE-1 using site-directed mutagenesis. These changes in the PEPCK promoter enhanced the binding affinity for C/EBP and CREB and increased responsiveness of the PEPCK-CAT gene to cAMP by more than 20-fold, as compared to 4-fold induction for the

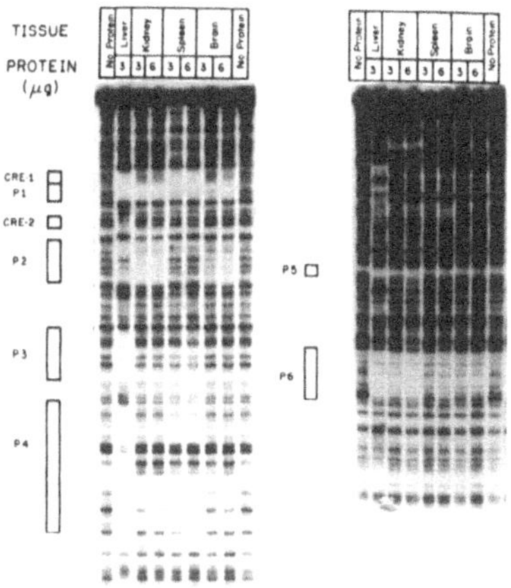

Fig. 3. The footprinting pattern of the PEPCK promoter with the nuclear proteins from different tissues. The Xbal-Bglll fragment in the PEPCK-CAT plasmid was isolated and the Xbal site was labeled with γ-^{32}P-ATP. This labeled DNA fragment was incubated with nuclear proteins isolated from different tissues, digested with DNasel and resolved in 8% polyacrylamide gels. Reprinted with permission from Roesler *et al.* [24].

Table 1. Summary of known proteins interacting with the PEPCK promoter

Binding sites	Locations	Sequence	Binding proteins	Function in transcription
TATA	– 21/– 28	TATTTAAA	TFIID	Basal
CRE-1	– 74/– 87	TTACGTCA	C/EBP, CREB, Fos/jun	Basal and cAMP
P1	– 87/– 123	TGGCTN3AGCCA	NF1/CTF	Basal
CRE-2	– 135/– 155	TTAGGTCA	C/EBP	cAMP
P2	– 164/– 200	ATTAAC	HNF-1	tissue-specific
P3 (I)	– 230/– 248	TTGTGTAAG	C/EBP	Basal, cAMP and tissue-specific
P3 (II)	– 249/– 260	TTAGTCA	Fos/Jun	Basal and cAMP
P4	– 269/– 320	ATCAGCAAC	C/EBP	Basal and cAMP
GR2	– 350/– 368	CATATGAAGTC	GR	Glucocorticoid
P5/GR1	– 375/– 390	ACACAAAATGTG	GR	Glucocorticoid
IRE/AF2	– 400/– 415	TGGTGTTTGACAAC	unknown	Glucocorticoid and insulin
P6/AF1	– 428/– 456	ATGACCTTTGGCCGT	unknown	Glucocorticoid

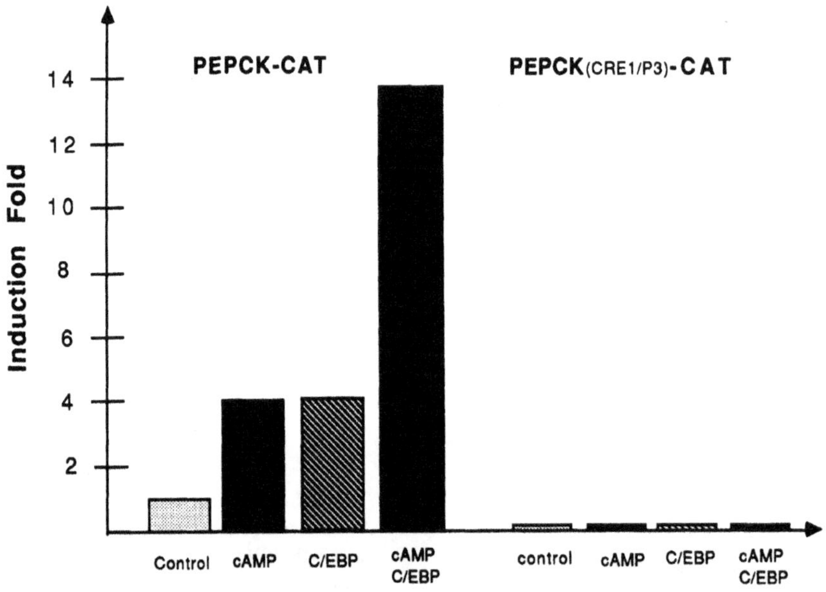

Fig. 4. Effect of mutations in the PEPCK promoter on the stimulation of transcription by cAMP and C/EBP. Mutation in the PEPCK promoter at CRE-1 and P3 were introduced by site-directed mutagenesis [57], and a double mutation at both sites was constructed by taking the advantage of an unique restriction site SauI at −200 between CRE-1 and P3 sites. The XbaI-SauI fragment containing the P3 mutation was switched to the XbaI-SauI site of the CRE-1 mutation to create the double site mutant. Transfection was carried out by using 10 μg of PEPCK-CAT and PEPCK(CRE-1/P3)-CAT together with 40 μg of MSV-C/EBP. Transcription from the PEPCK promoter was determined 12 hrs after the addition of 1 mM 8-Bromo-cAMP to HepG2 cells, as described in reference 31 and the legend of Fig. 2.

native PEPCK promoter (−490 to +73) [36]. These experiments suggest that the A at position −282 and the G at position −284 in P4 and G at position −140 of CRE-2 are involved in determining the affinity for the binding of transcription factors to the PEPCK promoter. It is possible that the ancestral promoter of the PEPCK gene contained three cAMP responsive elements, present at CRE-1, CRE-2 and P4 and that CRE-2 and P4 were altered during the course of evolution resulting in a decreased sensitivity of the PEPCK promoter to cAMP. Alterations in these ancesteral CREs by evolution could be responsible for the diverse pattern of transcriptional regulation of the PEPCK gene by hormones currently present in this gene in higher vertebrates.

Is C/EBP involved in cAMP regulation of PEPCK gene transcription?

The interaction of C/EBP with the cAMP responsive sites at CRE-1 and P3 in the PEPCK promoter suggests that C/EBP is involved in cAMP regulation of PEPCK gene transcription. However, other proteins have been shown to bind to CRE-like elements in mammalian promoters and to mediate cAMP effects on gene transcription. For example, CREB binds to the CRE in the promoter of somatostatin, another cAMP regulated gene and has been shown to be involved in the cAMP regulation of transcription [32]. CRE-1 of the PEPCK promoter is very similar (but not identical) to the CRE in the somatostatin gene, so that it is not surprising that CREB binds to CRE-1. CREB is phosphorylated on serine residues by a protein kinase A catalyzed reaction and this phosphorylation has been shown by Yamamoto *et al.* [32] to increase transcription from the somatostatin promoter *in vitro*.

On the other hand, C/EBP has not been previously shown to be involved in the cAMP responsiveness of specific gene, although such a possibility has been discussed [37]. C/EBP has been proposed to be a major factor in the transcription of genes coding for proteins having highly differentiated cellular functions, such as the enzymes of gluconeogenesis and lipogenesis [37]. In addition, C/EBP has a pattern of development in specific tissues of the mouse [38] which closely parallels that of PEPCK. We have noted that C/EBP can transactivate a chimeric PEPCK-CAT gene when an expression vector containing the cDNA for this transcription factor was co-transfected into hepatoma cells [31]. cAMP stimulates the transcription of PEPCK gene via two sites CRE-1 and P4 where the C/EBP used to transactivate the transcription of this gene is shown in Fig. 4. In the wild type promoter (PEPCK-CAT), cAMP and C/EBP cause a 4-fold induction of transcription from the PEPCK promoter and 13 fold when cAMP and C/EBP are introduced together. However, if the two sites required for the transactivation by C/EBP were simultaneously mutated (PEPCK(CRE-1/P3)-CAT), the promoter was unresponsive to stimulation by both cAMP or C/EBP. Since transcription factors which mediate the effect of cAMP on the PEPCK promoter recognize a sequence known to bind C/EBP, it is reasonable to assume that C/EBP is in some way involved in the action of cAMP or that other proteins involved in this effect have similar DNA binding properties.

Transcriptional control of the PEPCK promoter also includes negative regulation

One of the unique features of the PEPCK gene is that its transcription is so tightly controlled by negative regulatory signals. Insulin is the major negative factor in the control of PEPCK gene expression and is a dominant controlling signal, since it can inhibit PEPCK gene transcription even in the presence of cAMP [39]. The cis sequences necessary to confer negative regulation of PEPCK gene transcription appear to be contained in a region of the promoter from −455 to the start site of transcription, as demonstrated by a number of different studies [39, 28]. Recently, O'Brien *et al.* [27] reported that the sequences between −415 and −400 contain an insulin regulatory element (IRE) which can confer negative regulation by insulin to a heterologous promoter. This region of the promoter has been shown to be involved in the stimulatory effect of glucocorticoids on PEPCK gene transcription [27]. Since the negative effect of insulin was only apparent when glucocorticoids were used to stimulate transcription of PEPCK gene [27], it is possible that insulin interferes with the positive stimulation of transcription caused by glucocorticoids. However, the region of the promoter between −415 to −400 is not in itself sufficient to account for the total negative effect of insulin on PEPCK gene transcription, since sequences 3′ to this region of the PEPCK promoter are also involved in the response of the gene to insulin [27]. How the proposed IRE can inhibit cAMP, which acts at regions of the PEPCK promoter considerably downstream, remains to be determined. This negative regulation of PEPCK gene transcription by insulin is underlined by the observation by McGrane *et al.* [22] that feeding glucose to transgenic mice containing the PEPCK-bGH gene will reduce transcription from the PEPCK promoter to a level 5% of that noted in control, non-transgenic animals. The segment of the PEPCK promoter used in these studies was from −460 to +73 contained the proposed IRE. This suggests that a region of the PEPCK promoter which contains the upstream IRE was insulin responsive when linked to a heterologous structural gene and introduced into transgenic mice.

While insulin is the major physiological regulator which inhibits the transcription of the PEPCK gene in the liver, there are a number of other compounds which have been shown to also cause a marked reduction of PEPCK gene expression. These include vanadate [28] and phorbol esters [40], which block the cAMP induced stimulation in PEPCK gene transcription in hepatoma cells and glucose itself, which lowers the basal level of PEPCK mRNA by decreasing transcription of the endogenous PEPCK gene in hepatoma cells [41]. Metabolic acidosis also decreases the rate of transcription of the PEPCK gene in rat kidney [42]. The mechanism of action of these compounds on the expression of the PEPCK gene is currently not understood.

Model for the hormonal regulation of PEPCK gene transcription

It is generally accepted that peptide hormones exert their effects on gene expression by interacting with their corresponding receptors on the cell membrane which utilize the signal transduction pathways to turn on the expression of genes for specific transcription factors or to modify the structure of existing transcription factors. On the other hand, the steroid hormone receptors are themselves DNA binding proteins capable of moving from the cytoplasm to nucleus to bind to specific target sequences in the gene. The final result of hormonal activation is an alteration in the ability of specific transcription factors to interact with the transcription initiation complex, usually located at the TATA box. Since the hormonal regulation of PEPCK gene transcription is very complex, it is not currently possible to propose a general model which will summarize *all* of the regulatory features of this gene. We will present a model for the positive regulation of PEPCK gene transcription by cAMP based on the experimental evidence reviewed above and will speculate on how the negative regulation of this gene is achieved.

At least ten proteins which bind to CRE-like sequences in mammalian genes have been cloned and sequenced [43–47]. These transcription factors are related to the class of transcription factor AP-1 or Jun. These proteins have several interesting features; most notably is the leucine zipper structure, a structural domain just proposed for C/EBP by McKnight and colleagues [48] to be involved in the formation of transcriptionally active C/EBP homodimers. This leucine zipper structure can mediate formation of a dimer complex through parallel interaction of helical domains with leucine residues on the same side of helix. Heterodimerization can also occur via these leucine zippers linking together transcription factors such as Fos/Jun to form highly specific, functional dimers [49, 50]. Interactions of this type between the transcription factors add enormous scope and complexity for regulating gene expression.

A model for the hormonal regulation of gene expression must take into account several features involved in the selective interaction between proteins and DNA. First, each transcription factor has a specific DNA binding motif due to the primary structure of the protein. These include the so called 'zinc finger' domains in several transcription factors, helix-turn-helix motifs characteristic of bacterial proteins and the leucine zipper proteins [2]. Second, dimerization between transcription factors is not random, since only specific combinations of factors can form active dimers [46]. It is possible that covalent modification such or phosphorylation induced by hormones influences the dimerization of transcription factors. Third, the binding specificity of a transcription factor may be determined by the nature of the specific dimer formed. For example, Fos does not bind to AP-1 sites [50, 51] and the Jun/Jun homodimer has a low affinity for AP-1 sequences. However, the combination of Fos/Jun binds with a high affinity to these sites [50, 51]. Fourth, the effectiveness of a transcription factor to regulate gene transcription will depend not only on the binding sequence in the promoter, but also the sequences surrounding the regulatory element [52]. This high degree of selectivity of protein-DNA interaction may explain why a protein such as C/EBP can bind to seemingly diverse sets of sequences such as those in CRE-1 and P3 in the PEPCK promoter while CREB can only interact with CRE-1.

An example of a model for the regulated expression of the PEPCK gene is presented in Fig. 5. A series of binding proteins could simultaneously interact with each other or with the critical CRE-1 and P3 regions of the PEPCK promoter. These binding proteins could communicate either a positive or negative signal, with the extent of gene transcription depending on the balance between these signals. Although available evidence seems to favor C/EBP binding to CRE-1 and P3, other

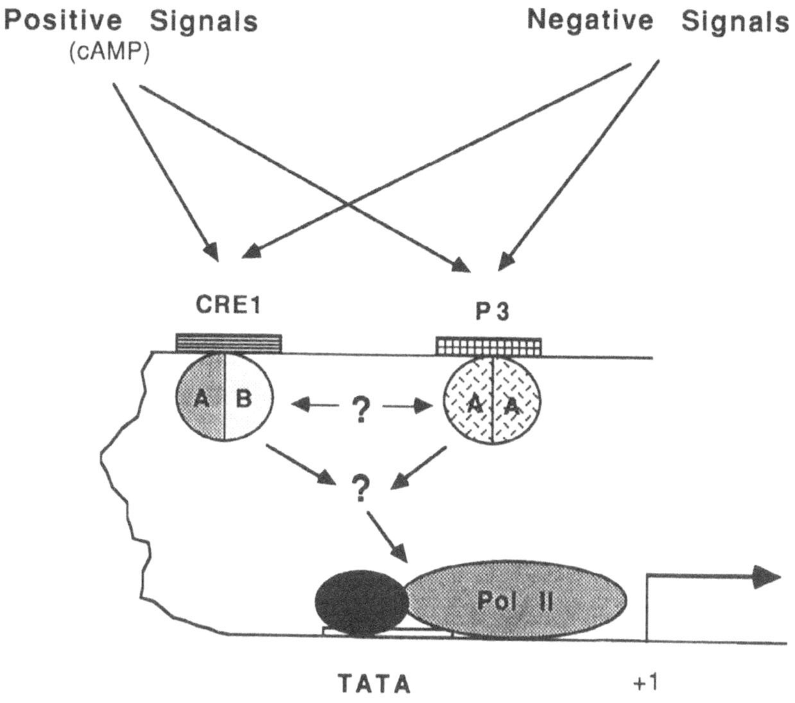

Fig. 5. A model for regulated transcription in the PEPCK gene. P3 and CRE-1 are sites critical for the positive regulation of PEPCK gene transcription. The binding protein designated as A, could form a homodimer with itself, as at P3 or could form a heterodimer with another transcription factor, designated B, as at CRE-1. In this model either combination of heterodimers or homodimers would be possible at both sites. RNA polymerase II is labelled as pol II and the TATA box binding factor, TFIID, is indicated in black. For the simplicity, other accessary transcription factors are not included in the model. These proteins could interact with each other or interact with some other factors at the TATA box to modulate the transcription.

transcription factors could theoretically bring about the same effect on PEPCK gene transcription. In our model, A and B indicate individual transcription factors which could form either homodimers (AA) or heterodimerize with each (AB). When the concentration of cAMP increases, the proteins at these sites could be phosphorylated, inducing an interaction with each other which will result in either a direct or indirect communication with the RNA polymerase II initiation complex at the TATA box. Indirect communication has been shown to involve an adaptor protein which can interact with proteins of the transcription initiation complex at the TATA box [53]. It must be emphasized that while these models for the hormonal regulation of PEPCK gene transcription are speculative, they are based on known mechanisms for the regulation of transcription of several of the less complex gene systems currently under investigation in this rapidly moving field.

This model also allows for negative regulation of transcription from the PEPCK promoter. First, the transcription factors involved in negative regulation could complete with proteins which stimulate PEPCK gene transcription by binding at key regulatory elements in the promoter. Those sites could interact with many different transcription factors as discussed above. Second, the protein(s) involved in negative regulation of transcription could dimerize with a positive factor(s) through leucine zipper domains, thereby inhibiting the binding of positive factors to CRE-1 or P3. We have found that Jun will markedly stimulate transcription from the PEPCK promoter when an expression vector containing Jun cDNA is co-transfected into hepatoma cells. Interestingly, Fos will completely block this effect of Jun and reduce expression of a chimeric PEPCK-CAT gene to basal level.[4] Third, recent evidence suggests that several transcription factors do not directly interact with the initiation complex, but rather act through an adaptor protein which contains a transcription activation domain which interacts with the transcription initiation complex (RNA polymerase II) at the TATA box. This type of negative regulation could be achieved by interfering with the interaction between the positive transcription factor and the adaptor or between the adaptor and initiation complex. If the negative factor is interfering the transactivation process by adaptor proteins, a specific DNA binding domain on the promoter would not be required for the negative effects on transcription. One of the best examples of this type of regulation is E1A, which is not itself a DNA binding protein but can activate or repress gene expression by interacting with specific transcription factors such as ATF or AP-1 [54, 55]. Regardless of the exact mechanism involved, the level of transcription from the PEPCK promoter is determined by the balance between these positive and negative signals.

Future directions

New transcription factors are being identified and characterized at a rapid pace, so that the control of gene transcription should be tested directly using regulatory proteins added to various promoters *in vitro*. Due to the complexity of the interaction of factors involved in the expression of RNA polymerase II transcribed genes, the development of a responsive, cell free system has been difficult. It will be necessary not only to identify the appropriate transcription factor(s) which binds to regulatory elements in a specific promoter, but also to understand the way these proteins interact with each other. Several cell-free transcription systems, which are responsive to purified transcription factors are currently available. Using a cell free transcription system, Klemm *et al.* [56] established that transcription from the PEPCK promoter can be stimulated by the addition of the catalytic subunit of protein kinase A. However, this transcription system is not totally reconstituted, since it relies on the addition of uncharacterized proteins extracted from rat liver nuclei. A complete understanding of the hormonal regulation of transcription of a gene as complex as PEPCK is clearly a long term undertaking which will rely on advances in the general field of gene transcription.

[4] Gurney, A. and Park, E.A., unpublished observations.

Acknowledgements

The authors wish to thank Drs. Edwards Park, Mary McGrane and Linda Brady for their helpful comments during the preparation of this manuscript.

This work was supported by grants DK 21859 and DK 24451 from the National Institutes of Health.

References

1. Mitchell PJ, Tjian R: Transcriptional regulation mammalian cells by sequence-specific DNA binding proteins. Science 245: 371–378, 1989
2. Johnson PF, McKnight SL: Eucaryotic transcriptional regulatory proteins. Annu Rev Biochem 799–839, 1989
3. Evans RM: The steroid and thyroid hormone receptor superfamily. Science 240: 889–895, 1988
4. Maniatis T, Goobourn S, Fisher JA: Regulation of inducible and tissue-specific gene expression. Science 236: 1237–1245, 1987
5. Utter MF, Kolenbrander HM: Formation of oxalacetate by CO_2 fixation on phosphoenolpyruvate. The enzymes 6: 117–168, 1974
6. Tilghman SM, Hanson RW, Ballard FJ: Hormonal regulation in gluconeogenesis: its regulation in mammalian species (Hanson RW, Mehlman MA ed.). John Wiley and Sons, New York, p 47
7. Garber AJ, Hanson RW: The interrelationships of the various pathways forming gluconeogenic precursor in guinea pig liver mitochondria. J Biol Chem 246: 589–598, 1971
8. Hod Y, Cook JS, Weldon SL, Short JM, Wynshaw-Boris, Hanson RW: Differential expression of the genes for the mitochondrial and cytosolic forms of phosphoenolpyruvate carboxykinase. Ann NY Acad Sci 478: 31–45
9. Hanson RW, Garber AJ: Phosphoenolpyruvate carboxyki-

nase: I. Its role in gluconeogenesis. American J Clin Nutr 25: 1010–1021, 1972
10. Ballard FJ, Hanson RW, Leverlle GA: Phosphoenolpyruvate carboxykinase and the synthesis of glyceride glycerol from pyruvate in adipose tissues. J Biol Chem 242: 2746–2750, 1967
11. Rognstad R: Rate-limiting step in metabolic pathway. J Biol Chem 245: 1875–1878, 1979
12. Ballard FJ, Hanson RW: P-enolpyruvate carboxykinase and pyruvate carboxylase in developing rat liver. Biochem J 104: 866–871, 1967
13. Mencher D, Cohen H, Benvenisty N, Meyuhas, Reshef L: Primary activation of cytosolic phosphoenolpyruvate carboxykinase gene in fatal rat liver and the biogenesis of its mRNA. Eur J Biochem 141: 199–203, 1984
14. Zargoli A, Turkenkopf IJ, Mueller VL: Gluconeogenesis in developing rat kidney cortex. Biochem J 111: 181–185, 1969
15. Lamers WH, Hanson RW, Meisner H: cAMP stimulates transcription of the gene for cytosolic phosphoenolpyruvate carboxykinase in rat liver nuclei. Proc Natl Acad Sci USA 79: 5237–5141, 1982
16. Granner DK, Andreone T, Sasaki K, Beale E: Inhibition of transcription of phosphoenolpyruvate carboxykinase gene by insulin. Nature 305: 549–551, 1983
17. Hod Y, Hanson RW: cAMP stabilizes the mRNA for P-enolpyruvate carboxykinase (GTP) against degradation. J Biol Chem 263: 7747–7752, 1988
18. Peterson DD, Koch SR, Granner DK: 3′ noncoding region of the phosphoenolpyruvate carboxykinase mRNA contains a glucocorticoid-responsive mRNA-stabilizing element. Proc Natl Acad Sci USA 86: 7800–7804, 1989
19. Quinn PG, Wong TW, Magnuson MA, Shabb JB, Granner DK: Identification of basal and cAMP regulatory elements in the promoter of the phosphoenolpyruvate carboxykinase gene. Mol Cell Biol 8: 3467–3475, 1988
20. Wynshaw-Boris A, Short JM, Loose DS, Hanson RW: Characterization of the phosphoenolpyruvate carboxykinase (GTP) promoter-regulatory region. I. multiple hormone regulatory elements and the effects of enhancers. J Biol Chem 261: 9414–9720, 1986
21. Short J, Wynshaw-Boris A, Short HP, Hanson RW: Characterization of the phosphoenolpyruvate carboxykinase (GTP) promoter-regulatory region, II. Identification of cAMP and glucocorticoid regulatory domains. J Biol Chem 261: 9721–9726, 1986
22. McGrane MM, deVente J, Yun J, Bloom J, Park EA, Wynshaw-Boris A, Wagner T, Rottman FM, Hanson RW: Tissue-specific expression and dietary regulation of a chimeric phosphoenolpyruvate carboxykinase/bovine growth hormone gene in transgenic mice. J Biol Chem 263: 11443–11451, 1988
23. McGrane MM, Yun JS, Moorman AFM, Lamers WH, Hendrick GK, Arafa BM, Park EA, Wagner TE, Hanson RW: Metabolic effects of developmental, tissue and cell specific expression of a chimeric phosphoenolpyruvate carboxykinase (GTP)/Bovine growth hormone gene in transgenic mice. J Biol Chem 245: 22371–22379, 1990
24. Roesler WJ, Vandenbark GR, Hanson RW: Identification of multiple proteins binding domains in the promoter-regulatory region of the phosphoenolpyruvate carboxykinase (GTP) gene. J Biol Chem 264: 9657–9664, 1989
25. Bokar JA, Roesler WJ, Vandenbark GR, Kaetzel DM, Hanson RW, Nilson JH: Characterization of the cAMP responsive elements from the genes for the α-subunit of glycoproteins hormones and phosphoenolpyruvate carboxykinase (GTP), conserved features of nuclear protein binding between tissues and species. J Biol Chem 263: 19740–19747, 1988
26. Imai E, Stromstedt P, Quinn PG, Carlstedt-Duke J, Gunstafsson J, Granner DK: Characterization of a complex glucocorticoid response unit in the phosphoenolpyruvate carboxykinase gene. Mol Cell Biol 10: 4712–4719, 1990
27. O'Brien RM, Lucas PC, Forest CD, Magnuson MA, Granner DK: Identification of a sequence in the PEPCK gene that mediates a negative effect of insulin on transcription. Science 249: 533–537, 1990
28. Bosch F, Hatzoglou M, Park EA, Hanson RW: Vanadate inhibits the expression of the gene for phosphoenolpyruvate carboxykinase (GTP) in rat heptoma cells. J Biol Chem 265: 13677–13682, 1990
29. Tony IPY, Poon D, Stone D, Granner DK, Chalkley R: Interaction of a liver-specific factor with an enhancer 4.8 kilobases upstream of the phosphoenolpyruvate carboxykinase gene. Mol Cell Biol 10: 3770–3781, 1990
30. Roesler WJ, Vandenbark GR, Hanson RW. cAMP and the induction of eucaryotic gene transcription. J Biol Chem 263: 9063–9066, 1988
31. Park EA, Roesler WJ, Liu J, Klemm DJ, Gurney AL, Thatcher JD, Shuman J, Friedman A, Hanson RW: The role of CCAAT/Enhancer binding protein in the transcriptional regulation of the gene for phosphoenolpyruvate carboxykinase (GTP) gene. Mol Cell Biol 10: 6264–6272, 1990
32. Yamamoto KK, Gonzalez GA, Briggs IIIWH, Montminy MR: Phosphorylation-induced binding and transcriptional efficacy of nuclear factor CREB. Nature 334: 494–498, 1988
33. Bohmann D, Bos TJ, Admon A, Sishimura T, Vogt PK, Tjian R: Human protooncogene c-jun encodes DNA binding proteins with structural and functional properties of transcriptional factors AP-1. Science 238: 1386–1392, 1987
34. MT Rauscher III FJ, Cohen DR, Curren T, Bos TJ, Vogt PK, Bohmann, Tjian R, Franza BR: Fos-associated proteins (p39) is the product of the Jun protooncogene. Science 240: 1010–1016, 1989
35. Chiu R, Boyle WJ, Meek J, Smeal T, Hunter T, Karin M: The c-fos protein interacts with c-Jun/AP-1 to stimulate transcription from AP-1 responsive genes. Cell 54: 541–552, 1988
36. Gurney A, Liu J, Park EL, Roesler WJ, Hanson RW: Single base substitutions within the protein binding domains of the promoter of the gene for phosphoenolpyruvate carboxykinase alter the effect of cAMP on transcription. Manuscript in preparation

37. McKnight SL, Lane MD, Gluecksohn-Waelsch S: Is CCAAT/enhancer binding protein a central regulator of energy metabolism? Genes Dev 3: 2012–2024, 1990
38. Birkenmeier EH, Gwynn B, Howard S, Jerry J, Gordon JI, Landschulz WH, McKnight SL: Tissue-specific expression, developmental regulation, and genetic mapping of the gene encoding CCAAT/enhancer binding protein. Genes Dev 3: 1146–1156, 1989
39. Magnuson MA, Quinn PG, Granner DK: Multihormonal regulation of PEPCK-CAT fusion genes: insulin's effects oppose those of cAMP and dexamethsone. J Biol Chem 262: 14917–14920, 1987
40. Chu DTW, Granner DK: The effect of phorbol esters and diacylglycerol on expression of the phosphoenolpyruvate carboxykinase gene in rat hepatoma H4IIE cells. J Biol Chem 261: 16848–16853, 1986
41. Kahn CR, Lauris V, Koch S, Crettaz M, Granner DK: Acute and chronic regulation of phosphoenolpyruvate carboxykinase mRNA by insulin and glucose. Molecular Endocrinology 3: 840–845, 1989
42. Meisner H, Loose DS, Hanson RW: Effect of hormones on transcription of the gene for cytosolic phosphoenolpyruvate carboxykinase in rat kidney. Biochemistry 24: 421–425, 1985
43. Hoeffler JP, Meyer TE, Yun Y, Jameson JL, Habener JF: Cyclic AMP-responsive DNA binding protein: structure determined from a cloned placental cDNA. Science 242: 749–752, 1988
44. Gonzalez GA, Yamamoto KK, Fischer WH, Karr K, Menzel P, Briggs III W, Vale WW, Montminy MR: A cluster of phosphorylation sites on the cyclic-AMP regulated nuclear factor CREB predicted by its sequence. Nature 337: 749–752, 1989
45. Maekawa T, Sakura H, Kanei-Ishii C, Sudo T, Yoshimura T, Fujisawa J, Yoshida M, Ishii S: Leucine zipper structure protein CRE-BP1 binding to the cyclic AMP responsive element in brain. EMBO J 8: 2023–2028, 1989
46. Hai T, Liu F, Coukos WJ, Green MR: Transcription factor ATF cDNA clones: an extensive family of leucine zipper proteins able to selectively form DNA-binding heterodimers. Genes Dev 3: 2083–2090, 1989
47. Habener JF: Cyclic AMP response element binding proteins: A cornucopia of transcription factors. Molecular Endocrinology 4: 1087–1094, 1990
48. Landschulz W, Johnson PF, McKnight SL: The leucine zipper: a Hypothetical structure common to a new class of DNA binding proteins. Science 240: 1759–1764, 1988
49. Gentz R, Rauscher III FJ, Abate C, Curren T: Parallel association of Fos and Jun leucine zippers juxtaposes DNA-binding domains. Science 243: 1695–1699, 1989
50. Smeal T, Angel P, Meek J, Karin M: Different requirement for formation of Jun : Jun and Jun : Fos complex. Genes Dev 3: 2091–2100, 1989
51. Cohen DR, Ferreira PCP, Gentz R, Franza BR, Curren: The product of fos-related gene, fra-1, binds cooperatively to the AP-1 site with Jun: transcription factor AP-1 is comprised of multiple protein complexes. Genes Dev 3: 173–184, 1989
52. Deutsch PJ, Hoeffler JP, Jameson JL, Habener JF: Structural determinants for transcriptional activation by cAMP-responsive elements. J Biol Chem 263: 18466–18472
53. Lewin B: Commitment and activation at Pol II promoters: A tail of protein-protein interactions. Cell 61: 1161–1164, 1990
54. Martin KJ, Lillie JW, Green M: Evidence for interaction of different eucaryotic transcription activators with distinct cellular targets. Nature 346: 147–152
55. Liu F, Green M: A specific member of the ATF transcription factor family can mediate transcription activation by the adenovirus E1a protein. Cell 61: 1217–1224, 1990
56. Klemm DJ, Roesler WJ, Liu J, Park EA, Hanson RW: *In vitro* analysis of promoter elements regulating transcription of the phosphoenolpyruvate carboxykinase (GTP) gene. Mol Cell Biol 10: 480–485, 1990
57. Liu J, Roesler WJ, Hanson RW: An Efficient method to introduce block mutations into specific regions of a gene. Biotechniques 4: 738–742, 1990

Address for offprints: R.W. Hanson, Department of Biochemistry, Case Western Reserve University School of Medicine, Cleveland, OH 44106, USA

Molecular and Cellular Biochemistry **104**: 101–107, 1991.

Regulation of myosin heavy chain and actin isogenes expression during cardiac growth

K.R. Boheler, L. Carrier, C. Chassagne, D. de la Bastie, J.J. Mercadier and K. Schwartz
I.N.S.E.R.M. Unité 127, Hôpital Lariboisiére, 41 Boulevard de la Chapelle, 75010 Paris, France

Key words: α-skeletal and α-cardiac actins, α- and β-myosin heavy chains, mRNA accumulations and nuclear run-on assay, development and aging, hypothyroid development, rat and human myocardium

Abstract

The cardiac ventricular myosin heavy chain phenotype is developmentally and hormonally regulated, but less is known concerning the actin phenotype. In this study, the levels of accumulation of α-skeletal and α-cardiac actin mRNAs were investigated in rat and human ventricles by primer extension assays. In rat, the two iso-mRNAs are present in approximately equal amounts from birth until 15 days of age and the cardiac form is predominant in adult and senescent hearts. Hypothyroid development has no effect, at least during the first two weeks of age. In man, the two isoactins are co-expressed to similar ratios in one control heart and in one failing heart. It therefore appears that myosin heavy chain and actin multigene families are both expressed in a species specific fashion but are independently regulated within a species. Preliminary results from nuclear run-on assays are presented that indicate differences in the level of transcription of the α-actin and β-myosin heavy chain isogenes in the rat heart.

The molecular heterogeneity of skeletal muscles and their dynamic properties were recognized more than 20 years ago when Barany showed that the differences in the ATPase activities of various skeletal myosins correlated with differences in maximum shortening velocities from the corresponding muscles [1], and when Buller *et al.* [2] showed by cross-innervation of a fast muscle with a slow nerve, that a nerve could modulate the properties of a muscle, changing both their contractility and myosin ATPase activities. Since that time, and more specifically during the last decade, many studies have been conducted which have led to the elucidation of some of the mechanisms controlling the expression of structurally distinct and developmentally regulated protein isoforms in fast and slow skeletal muscles. Much less is known, however, concerning the heterogeneity of cardiac muscles and the plasticity of the differentiated phenotype of cardiomyocytes and as H. Morgan and his colleagues wrote only 3 years ago 'The control of gene expression in heart is largely unexplored . . .' [3].

The purpose of this paper is to present our recent contributions to the identification of some of the mechanisms by which contraction is governed during cardiac growth and atrophy. Our studies have concerned three specific issues: 1) to determine if the expressions of myosin heavy chain and actin multigene families are independently or dependently regulated, 2) to establish whether the isoactin transitions described in the hypertrophied rat heart also occured in the human heart, and 3) to determine at what level these genes are regulated – transcriptional or post-transcriptional. For this, we have analyzed by primer extension assays the respective mRNA levels of α-skeletal and α-cardiac actins in rat and human ventricles submitted to several physiological and pathological conditions and then compared these results with those previously reported for α- and β-myosin heavy chains. A novel nuclear run-on assay has also been established to measure the transcriptional activity of these isogenes.

Methods

Animals

The time course of left ventricular isoactin mRNA accumulations during normal cardiac development was studied in male rats of 15 days (n = 2) and 24 months (n = 2) of age. These time points were chosen because the actin composition of neonatal and 2 month old adult rat hearts have previously been reported [4, 5]. Hypotrophic development was produced by severe hypothyroidism, as in [6] and the accumulations were analyzed at 15 days of age. For this, a pregnant female was fed a low iodine diet (ICN Life Science) and treated from the 15th day of gestation throughout the period of lactation with 0.1% 6-propyl-2-thiouracil (Sigma). Under these conditions, ventricular weights of the pups are 2.5 fold less than those of euthyroid animals. All hearts used in these studies were rapidly removed after sacrifice of the animals and the ventricles frozen in liquid N_2 and stored at − 70° C.

The run-on assays were performed on nuclei isolated from 23 day-old female rats, because at this age the level of transcription is in theory still sufficiently high to yield detectable transcripts whereas myocyte division has virtually disappeared. The nuclei were isolated essentially as described by Liew *et al.* [7] and the exact composition of each solution can be found there with the exception that β-mercaptoethanol was substituted by dithiothreitol to a final concentration of 2.8 mM. In brief, the animals were sacrificied, the hearts removed and rinsed, followed by homogenization with a polytron. This and all subsequent steps were carried out at 4° C. After centrifugation, the pellet was resuspended, further homogenized with a mechanical pestle, and the homogenate filtered through two stainless steel sieves (200 and 350 mesh). This fraction was again pelleted and the cells lysed in a 0.5% Triton X-100/buffer. After centrifugation and resuspension, this fraction was loaded onto a sucrose gradient and centrifuged at 25000 rpm's in a SW28 Beckman rotor for two hours. The relatively pure nuclei were located at the interface of the 2.4 and 2.7 M sucrose beds. These nuclei were isolated, rinsed to remove excess sucrose, resuspended in Keller Storage Buffer [8], frozen in liquid N_2, and stored at − 70° C.

Patients

Hearts from 1 patient without heart failure (C) and 1 patient undergoing cardiac transplantation (NYHA class IV) (HF) were investigated. The patient without heart failure was an accident victim maintained under intensive care. The other patient suffered from chronic end-stage heart failure due to ischemic heart disease, and the heart was hypertrophied (heart weight = 500 g). In addition, a muscle specimen, excised < 12 hours after death from an autopsy patient, was used as a control of skeletal muscle. Tissue specimens were obtained as approved by the committees for the protection of human subjects at the respective institutes.

Primer extension

This type of assay allows simultaneous unambiguous identification and quantification of each actin iso-mRNA. It was performed in the rat heart as previously described [4, 5]. Briefly, total RNA was prepared by extraction in guanidinium isothiocyanate and isolated by sedimentation in cesium chloride followed by precipitation twice with acetic acid and ethanol. It was then hybridized with a ^{32}P labeled 18-base oligonucleotide whose sequence is complementary to codons 31–37 (in exon 2) which is identical in the two forms of rat sarcomeric α-actin mRNAs. Extension was initiated by the addition of reverse transcriptase (Bethesda Research Laboratories) and the extended fragments were separated on a 6% denaturing polyacrylamide/urea gel. For the human hearts, another oligomer was synthesized, complementary to a sequence identical to both human iso-mRNAs and located as for the rat in exon 2. The ability of this technique to resolve and precisely quantitate the two human iso-actins is described in detail elsewhere (Boheler *et al.* in preparation). The lengths of the extended fragments are in the rat 186 bases for α-skeletal actin and 195 bases for α-cardiac actin, and in man 222 and 174 bases, respectively. The data are reported as the ratio of skeletal to cardiac mRNA in each sample.

Nuclear run-on assay

Nuclear run-on assays were performed essentially as described by Konieczny and Emerson [8]. Each reaction was performed in the presence of 18×10^6 isolated cardiac nuclei, 500 μCi ^{32}P UTP (800 Ci/mmol), 312 μM CTP and GTP, and 625 μM ATP, and in the presence or absence of 2.0 μg/ml α-amanitin (Sigma). After the initial 40 minute incubation, the endogenous DNA was degraded by RQ-1 DNase 1 (Promega). Proteins were eliminated by incubation with proteinase K and SDS followed by phenol/chloroform extractions, and the transcripts were isolated from free nucleotides by 3 consecutive ethanol precipitations. The labeled transcripts were then hybridized [9] to single stranded probes fixed to Nylon membranes by UV illumination and prepared from M13 phage. The single stranded M13 probes fixed to the membranes were specific to the + and − strands of the corresponding genes of α-cardiac and α-skeletal actin [10], β-myosin heavy chain [11], and fibronectin (Boheler *et al.* in preparation). Membrane washes were as in [12] and were sufficient to remove all non-specific binding. Background was determined by hybridization to the M13 vector or to the + strand of the respective genes, and positive signals were considered as any signal 2 × greater than background.

Table 1.

Age	α-sk/α-card Actin mRNA	Reference
1 day	1.0	Bishopric *et al.* [4]
15 day	0.89	This study
	1.2	
2 months	0.036	Winegrad *et al.* [5]
24 months	0.0	This study
	0.01	
15 day	1.04	This study
hypothyroid	1.29	

Results

Table 1 shows that both isoactin mRNAs are present in the hearts of 15 day-old rats in approximately equal amounts, and that in hypothyroid animals, the ratio has the same range as that of euthyroid ones. In 24 month old animals by contrast, α-skeletal mRNA is hardly detectable and total sarcomeric actin is almost entirely composed of the cardiac form. In neonate ventricles, α-cardiac actin is ~ 50%, which corresponds to a ratio of skeletal to cardiac mRNA of around 1 [4], and in 2 month-old rats, we previously showed that the ratio dramatically dropped to 0.036 [5]. All these data indicate that α-skeletal mRNA is down-regulated in rats during the first two months after birth and that it is not reinduced with aging. Moreover, hypothyroid development does not seem to affect the relative proportions of the two actins, at least during the first two weeks of age.

In man, the two run-off products corresponding to α-skeletal and α-cardiac iso-mRNAs were of the expected length, and almost identical results were obtained by duplicate analysis of each RNA (Fig. 1). This demonstrated that the human RNAs were not degraded and that this method could be used to precisely quantitate the relative proportions of the two iso-mRNAs. Both sarcomeric α-actin transcripts were present in the two samples, and the ratios, α-skeletal to α-cardiac, were equal to 0.5 and 1.5 for the normal (C) and the failing heart (HF), respectively. It therefore appears that the two actin isogenes are co-expressed in adult human ventricles, regardless of the physiological or pathological condition.

Preliminary results using the run-on assays are

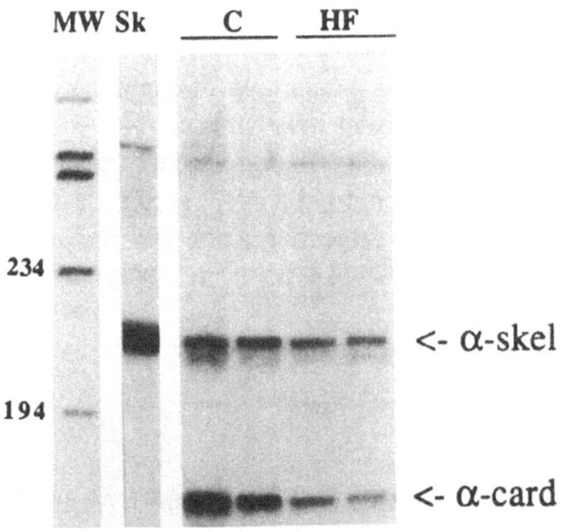

Fig. 1. Autoradiogram from a denaturing gel. The results of a primer extension analysis using a synthetic 18-base oligonucleotide with RNA from human skeletal muscle (Sk), and control (C) and failing (HF) left ventricles. Molecular weight markers (MW) are shown in the first lane indicating that the sizes of the reversed transcribed α-skeletal and α-cardiac actins migrate as predicted (222 and 174 nucleotides, respectively).

shown in Fig. 2. In this figure, it is important to note that positive signals are seen only with the probes specific for the two α-actin mRNAs, and in the presence of α-amanitin, these signals disappear. This implies that the genes for the α-actins are being actively transcribed in the nuclei of three week old rat hearts by RNA polymerase II, the enzyme responsible for mRNA transcription. No signal could be detected for β-MHC suggesting that this gene is not being actively transcribed. Additionally, no signal is detected from the fibronectin probe, and since fibronectin would only be expected to be expressed in non-myocytes, this indicates that the nuclei used in our experiments consist predominantly of myocytic nuclei. Thus at three weeks of age, the sarcomeric α-actin and β-myosin heavy chain genes appear to be independently regulated at the transcriptional level in the rat heart.

Discussion

Myosin heavy chain and actin multigene families are independently regulated

The results presented here, together with those of Bishopric *et al.* [4] and Winegrad *et al.* [5], indicate that the two α-actin genes are coexpressed in the rat heart at similar levels during at least the first two weeks of age and that the cardiac form becomes and remains virtually unique after 2 months. The values presented here from 15 day old rat hearts are somewhat higher than those we and others previously reported (inferior or equal to 10% in 8 day-old and 15 day-old animals) [13, 14]. Most probably, those lower values can be attributed to the potential ambiguity in interpreting signals obtained from the use of two different probes hybridized to dot and northern blots. The precise timing of α-skeletal actin down regulation after 15 days remains undefined, but these and previously reported results for α-cardiac actin and the myosin heavy chains suggest that the two main contractile proteins follow independent regulatory pathways during ontogenic development. We have summarized in Table 2 the time course of the expression of the two myosin heavy chains isoforms, α and β [15, 16]. Whereas the mRNAs of both species accumulate in almost equivalent amounts at birth, the β isoform decreases rapidly (the ratio β/α at 15 days is approximately half of that in neonates). β myosin heavy chain mRNA has almost disappeared in the 2 month-old animals, but it increases to represent almost 80 to 90% of total myosin heavy chain mRNA in aged rat hearts [16]. Thus the β myosin heavy chain gene is repressed earlier during development than the α-skeletal actin gene and is reinduced during senescence whereas α-skeletal actin is not. Moreover, the two multigene families respond differently with hypothyroid development. At 15 days, the pattern of isomyosins is drastically different from that of the euthyroid rats [6, 17], but there is apparently no difference in the isoactin pattern, at least in the two animals that we have studied (Table 1). In adult rats, hypothyroidism seems to de-induce the α-skeletal actin gene [5] and consequently, it is of interest to determine if some

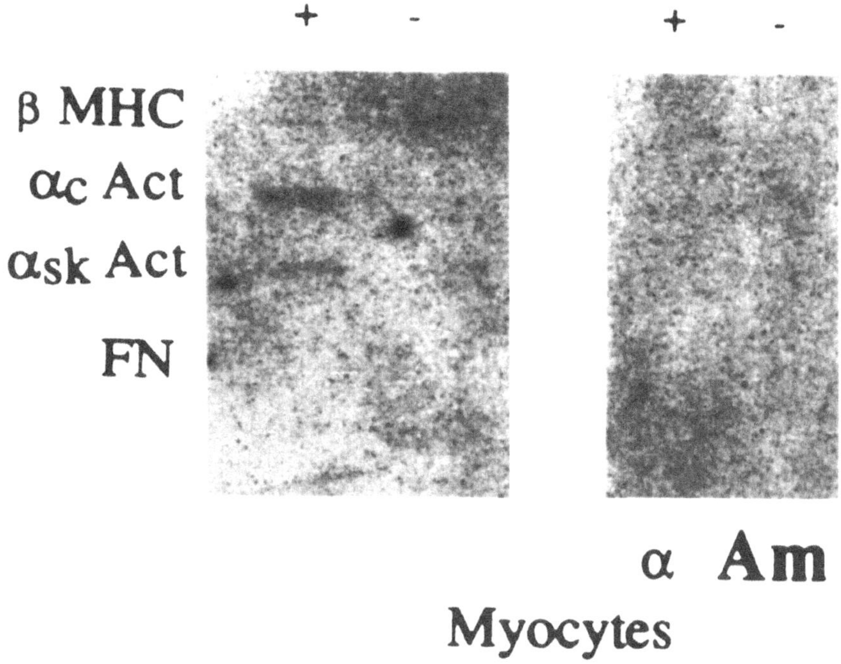

Fig. 2. Autoradiogram from the nuclear run-on transcription reactions. The isolated ^{32}P-labeled RNA from two run-on reactions was hybridized to nylon membranes containing 20 μg of the indicated single stranded probes derived from M13. 36×10^6 nuclei from three nuclei preparations were mixed on ice and equally divided into different tubes, one of which contained 2 μg/ml α-amanitin. β MHC, β myosin heavy chain; α_c and α_s Act, α-cardiac and α-skeletal actin; FN, fibronectin; +, single stranded probes that are complementary to the indicated RNA: −, single stranded probes that are identical to the indicated RNA; α AM, α-amanitin.

thyroid associated changes take place later during hypothyroid development.

In the near future, the nuclear run-on assay that we have established (Fig. 2) should allow us to distinguish between transcriptional and post-transcriptional events in the regulation of myosin heavy chain and actin isomRNA accumulations. Myosin heavy chain switches during development and in response to thyroid hormone or pressure overload depend upon pretranslational mechanisms [15, 18], but it has until now been difficult to study the transcriptional events because a reliable run-on assay on myocytes from juvenile and adult rats was not available. There are multiple reasons for this, the main ones being the relatively low level of cardiac transcription that rapidly declines between the 5th and 20th days after birth [19] and the difficulty of obtaining intact nuclei from myocytes containing a strong sarcomeric architecture and which are surrounded by an extensive extracellular matrix. We have, however, been able to obtain a high level of total incorporated radioactivity, which is only 5 fold less than that from freshly cultured neonatal myocytes [12] and the isolated transcripts are of relatively high molecular weight [20]. Moreover, the pattern of transcription that we observed

Table 2.

Age	β/α MHC mRNA	Reference
1 day	1.0	Lompre *et al.* [15]
15 day	0.5	Lompre *et al.*
[15]		
2 months	0.05	Izumo *et al.* [18]
24 months	9.0	O'Neill *et al* [16]
15 day hypothyroid	10.0*	Samuel *et al.* [6]

* This value was determined from the protein and not the mRNA accumulation levels.

(Fig. 2) is consistent with what is known concerning the levels of accumulation for the corresponding mRNAs. We nevertheless have very faint signals, most probably because the probes used here were very short, 68 nucleotides for β myosin heavy chain, and ~ 50 nucleotides for both α-skeletal and α-cardiac actin. By comparison, the actin probes used to study the effects of α 1-adrenergic stimulation on the rate of transcription from the actin genes in cultured neonate myocytes were approximately ten times longer than those used here, and even with those relatively long probes the signals were weak in the basal state [12]. We are now preparing longer intronic and exonic probes, and with these hope to explore in detail whether transcription is the major regulatory step in the modulation of contractile protein isogenes during cardiac development.

Both sarcomeric actins are expressed during end-stage heart failure in man

The iso-mRNA pattern of the human hearts that we studied (Fig. 1) is consistent with the two previous values reported in the literature. By northern blot analysis, Gunning *et al.* [21] found in a patient suffering from hypertrophic obstructive cardiomyopathy a ratio of α-skeletal to α-cardiac actin of approximately 1. Bennets *et al.* [22], by dot-blot analysis, reported in a patient who had fatal head injuries and a priori no cardiovascular disease, a lower ratio, 0.25. We found by primer extension, 0.5 in a control heart and 1.5 in an hypertrophied heart. Ratios around 0.25 were observed at the protein level in a control heart and in the heart of a patient suffering hypertrophic obstructive cardiomyopathy [23]. Thus it is not yet clear whether hypertrophy in man is accompanied by an isoactin switch. Analysis of more samples is necessary to answer this question. There can be no doubt, however, that the pattern of expression of sarcomeric actins is species-specific: α-cardiac is almost the exclusive form of adult rodent hearts, whereas both isogenes are expressed in human, pig and bovine hearts [23].

In conclusion, the expression of the multigene families encoding the two main contractile proteins is tissue and species dependent and follows independent time courses. These results provide further insights into the search for the regulatory elements that control these isogenes and underline the great complexity of sarcomere formation during myocardial growth.

Acknowledgements

We authors would like to thank Drs. P. Allen and M. Komajda (Coordonateur du réseau I.N.S.E.R.M. CAR 487018) for the human heart samples and Dr. E. Lakatta for the 24 month-old rat hearts. We would also like to thank Drs. Margaret Buckingham and Roger Cox for helping us establish the *in vitro* transcription assay and Ms. Claudine Wisnewsky for preparing the RNAs. This study was supported by I.N.S.E.R.M. and a grant from the Association Française contre les Myopathies. Dr. K. Boheler is the recipient of a fellowship from the Fondation de la Recherche Medicale in France and Dr. L. Carrier is a recipient of a fellowship from the Association Française contre les Myopathies.

References

1. Barany M: ATPase activity of myosin correlated with speed of muscle shortening. J Gen Physiol 50: 197–216, 1967
2. Buller AJ, Mommaerts W, Seraydarian K: Enzymic properties of myosin in fast and slow twitch muscles of the cat following cross-innervation. J Physiol 205: 581–597, 1969
3. Morgan HE, Gordon EE, Kira Y, Chua BHL, Russo LA, Peterson CJ, McDermott P, Watson PA: Biochemical mechanisms of cardiac hypertrophy. Ann Rev Physiol 49: 533–543, 1987
4. Bishopric NH, Simpson PC, Ordhal CP: Induction of the skeletal α-actin gene in α 1-adrenoreceptor-mediated hypertrophy of rat cardiac myocytes. J Clin Invest 80: 1194–1199, 1987
5. Winegrad S, Wisnewsky C, Schwartz K: Effect of thyroid hormone on the accumulation of mRNA for skeletal and cardiac α-actin in hearts from normal and hypophysectomised rats. Proc Natl Acad Sci USA 87: 2456–2460, 1990
6. Samuel JL, Rappaport L, Syrovy I, Wisnewsky C, Marotte F, Whalen RG, Schwartz K: Differential effect of thyroxine on atrial and ventricular isomyosins in rats. Amer J Physiol 250: H333–H341, 1986

7. Liew CC, Jackowski G, Ma T, Sole MJ: Nonenzymatic separation of myocardial cell nuclei from whole heart tissue. Am J Physiol 244: C3–C10, 1983
8. Konieczny SF, Emerson CP: Differentiation, not determination, regulates muscle gene activation: transfection of troponin I genes into multipotential and muscle lineages of 10T1/2 cells. Mol Cell Biol 5: 2423–2432, 1985
9. Marzluff WF, Huang RCC: Transcription of RNA in Isolated Nuclei. In: BD Hames, SJ Higgins (eds) Transcription and Translation. IRL Press, Washington, 1984, pp 89–129
10. Sassoon DA, Garner I, Buckingham M: Transcripts of α-cardiac and α-skeletal actins are early markers for myogenesis in the mouse embryo. Development 104: 155–164, 1988
11. Schiaffino S, Samuel JL, Sassoon D, Lompré AM, Garner I, Marotte F, Buckingham M, Rappaport L, Schwartz K: Nonsynchronous accumulation of α-skeletal actin and β-myosin heavy chain mRNAs during early stages of pressure-overload- induced cardiac hypertrophy demonstrated by *in situ* hybridization. Circ Res 64: 937–948, 1989
12. Long CS, Ordhal CP, Simpson PC: α 1-Adrenergic receptor stimulation of sarcomeric actin isogene transcription in hypertrophy of cultured rat heart muscle cells. J Clin Invest 83: 1078–1082, 1989
13. Schwartz K, de la Bastie D, Bouveret P, Oliviéro P, Alonso S, Buckingham M: α-skeletal actin mRNA's accumulate in hypertrophied adult rat hearts. Circ Res 59: 551–555, 1986
14. Mayer Y, Czosnek H, Zeelon P, Yaffe D, Nudel U: Expression of the genes coding for the skeletal muscle and cardiac actins in the heart. Nucleic Acids Res 12: 1087–2000, 1984
15. Lompré AM, Nadal-Ginard B, Mahdavi V: Expression of the cardiac ventricular α- and β- myosin heavy chain genes is developmentally and hormononally regulated. J Biol Chem 259: 6437–6446, 1984
16. O'Neill L, Holbrook N, Lakatta EG: Progressive changes in steady state mRNA levels of rat cardiac myosin heavy chain genes from young age to senescence. J Mol Cell Cardiol 22, Suppl I: S34, 1990
17. Chizzonite RA, Zak R: Regulation of myosin isoenzyme composition in fetal and neonatal rat ventricle by endogenous thyroid hormones. J Biol Chem 259: 12628–12632, 1984
18. Izumo S, Lompré AM, Matsuoka R, Koren G, Schwartz K, Nadal-Ginard B, Mahdavi V: Myosin heavy chain messenger RNA and protein isoform transitions during cardiac hypertrophy. J Clin Invest 79: 970–977, 1987
19. McCully JD, Liew CC: RNA transcription in myocardial cell nuclei during post-natal development. A study establishing an assay system for transcription *in vitro*. Biochem J 256: 441–445, 1988
20. Chassagne C, Boheler KR, Schwartz K: Description of an *in vitro* transcription assay in nuclei isolated from control and hemodynamically overloaded rat cardiac myocytes. CR Acad Sci Paris 312(III): 7–12, 1991
21. Gunning P, Ponte P, Blau H, Kedes L: α-skeletal and α-cardiac actin genes are coexpressed in adult human skeletal muscle and heart. Mol Cell Biol 3: 1449–1452, 1984
22. Bennetts BH, Burnett L, Dos Remedios CG: Differential co-expression of α-actin genes with the human heart. J Mol Cell Cardiol 18: 993–996, 1986
23. Vanderkerckove J, Bugaisky G, Buckingham M: Simultaneous expression of skeletal muscle and heart actin proteins in various striated muscle tissues and cells. J Biol Chem 261:1838–1843, 1986

Address for offprints: K. Schwartz, I.N.S.E.R.M. Unité 127, Hôpital Lariboisiére, 41 Blvd de la Chapelle, 75010 Paris, France

Molecular and Cellular Biochemistry **104**: 109–116, 1991.

Post-translational modification of proteins by 15-carbon and 20-carbon isoprenoids in three mammalian cell lines

James H. Reese and William A. Maltese
Weis Center for Research, Geisinger Clinic, Danville, PA 17822, USA

Key words: isoprenylation, farnesyl, geranylgeranyl, mevalonate, post-translational modification

Summary

A number of cellular proteins, including p21^{ras}, lamin B, and the G-protein γ subunits, undergo post-translational modification by 15-carbon farnesyl or 20-carbon geranylgeranyl isoprenoid moieties derived from pyrophosphate intermediates of the cholesterol biosynthetic pathway. In this study, isoprenylated proteins in three mammalian cell lines (Hela cells, Rat-6 fibroblasts and COS cells) were radiolabeled with an isoprenoid precursor, [^{3}H]mevalonate, and resolved by SDS gel electrophoresis. Groups of proteins with different molecular masses were eluted from the gels and the chain-lengths of the radiolabeled isoprenyl groups, released from the proteins by Raney-nickel-catalyzed desulfurization, were established by gel permeation chromatography. 15-Carbon and 20-carbon isoprenyl groups were found in separate classes of proteins within each cell line. With the exception of p21^{ras}, which incorporated a 15-carbon group when expressed in COS cells, the proteins in the region of the 21–28 kDa *ras*-related GTP binding proteins contained mostly 20-carbon isoprenyl chains. In contrast, proteins belonging to the 66–72 kDa nuclear lamin family, as well as unidentified proteins with molecular masses of 41–46 kDa and 53–55 kDa, contained predominantly 15-carbon isoprenyl chains. The chain-lengths of the isoprenoids associated with particular classes of proteins did not vary from one cell line to another, suggesting that the nature of the isoprenoid modification (farnesyl versus geranylgeranyl) is determined by intrinsic structural features of the proteins, rather than the cell type in which the proteins are expressed.

Abbreviations: MVA – Mevalonolactone; SDS – Sodium Dodecyl Sulfate; PAGE – Polyacrylamide Gel Electrophoresis.

Introduction

Isoprenylation is a post-translational modification wherein an isoprenoid moiety derived from mevalonate is attached to a protein via a cysteine thioether linkage [reviewed in refs. 1–2]. Thus far two types of protein-associated isoprenyl groups have been identified. Farnesyl groups (15-carbon) have been found in extracts of total protein from erythroleukemia cells [3] and are known to be incorporated into specific proteins such as p21^{ras} [4–6], lamin B [7] and the γ subunit of transducin [8]. Geranylgeranyl groups (20-carbon) are the predominant isoprenoids in total protein extracts from Hela cells [9] and Chinese hamster ovary cells [10], and in the γ subunits of brain G-proteins [11, 12]. The structural motif that marks proteins as substrates for farnesylation is thought to include the last four amino acids of the carboxyl-terminal. This motif, known as a '*Caax*' box, contains the target cysteine followed by two aliphatic amino acids and a variable amino acid [4–6, 13–15]. Geranylgeranylated proteins, such as the γ subunits of brain G-proteins [11, 12, 16], also contain carboxyl-terminal sequences that fit this general pattern.

It is not yet clear why some proteins are modified by farnesyl groups while others are modified by geranylgeranyl groups. One possibility is that proteins with a *Caax* box may serve as acceptors for either farnesyl or geranylgeranyl groups, with the overall proportion of proteins modified by 15-carbon versus 20-carbon isoprenyl groups reflecting the relative sizes of the isoprenyl pyrophosphate donor pools (farnesyl-PP versus geranylgeranyl-PP) produced by the isoprenoid biosynthetic pathway in a particular tissue or cell type. A second possibility is that cells contain separate enzyme systems which catalyze each type of isoprenoid modification, and that specific proteins can serve as acceptors for only one type of isoprenyl group. The present study was initiated with the aim of distinguishing between these possibilities.

Materials and methods

Materials

R,S-[^{3}H]Mevalonolactone was prepared by reduction of mevaldic acid precursor with sodium boro[^{3}H]hydride (Amersham) as described by Keller [17]. Dulbecco's modified Eagle medium and fetal bovine serum were from GIBCO/BRL. Electrophoresis reagents and protein determination kits were from Bio-Rad Laboratories. Tissue culture plasticware (Corning), platinum IV oxide hydrate (Aldrich), HPLC-grade organic solvents (Fisher), ReadySafe scintillation cocktail (Beckman), and molecular mass standards (Amersham) were obtained from the designated sources. 2,6,10-Trimethyldodecane (farnesane) was purchased from Wiley Organics and 2,6,10,14-tetramethylhexadecane (phytane) was custom-synthesized by Farchan Laboratories. Lovastatin was a gift from Alfred W. Alberts of Merck, Sharp & Dohme Research Laboratories. Raney nickel activated catalyst and all other chemicals were from Sigma Chemical Co.

Cell culture and metabolic labeling of isoprenoids

Hela(S3) cells were obtained from the American Type Culture Collection. COS-M6 cells were a gift from Dr. David Russell (University of Texas Southwestern Medical Center), and H-*ras*-transformed Rat-6 fibroblasts were generously provided by Drs. Wendy Hsiao and I. Bernard Weinstein (Colombia University Cancer Center, NY). All cells were grown in monolayer culture in Dulbecco's modified Eagle medium (high glucose formula), supplemented with 10% (v/v) fetal calf serum, as described previously [18, 19].

To obtain cells that were capable of transiently overexpressing p21$^{\text{H-ras}}$, COS cells that had attained 50% confluency were transfected with the pCMV5 plasmid containing the H-*ras* (wt) insert as described previously [16, 20], except that the medium was supplemented with 2% (v/v) Low-Protein Serum Replacement (Sigma) instead of fetal calf serum during the initial incubation of cells with plasmid DNA and DEAE-dextran.

For labeling of cellular isoprenoids with [^{3}H]MVA, cells were grown in 75 cm^2 culture flasks until they were 70–80% confluent. Fresh medium containing 25 μM lovastatin and [^{3}H]MVA (3.15 Ci/mmol, 200 μCi/ml of medium) was added and the cells were incubated for 18 h at 37° C. Cells were harvested as described previously [18, 19] and protein was quantitated by means of the Bio-Rad assay kit.

Electrophoresis and elution of isoprenylated proteins

Proteins from whole cells were solubilized in SDS-electrophoresis sample buffer containing 8 M urea and subjected to one-dimensional SDS-PAGE on 8%–17% polyacrylamide gradient gels as described previously [19, 20]. For each cell line, approximately 1.3 mg of protein was distributed among 13 lanes of a gel. In the cases of the H-*ras*-transfected COS cells (Fig. 4), the total protein loaded on the gel was approximately 400 μg (100 μg/lane). Upon completion of electrophoresis one lane of each gel was dried, permeated with Amplify (Amersham) and exposed to X-Omat AR (Kodak) film for 3–5 days to create a fluorographic record of the proteins

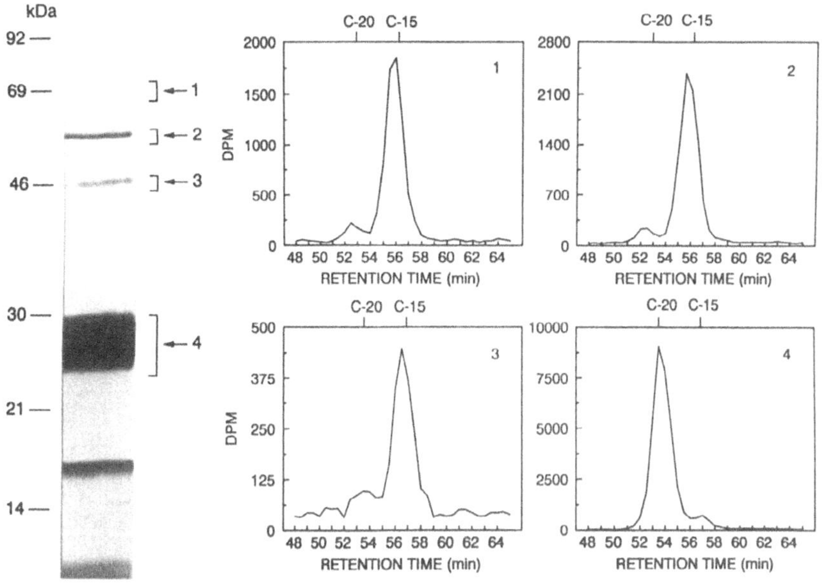

Fig. 1. Determination of the chain-lengths of radiolabeled isoprenyl groups derived from Hela cell proteins. Cells were incubated for 18 h with [^{3}H]MVA. Isoprenylated proteins were separated by SDS-PAGE and visualized by fluorography (positions of the molecular mass standards are indicated on the left). Specific zones indicated by the numbered arrows were excised from adjacent sections of the non-fluorographed gel, and the proteins were electroeluted. Radiolabeled isoprenyl groups were released from the proteins by Raney-nickel cleavage, hydrogenated over platinum, and characterized by gel permeation chromatography as described in *Materials and methods.* The chromatograms of the radiolabeled hydrocarbons derived from the isoprenylated proteins are numbered to correspond to the zones of the gel from which they were eluted. The retention times of the 15-carbon farnesane and 20-carbon phytane standards are shown at the top of each chromatogram.

containing [^{3}H]MVA-derived isoprenyl groups. The remaining portion of each gel was frozen and stored at − 80° C. Following exposure of the fluorographs, R_f values were calculated for various radiolabeled protein bands and these values were used to locate the corresponding bands on the frozen gels. Individual bands were excised and the proteins were electroeluted in a Bio-Rad apparatus (10 mA/elution tube) for 5 h at room temperature in a buffer consisting of 25 mM Tris base, 192 mM glycine and 0.1% (w/v) SDS.

Removal of radiolabeled isoprenyl groups from proteins

Bovine serum albumin (final concentration, 0.5 mg/ml) was added to each eluted protein sample and SDS was removed by precipitation with potassium phosphate [21]. Proteins were collected by precipitation with ice-cold acetone and dissolved in 0.4 ml 8 M guanidine-HCl, 200 mM sodium phosphate, pH 7.0. The tritium-labeled isoprenyl groups were released from the proteins by reaction with Raney nickel-activated catalyst under nitrogen at 100° C for 16 h, essentially as described by Farnsworth *et al.* [7].

Estimation of the chain-length of radiolabeled isoprenyl groups

The radiolabeled isoprenyl groups released by Raney nickel cleavage were extracted into pentane and hydrogenated over platinum oxide (Adams' catalyst) for 2 h [7]. The chain-lengths of the radiolabeled saturated hydrocarbons in a 50 μl aliquot of the pentane extract were determined by gel per-

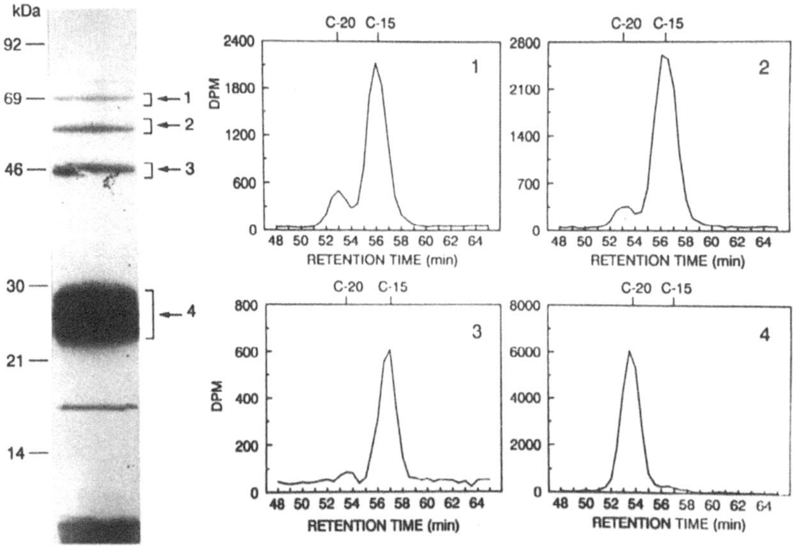

Fig. 2. Determination of the chain-lengths of radiolabeled isoprenyl groups derived from proteins of H-*ras*-transformed Rat-6 fibroblasts. Cells were incubated for 18 h with [^{3}H]MVA. Isoprenylated proteins were separated by SDS-PAGE and visualized by fluorography (positions of the molecular mass standards are indicated on the left). Specific zones indicated by the numbered arrows were excised from adjacent sections of the non-fluorographed gel, and the proteins were electroeluted. Radiolabeled isoprenyl groups were released from the proteins by Raney-nickel cleavage, hydrogenated over platinum, and characterized by gel permeation chromatography as described in *Materials and methods.* The chromatograms of the radiolabeled hydrocarbons derived from the isoprenylated proteins are numbered to correspond to the zones of the gel from which they were eluted. The retention times of the 15-carbon farnesane and 20-carbon phytane standards are shown at the top of each chromatogram.

meation chromatography. HPLC was performed on a Beckman System Gold chromatograph, using two tandem 300 mm × 7.8 mm columns of 5 micron Phenogel (Phenomenex). The first column had a pore size of 50 Å and the second had a pore size of 100 Å. Hydrocarbons were eluted with tetrahydrofuran at a flow rate of 0.25 ml/min. Fractions were collected at 0.5 min intervals, mixed with ReadySafe, and counted for tritium in a liquid scintillation spectrometer. Prior to injection, the radiolabeled material was mixed with 2.5 μl of 2,6,10-trimethyldodecane (farnesane) and 2,5 μl 2,6,10,14,-tetramethylhexadecane (phytane). The elution positions of the 15-carbon and 20-carbon standards in each run were determined with a Beckman model 156 refractive index detector.

Results

The electrophoretic profiles of isoprenylated (i.e., [^{3}H]MVA-labeled) proteins are generally similar in various types of cells (1; see Figs. 1–4). Based on their molecular masses, calculated from mobilities in SDS gels, the isoprenylated proteins can be grouped into several distinct categories. Most of the radiolabeled MVA is incorporated into a group of 21–28 kDa proteins that includes p21^{ras} and many other *ras*-related GTP-binding proteins [4, 15, 19, 20]. The labeled proteins at 66–72 kDa consist primarily of the nuclear lamins [7, 22, 23], whereas those at 41-46 kDa, 53–55 kDa and 14–17 kDa are currently unidentified. Although the isoprenylated G-protein γ subunits (5–8 kDa) normally migrate in the heavily labeled region near the dye front, most of the radiolabeled material in this region appears not to be covalently associated with protein, since it can be extracted by treatment of the

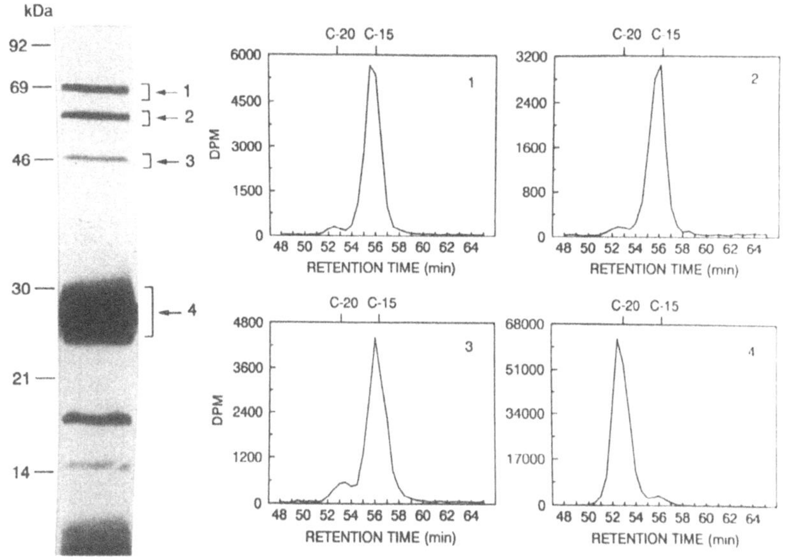

Fig. 3. Determination of the chain-lengths of the radiolabeled isoprenyl groups derived from simian COS cell proteins. Cells were incubated for 18 h with [^{3}H]MVA. Isoprenylated proteins were separated by SDS-PAGE and visualized by fluorography (positions of the molecular mass standards are indicated on the left). Specific zones indicated by the numbered arrows were excised from adjacent sections of the non-fluorographed gel, and the proteins were electroeluted. Radiolabeled isoprenyl groups were released from the proteins by Raney-nickel cleavage, hydrogenated over platinum, and characterized by gel permeation chromatography as described in *Materials and methods*. The chromatograms of the radiolabeled hydrocarbons derived from the isoprenylated proteins are numbered to correspond to the zones of the gel from which they were eluted. The retention times of the 15-carbon farnesane and 20-carbon phytane standards are shown at the top of each chromatogram.

cell lysates with organic solvents prior to electrophoresis.

Recent studies have shown that lamin B isolated from Hela cells is modified by a 15-carbon farnesyl moiety [7], whereas peptide fragments derived from digests of total Hela cell protein contain mostly 20-carbon geranylgeranyl groups [9]. To determine which classes of isoprenylated proteins in Hela cells might be modified by 15-carbon versus 20-carbon groups, [^{3}H]MVA-labeled proteins were excised from defined zones of SDS gels, and the radiolabeled hydrocarbon chains derived from isoprenyl groups released by Raney-nickel-catalyzed cleavage were separated according to size, using gel permeation chromatography (Fig. 1). Consistent with the known farnesyl modification of lamin B in Hela cells, the radiolabeled hydrocarbon released from the 66–72 kDa proteins in this cell line migrated with the 15-carbon farnesane standard. Similarly, the radiolabeled material released from the unidentified isoprenylated proteins with molecular masses of 53–55 kDa and 41–46 kDa also appeared to consist primarily of 15-carbon isoprenoids. In contrast to the results obtained with the high-molecular-mass proteins, the radiolabeled hydrocarbons released from the 21–28 kDa proteins migrated with the 20-carbon phytane standard, suggesting that the geranylgeranyl groups observed in previous studies of Hela cell peptides [9] arose primarily from the low molecular mass GTP-binding proteins.

Yields of pentane-extractable radiolabeled isoprenoid after incubation with Raney-nickel activated catalyst generally were 80–95% for proteins eluted from the 21–28 kDa, 41–46 kDa and 66–72 kDa regions of the gel, and 50–58% for proteins in the 53–58 kDa region. Therefore, the primary mode of association between the isoprenyl groups and these proteins appeared to be via a thioether bond. Analysis of the [^{3}H]MVA-derived material

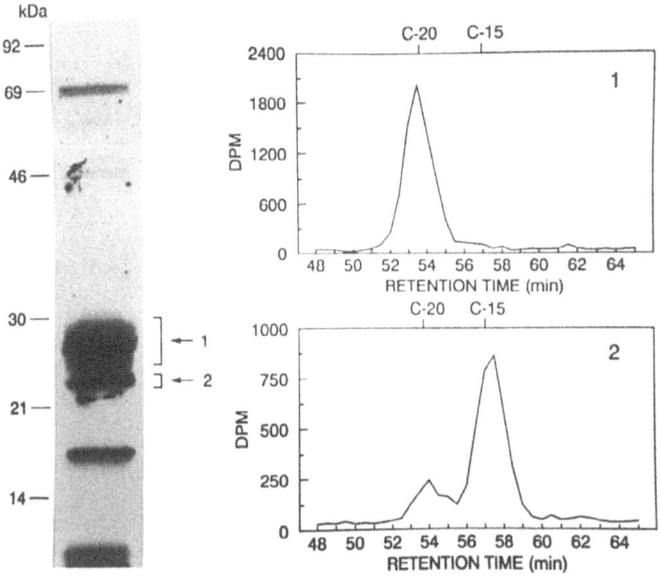

Fig. 4. Determination of the chain-lengths of the radiolabeled isoprenyl groups derived from proteins of COS cells transfected with H-*ras* and transiently expressing p21$^{H\text{-}ras}$ protein. Cells were incubated for 18 h with [^{3}H]MVA. Isoprenylated proteins were separated by SDS-PAGE and visualized by fluorography (positions of the molecular mass standards are indicated on the left). Specific zones indicated by the numbered arrows were excised from adjacent sections of the non-fluorographed gel, and the proteins were electroeluted. Radiolabeled isoprenyl groups were released from the proteins by Raney-nickel cleavage, hydrogenated over platinum, and characterized by gel permeation chromatography as described in *Materials and methods*. The chromatograms of the radiolabeled hydrocarbons derived from the isoprenylated proteins are numbered to correspond to the zones of the gel from which they were eluted. The retention times of the 15-carbon farnesane and 20-carbon phytane standards are shown at the top of each chromatogram.

associated with the proteins at approximately 17 kDa was complicated by poor yields of pentane-extractable material (< 10%) following Raney-nickel cleavage. Incubation of these gel-eluted proteins under conditions designed to break thioester bonds (i.e., 0.1 M methanolic KOH, 1 N hydroxylamine) also failed to release significant amounts of radiolabeled material. For this reason, these proteins were not included in the subsequent studies described below.

Having established which groups of proteins were modified by thioether-linked 15-carbon or 20-carbon isoprenyl chains in Hela cells, we next asked whether proteins in the same molecular mass categories were modified by isoprenoids of the same chain-length in two additional cell lines; H-*ras*-Rat-6, a rat fibroblast line stably transformed by H-*ras* (Fig. 2), and simian COS cells (Fig. 3). In each case the distribution of 15-carbon versus 20-carbon isoprenyl groups among the major isoprenylated proteins, classified according to molecular mass, was the same as that observed in the Hela cells.

As mentioned earlier, several studies have provided evidence that the H-, K- and N-p21^{ras} proteins are farnesylated [4–6]. The fact that the predominant chain-length of the isoprenyl groups extracted from the 21–28 kDa proteins of the H-*ras*-Rat-6 cells was 20 carbon (see Fig. 2) was not inconsistent with this finding, since p21^{ras} actually accounts for only a small proportion of the [^{3}H]MVA-derived radioactivity incorporated into the 21–28 kDa protein fraction in this cell line [19]. The ability of the gel permeation technique to detect the 15-carbon modification of p21^{ras} was verified in a separate study in which the isoprenoid chain-length analysis originally performed on proteins from non-transfected COS cells (Fig. 3) was repeated with cells that were transiently overproducing p21$^{H\text{-}ras}$ after transfection with the H-*ras* gene (Fig. 4). Enhanced incorporation of [^{3}H]MVA into a protein at approximately 22 kDa was readily detected in the fluorograph of the proteins from the H-*ras* tranfected COS cells (Fig. 4). Previous studies have shown that this protein is recognized by antibodies to p21^{ras} [16, 20]. When this protein was eluted and its isoprenyl group was subjected to chain-length analysis, the radiolabeled hydrocarbon migrated with the 15-carbon standard, in marked contrast to the other proteins of this molecular mass category, which contained 20-carbon isoprenyl chains.

Discussion

Farnesyl (15-carbon) and geranylgeranyl (20-carbon) are the two types of isoprenyl groups thus far identified as being linked to proteins in mammalian cells. Our findings extend the previous work of Farnsworth *et al.* [7, 9], insofar as they confirm that both 15-carbon and 20-carbon isoprenylated proteins can be generated within a single cell line. However, the present study yields several new insights regarding protein isoprenylation. The data indicate that with the exception of the p21^{ras} modification, which involves a 15-carbon farnesyl group, the most prevalent isoprenoid modification of the low molecular mass GTP-binding protein family (21–28 kDa) involves the 20-carbon geranylgeranyl moiety. The analyses of the isoprenylated proteins with molecular masses of 41–46 kDa and 53–55 kDa also provide the first evidence that these unidentified proteins, like the members of the 66–72 kDa lamin family, are modified predominantly by 15-carbon isoprenyl chains. Most importantly, the observation that particular classes of proteins were always modified by isoprenyl groups of a specific chain-length in three different cell lines suggests that the nature of the isoprenoid modification (15-carbon versus 20-carbon) is determined by intrinsic characteristics of the proteins, and is therefore protein-specific rather than cell-type specific.

The exact structural features that serve to identify proteins as acceptors for 15-carbon or 20-carbon isoprenyl groups remain to be delineated, but recent studies conducted *in vitro* with p21^{ras} farnesyl : protein transferase suggest that the amino acid composition and sequence of the *Caax* box at the carboxyl terminus of the acceptor protein may play a critical role in the recognition process [5, 6]. Another major unanswered question is whether the farnesyl : protein transferase can utilize geranylgeranyl-PP as an alternate substrate and thus catalyze both the 15-carbon and the 20-carbon modifications, or whether the transfer of geranylgeranyl groups to proteins is catalyzed by a separate enzyme.

The function of protein isoprenylation represents a major area for future investigation. Studies of the *ras* proteins indicate that farnesylation is required for proteolytic processing and subsequent carboxylmethylation of the carboxyl-terminal cysteine [15]. Moreover, prevention of isoprenylation of oncogenic *ras* proteins by site-directed mutagenesis or pharmacological blockade of isoprenoid biosynthesis abolishes their transforming activity, presumably by interfering with membrane localization of the proteins [4, 15, 24]. A role for farnesylation in the localization of the lamins to the nuclear envelope has also been suggested [25–27]. However, the concept of the isoprenyl chain as a simple membrane anchor may need to be modified in light of recent evidence that some isoprenylated low molecular mass GTP-binding proteins exist primarily as soluble entities in several cell lines [19, 20]. It will be of particular interest to determine whether the 15-carbon farnesyl and 20-carbon geranylgeranyl post-translational modifications are functionally equivalent, or whether each type of modification triggers a different set of protein processing and localization events.

Acknowledgements

This work was supported in part by United States Public Health Service Grant CA 34569.

References

1. Maltese, WA: Post-translational modification of proteins by isoprenoids in mammalian cells. FASEB J. 4: 3319–3328, 1990
2. Glomset JA, Gelb MH, Farnsworth CC: Prenyl proteins in eukaryotic cells: a new type of membrane anchor. TIBS 15: 139–142, 1990
3. Maltese WA, Erdman RA: Characterization of isoprenoid involved in the post-translational modification of mammalian cell proteins. J Biol Chem 264: 18168–18172, 1989
4. Casey PJ, Solski PA, Der CJ, Buss JE: $p21^{ras}$ is modified by a farnesyl isoprenoid. Proc Natl Acad Sci USA 86: 8323–8327, 1989
5. Reiss Y, Goldstein JL, Seabra MC, Casey PJ, Brown MS: Inhibition of purified $p21^{ras}$ farnesyl:protein transferase by Cys-AAX tetrapeptides. Cell 62: 81–88, 1990
6. Schaber MD, O'Hara MB, Garsky VM, Mosser SD, Bergstrom JD, Moores SL, Marshall MS, Friedman PA, Dixon RAF, Gibbs JB: Polyisoprenylation of ras *in vitro* by a farnesyl-protein transferase. J Biol Chem 265: 14701–14704, 1990
7. Farnsworth CC, Wolda S, Gelb MH, Glomset JA: Human lamin B contains a farnesylated cysteine residue. J Biol Chem 264: 20422–20429, 1989
8. Fukada Y, Takao T, Ohguro H, Yoshizawa T, Akino T, Shimonishi Y: Farnesylated γ subunit of photoreceptor G protein indispensable for GTP-binding. Nature 346: 658–660, 1990
9. Farnsworth CC, Gelb MH, Glomset JA: Identification of geranylgeranyl-modified proteins in Hela cells. Science 247: 320–322, 1990
10. Rilling HC, Breunger E, Epstein WW, Crain PF: Prenylated proteins: the structure of the isoprenoid group. Science 247: 318–319, 1990
11. Yamane HK, Farnsworth CC, Xie H, Howald W, Fung BKK, Clarks S, Gelb MH, Glomset JA: Brain G protein γ subunits contain an all-trans-geranylgeranyl-cysteine methyl ester at their carboxyl termini. Proc Natl Acad Sci USA 87: 5868–5872, 1990
12. Mumby SM, Casey PJ, Gilman AG, Gutowski S, Sternweis PC: G protein γ subunits contain a 20-carbon isoprenoid. Proc Natl Acad Sci USA 87: 5873–5877, 1990
13. Anderegg RJ, Betz R, Carr SA, Crabb JW, Duntze W: Structure of *Saccharomyces cerevisiae* mating hormone a-factor: identification of S-farnesyl cysteine as a structural component. J Biol Chem 263: 18236–18240, 1988
14. Clarke S, Vogel JP, Deschenes RJ, Stock J: Post-translational modification of the Ha-ras oncogene protein: evidence for a third class of protein carboxyl methyltransferases. Proc Natl Acad Sci USA 85: 4643–4647, 1988
15. Hancock JT, Magee AI, Childs JE, Marshall CJ: All ras proteins are polyisoprenylated but only some are palmitoylated. Cell 57: 1167–1177, 1989
16. Maltese WA, Robishaw JD: Isoprenylation of C-terminal cysteine in a G-protein γ subunit. J Biol Chem 265: 18071–18074, 1990
17. Keller RK: The mechanism and regulation of dolichyl phosphate biosynthesis in rat liver. J Biol Chem 261: 12053–12059, 1986
18. Maltese WA, Sheridan KM: Isoprenylated proteins in cultured cells: subcellular distribution and changes related to altered morphology and growth arrest induced by mevalonate deprivation. J Cell Physiol 133: 471–481, 1987
19. Maltese WA, Sheridan KM, Repko EM, Erdman RA: Post-translational modification of low molecular mass GTP-binding proteins by isoprenoid. J Biol Chem 265: 2148–2155, 1990
20. Maltese WA, Sheridan KM: Isoprenoid modification of G25K (G_p), a low molecular mass GTP-binding protein distinct from $p21^{ras}$. J Biol Chem 265: 17883–17890, 1990
21. Suzuki H, Terada T: Removal of dodecyl sulfate from protein solution. Anal Biochem 172: 259–263, 1988
22. Wolda SL, Glomset JA: Evidence for modification of lamin B by a product of mevalonic acid. J Biol Chem 263: 5977–6000, 1988
23. Beck LA, Hosick TJ, Sinensky M: Incorporation of a product of mevalonic acid metabolism into proteins of Chinese hamster ovary cell nuclei. J Cell Biol 107: 1307–1316, 1988
24. Schafer WR, Kim R, Sterne R, Thorner J, Kom SH, Rine J: Genetic and pharmacologic suppression of oncogenic mutations in RAS gene of yeast and humans. Science 245: 379–284, 1989
25. Holtz D, Tanaka RA, Hartwig J, McKeon F: The CAAX motif of lamin A functions in conjunction with the nuclear localization signal to target assembly to the nuclear envelope. Cell 59: 969–977, 1989
26. Krohne G, Waizenegger I, Hoger TH: The conserved carboxyl-terminal cysteine of nuclear lamins is essential for lamin association with the nuclear envelope. J Cell Biol 109: 2003–2011, 1989
27. Beck LA, Hosick TJ, Sinensky M: Isoprenylation is required for the processing of the lamin A precursor. J Cell Biol 110: 1489–1499, 1990

Address for offprints: W.A. Maltese, Weis Center for Research, Geisinger Clinic, 26–16, North Academy Ave. Danville, PA 17822, USA

RIBOSOME BIOGENESIS: TRANSCRIPTION OF rDNA

Molecular and Cellular Biochemistry **104**: 119–126, 1991.

Initiation and regulation mechanisms of ribosomal RNA transcription in the eukaryote Acanthamoeba castellanii*

Marvin R. Paule, Erik Bateman[2], Laura Hoffman[3], Calvin Iida, Martin Imboden, William Kubaska, Preecha Kownin, Han Li, Alan Lofquist, Peggy Risi, Qin Yang and Michael Zwick[4]
Department of Biochemistry and the Cell and Molecular Biology Program, Colorado State University, Fort Collins, CO 80523, USA; [2] present address; Department of Medical Biology, College of Medicine, University of Vermont, Burlington, VT 05405, USA; [3] present address: Department of Biochemistry, School of Medicine, University of Utah, Salt Lake City, UT 84132, USA; [4] present address: Department of Biochemistry, University of Missouri, Columbia, MO 65211, USA

Key words: DNA melting, transcription initiation, footprinting, RNA polymerase I, posttranslational modification

Abstract

Acanthamoeba rRNA transcription involves the binding of a transcription initiation factor (TIF) to the core promoter of rDNA to form the preinitiation complex. This complex is formed in the absence of RNA polymerase I, and persists for multiple rounds of initiation. Polymerase I next binds to form the initiation complex. This binding is DNA sequence-independent, and is directed by protein-protein contacts with TIF. DNA melting occurs in a separate step. In contrast to most prokaryotic transcription, melting occurs only following nucleotide addition and β-γ hydrolysis of ATP is not required as for polymerase II. Growth-dependent regulation of rRNA transcription is accomplished by modification of RNA polymerase I. The inactive form of polymerase (Pol_E) is unable to bind to the promoter and has altered heat stability. Pol_E is still active in elongation; thus, the modification affects the polymerase site involved in TIF contact. Modification of a polymerases I and III common subunit has been detected leading to the suggestion that transcription of stable RNAs of the ribosome might be co-regulated by this mechanism.

Introduction

Eukaryotic transcription requires one or several proteins in addition to RNA polymerase for specific initiation to occur. These factors function by binding to specific sequences in the DNA, or to previously DNA-bound proteins, on what may be thought of as an extended promoter. For transcription of rRNA precursor genes by RNA polymerase I in *Acanthamoeba*, only a single protein is required, in contrast to vertebrate systems. Preliminary data which may reconcile this apparent difference between the species, plus information which unifies the apparently smaller promoter of the *Acanthamoeba* rRNA gene are discussed below. Until recently, it had not been clear how the RNA polymerase itself found the promoter, nor what subsequent steps occurred to produce an elongation complex. Here, we review experiments carried out in our laboratory which demonstrate that RNA polymerase I, and presumably all three eukaryotic RNA polymerases, locates the promoter by direct

* Dedicated to the memory of Preecha Kownin

interaction with the transcription factor(s) and not by sequence-specific interactions with the DNA. Further, we show that DNA melting occurs as a distinct subsequent step requiring nucleoside triphosphates.

Studies in this laboratory and others have revealed that several mechanisms are used by eukaryotes to regulate ribosomal RNA transcription [1–4]. In *Acanthamoeba,* covalent modification of RNA polymerase I, or a transcription factor tightly associated with it, preventing it from binding to the promoter was the first demonstrated mechanism used to regulate rRNA synthesis in response to cellular growth rate [1], and the data leading to this conclusion is reviewed. In many cell types including *Acanthamoeba,* 5S RNA transcription is coregulated in response to growth, and we have detected modification of a subunit common to the two polymerases responsible for transcription of these two genes. This subunit appears to be the homolog of the *E. coli* α-subunit [5]. A mechanism for coordinate regulation of both transcription units is proposed.

The ribosomal RNA transcription initiation process

Formation of the preinitiation complex

RNA polymerase I is unable to faithfully initiate transcription *in vitro* in the absence of an additional protein or proteins which we have termed *t*ranscription *i*nitiation *f*actor(s) [TIF]. In the eukaryotic protist, *Acanthamoeba castellanii,* only a single TIF appears to be required. This contrasts distinctly with results obtained in several vertebrate systems [6–9], and has been a subject of controversy. In these higher organisms, a minimum of two *trans*-acting transcription factors have been identified; UBF [also called SF-1[1], factor D, TFID] and SL-1 [also called TIF-IB or factor D, the latter may be an unresolved mixture of two or more factors]. Of these, only UBF, which is present in amounts 100 fold higher than the other, has been purified to homogeneity from several organisms [10–12]. The purified protein usually demonstrates a pair of closely spaced 'bands' on SDS-PAGE, with an average size ranging from 88 Kd to 98 Kd. A clone of this protein suggests it to be a single gene product [10]. The remaining protein has proven difficult to purify, partly because it is present in extremely low amounts even in rapidly growing cells. The subunit composition of TF partially purified from *Acanthamoeba* suggests it to consist of two heterologous subunits, one of which has a molecular weight and electrophoretic characteristics similar to human UBF. The second, larger *Acanthamoeba* TIF subunit is presumably the cognate of SL-1.

We have shown by using a template commitment assay and DNA footprinting that one role of TIF is to form a stable complex with the DNA template, and that this occurs in the absence of RNA polymerase [13]. In the template commitment assay, a promoter-containing DNA template is added to a limiting amount of one or more transcriptional proteins, given the opportunity to form a complex with the protein, and then challenged with a second DNA whose transcription product is distinguishable by length (in a runoff assay) from the first. Exclusion of second-DNA transcription takes place whenever a stable complex forms with the first DNA during the preincubation. Our results indicated that TIF forms a stable *preinitiation complex* with the template. Even TIF purified to near homogeneity forms preinitiation complexes. These complexes remain completely intact for at least 16 hours in the absence of RNA polymerase I and during multiple rounds of transcription (see below). RNA polymerase alone, however, is not able to recognize or to bind the promoter in the absence of TIF. These conclusions have been verified by footprinting of the various complexes (see below).

[1] The nomenclature used by different laboratories for the required *trans*-acting transcription factors remains confused, partly because most laboratories are working with fractions, rather than pure components. Until all factors have been purified to homogeneity, cloned, and their relationships at the sequence level reconciled, this situation is likely to continue.

Formation of the initiation complex

Subsequent to the formation of the preinitiation complex, polymerase I binds to form the *initiation complex.* This temporal order of binding can be inferred from the independent binding of TIF, but not of polymerase, to the DNA template in the template commitment assay. In addition, the use of a DNA footprinting assay provides direct evidence that polymerase binds in a separate step [14]. This assay involves determination of the regions of the DNA protected by bound proteins from cleavage reagents [DNase I or methidiumpropylEDTA : Fe (II) (MPE)] as revealed by gaps in the ladder of bands seen on a sequencing gel [15].

Bound TIF protects a region from − 17 to − 67. When polymerase is added to the preinitiation complex, the footprint is extended from the border of the TIF footprint to + 18. No footprint appears when polymerase is added to DNA in the absence of TIF. The change in the protected region upon addition of polymerase is due to the binding of polymerase, as opposed to an induced conformational change in TIF, since the putative polymerase I footprint can be made to translocate down the template in response to the addition of nucleoside triphosphates [2, 16]. Incubation with all four NTPs eliminates the polymerase footprint entirely while incubation with ATP, GTP and 3′-O-methyl-CTP, which allows the synthesis of an octamer product, results in translocation of the footprint approximately 8 base pairs.

It has recently been reported (see Sollmer-Webb, this issue) that in a relatively crude *in vitro* transcription system from mouse, RNA polymerase I is only capable of catalyzing a single round of transcription. Polymerase recovered after synthesis of a 53 nucleotide product is inactive, while after elongating a 10 nucleotide product it is active. This is an exciting result in relation to regulation (see below), but does not appear to be fundamental to the initiation/elongation process per se. In the *Acanthamoeba* system, using more purified components than the mouse system, we find that the polymerase promoter-footprint is not completely lost unless the enzyme is given very long runoff template (approaching 4 Kb). With more truncated templates (ca. 700 bp), the polymerase which has run off returns and rebinds to the promoter, sitting in the initiation complex before initiation long enough to partially protect the DNA.

Persistence of the preinitiation complex following

When initiation is allowed to occur, either to form an octamer or a complete transcript, the TIF footprint remains unaltered [2]. In addition, in template commitment assays, one never observes significant transcription of the second template despite multiple rounds of transcription [13]. From these results, we infer that the preinitiation complex of TIF and DNA remains stable in the cell through many rounds of transcription, acting in a catalytic fashion to guide multiple polymerase molecules to the promoter. This observation also has implications for regulatory mechanisms of rRNA transcription (see below).

The promoter

A knowledge of the required promoter sequence elements is necessary to understand detailed aspects of the initiation mechanism. We have studied the *cis*-acting sequences involved in mediating the above steps in initiation by directed mutation of the promoter [13, 17, 18]. A series of deletion mutations of the DNA template were analyzed by *in vitro* runoff transcription and template commitment assays. This study delineated the borders of the promoter as − 47 and + 8. The region between − 17 and − 31 is required for TIF binding, and the − 32 to − 47 region stabilizes the TIF-promoter complex. The sequence between − 17 and the transcription initiation point is required for other steps in initiation (see below).

The interaction of TIF and polymerase with the promoter was further dissected by the production of a series of single point mutations throughout the region [18]. Fifty mutants were analyzed over the − 55 to + 8 region. It was found that point mutants with the most severe deleterious effects on TIF binding and transcription are clustered around − 29 [− 30, − 29, − 28] and − 40 [− 41, − 39], with slightly less harmful sites upstream [− 45, − 31, − 32] and downstream [− 24, − 19, − 17]. All the mutant data delineates the TIF sequence-specific binding region as − 17 to − 45.

DNase I and MPE footprinting studies indicate additional tight contacts between TIF and the DNA region between − 51 and − 63 [14, 19]. This is interesting since there are apparently no significant sequence-specific interactions in this region, based upon the mutation studies discussed above. MPE footprinting reveals a single unprotected site at − 50 available for MPE intercalation which separates the sequence specific region from the upstream region [19, 20]. It is also interesting to speculate that this site reveals a cleft between the two putative TIF subunits.

Mechanism of promoter recognition by polymerase: protein-protein contacts with TIF

All the remaining down mutations map within the region protected from DNase I digestion by RNA polymerase I [18], but as discussed below, these are not involved in sequence-specific interactions with the polymerase during the binding step. We have dubbed this region the start motif.

Point mutation studies revealed a number of critical base pairs [− 11, − 12, − 13 and − 4] in the start motif [18]. Since the footprinting studies had shown that polymerase was unable to bind the promoter without TIF, it was important to determine whether these represented sequence-specific contact points between polymerase and the DNA. DNase I footprinting of the − 4 down mutant (69% relative activity) showed no diminished protection by polymerase I, while the − 30 mutant (68% relative activity) showed significant loss of protection by TIF [18]. Further, when assayed in competition with wild type DNA, the activity of the − 30 mutant dropped to 24%, but the − 4 mutant retained 68% relative activity suggesting that there is no competition for trans-acting factors which leads to the lower activity of the latter mutant. To further investigate this phenomenon, a series of spacer mutants were constructed [19] which have additional sequence inserted into or deleted from a site (at − 14) between the TIF binding site and the start motif. Such constructs not only result in a change in the normal spacing between the wild type transcription start site sequence and the TIF binding site, but also cause the equivalent of many clustered point mutations in the region downstream of the TIF site (where the start motif normally resides). *In vitro* transcription studies showed that as these constructs diverged from the wild type sequence, either by insertion or deletion of additional sequence, the efficiency of transcription decreased. Therefore, as suggested by the single-point mutation studies, there is some critical requirement for the sequence of the start motif (see below).

Surprisingly, DNase I footprinting studies of each of the constructs showed that polymerase was still able to bind to form the initiation complex, regardless of whether the construct was or was not transcriptionally active. Polymerase was bound normally even to those constructs in which the start motif sequence was completely different from wild type. An examination of MPE footprints of bound polymerase verified that there is almost no contact between the enzyme and the DNA in this region. Polymerase only protects a small region (3 bp) next to TIF and surrounding the nucleotide pairs encoding the 5′-end of the RNA (3 bp). Finally, in those constructs which are transcriptionally active, the initiating nucleotide for transcription is a constant distance downstream of the TIF binding site suggesting that polymerase acts as a molecular tape measure, reaching down 17 bp to find the initiating nucleotide, usually a purine. We have concluded from these results that polymerase binds to the promoter in a DNA sequence-independent fashion, and that the binding is directed solely by protein-protein contact with transcription initiation factor. (It should be noted that general DNA-polymerase interactions, eg. electrostatic, are not completely ruled out by these results, although the MPE footprinting suggests them to be weak or absent.)

Transcriptional enhancers in *Acanthamoeba*

In vertebrates and insects, the intergenic spacer contains repeated elements which serve as transcriptional enhancers. A similar element, though not repeated, is found in yeast [reviewed in [21]]. Since enhancers are found over such a wide phylogenic range, we wondered whether the *Acanthamoeba* gene also had similar modulatory elements. To investigate this possibility, the entire 2.5 Kbp intergenic spacer was sequenced and examined for repeated sequences. In an approximately 1 Kbp region downstream of the end of the 26S coding sequence, we found 7 copies of an inexact repeat sequence. When this entire element is used in competition with the core promoter in an *in vitro* transcription system, it inhibits transcription, in analogy to similar results found with other enhancer elements. When this repeat element is cloned just upstream of the core promoter, it stimulates transcription initiation from the core when partially purified TIF preparations are used, but stimulation is only about 2 fold. It does not stimulate, nor does it cause termination, when placed downstream of the core promoter. Band shift assays have been used to examine the possibility that these repeats bind TIF. Surprisingly, the repeats and the core promoter bind distinct proteins; TIF binds only to the core and another protein, dubbed spacer binding factor (SBF), binds to the repeats. Core sequence oligonucleotides compete TIF, but not SBF, and repeats compete SBF and not TIF. TIF and SBF can be partially separated on a Mono Q column. Is SBF another *trans*-acting factor? Why do the repeats compete transcription driven by the core promoter, but do not bind TIF? Is transcriptional enhancement mediated by SBF? Is SBF related to UBF, or to p16 (21a) (the band shift experiments suggest it to be much larger than 16 Kd)? These questions will have to await further experiments, but it appears that the presence of transcriptional enhancers for the rRNA gene is widespread, or even universal, in eukaryotes, as suggested by Reeder [21].

Transition from preinitiation to postinitiation: DNA melting

An examination of the structure of the post-initiation complex may have given us an insight into further differences between eukaryotic and prokaryotic transcription [16]. If the initiation complex is incubated with ATP, GTP and 3′-O-methyl CTP, then an RNA chain of 8 nucleotides can be produced from the wild type DNA before elongation is blocked by the incorporation of the 3′-O-methyl derivative. Footprinting this complex with MPE revealed that a significant conformational change occurred in the DNA-polymerase complex [16]. Instead of only two small regions of protection as before initiation, an extended region became insensitive to MPE cleavage. The post-initiation complex was treated with diethyl-pyrocarbonate, a reagent which can modify purines in melted DNA making them susceptible to piperidine cleavage [22]. The results showed that the conformational change was due to melting of at least 10 base pairs of DNA. The finding that DNA melting only occurs following the addition of nucleoside triphosphates was confirmed by an examination of the topology of a covalently closed circular DNA template during each step of initiation [16]. In this assay, DNA or DNA-protein complexes were treated first with topoisomerase I and then with sodium dodecylsulfate. This treatment results in an assayable change in the linking number of the circular DNA if regions are melted within the DNA-protein complex. The appearance of new, underwound topoisomers in the postinitiation complex, but not in the preinitiation complex, verified that melting only occurs AFTER nucleotide addition during transcription by polymerase I. Substitution of β-γ-imido ATP had no effect on transcription, therefore, there does not appear to be a need for protein phosphorylation or energy as in polymerase II-mediated transcription [23]. It is this melting step which is a prime candidate for the step which is affected by the point mutations in the start motif.

Regulation of ribosomal RNA transcription

When starved for essential nutrients, *Acanthamoeba* ceases cellular division and differentiates into a dormant cyst. In the growing cell, rRNA synthesis represents 75% of the transcription of the cell, but by 10–12 hours post-starvation, rRNA transcription has ceased [1]. This regulation can be mimicked in an *in vitro* system using crude S100 extracts as the source of both TIF and polymerase [1].

Since both TIF and RNA polymerase I are solely dedicated to the transcription of rRNA, this transcription unit could be regulated by altering the level or activity of either of these components. This notion was tested by comparing S100 extracts from various stages of encystment, assaying either in the presence or the absence of added RNA polymerase I purified from actively transcribing cells. Extracts from inactive cells were fully capable of transcription initiation when supplemented with polymerase, indicating that the level and activity of TIF is unaltered during transcriptional shutdown. Recent results indicate that identical activities are exhibited by TIF whether purified to near homogeneity from active or inactice cells. However, extracts from starved cells are *not* active in faithful initiation in the absence of added polymerase, indicating a change in the activity of the enzyme. The level of polymerase in the cells is unaltered since the extracts all contain the same number of polymerase molecules functional in elongation once artificially initiated [1, 24]. This is estimated by using damaged DNA templates (commercial calf thymus DNA or poly dl:poly dC) which allow nonspecific initiation at nicks, single stranded gaps and at DNA ends. Thus, it appears that RNA polymerase I is modified during encystment so that it becomes incapable of specific transcription.

This inability of polymerase I from cysts to correctly initiate at the promoter is also retained in the purified enzyme. Using enzyme purified to homogeneity from vegetative (growing) cells (*A*ctive or Pol_a) or cysts (*E*ncystment or Pol_E), one can demonstrate an approximately 5 to 7 fold difference in the ability of the enzyme to correctly initiate at the promoter [1, 2]. The two forms of the polymerase also have different heat stabilities, indicating that the enzyme is structurally, as well as functionally dimorphic [1]. Using the heat stability difference, it can be demonstrated that the residual activity in the Pol_E preparation is due to contamination with vegetative-type enzyme [1].

The effect of polymerase modification is on the initiation step because, in nonspecific assays, the specific activities of the two purified enzyme forms are the same. Indeed, footprinting assays of the cyst and vegetative polymerase show that the cyst enzyme is unable to bind to the promoter [2]. Therefore, it is the formation of the initiation complex which is altered to regulate rRNA transcription.

Thus, regulation or rRNA transcription in eukaryotic cells in response to changes in cellular growth rate is accomplished by altering the ratio of two dimorphic forms of RNA polymerase I. The TIF presumably remains bound to the promoter, keeping the gene primed for resumption of transcription as soon as active polymerase is available.

The exact nature of the polymerase modification is unknown. *Acanthamoeba* polymerase I contains 11 distinct polypeptide subunits [25], and comparison of the Pol_E and Pol_A enzymes by SDS polyacrylamide gel electrophoresis shows no gain or loss of subunits [1]. Analysis by gel electrophoresis in the presence of urea under basic conditions shows no change in the charge of the subunits with molecular weights 13.3 kD, 15.5 kD (subunits common to all three polymerases) or 17.5 kD (common to polymerases I and III). However, the 39 Kd subunit becomes modified in a temporally consistent manner during encystment, so that it is more acidic in charge and has an increased electrophoretic mobility in SDS-PAGE. The 39 Kd subunit is common to polymerase I and III [26] and is homologous to the *E. coli* α-subunit [5]. Intriguingly, 5S RNA transcription is co-regulated with the rRNA gene during encystment of *Acanthamoeba*, and the 39 Kd subunit of polymerase III is similarly modified. Therefore, co-regulation of these two transcription units could be mediated by a common polymerase modification mechanism. A homologous 5S RNA transcription system has been developed from

Acanthamoeba which mimics regulation. The mechanism of this regulation is currently under study.

Acknowledgements

Supported by U.S. Public Health Service grants GM22580, GM26059 and GM39901 and American Cancer Society Grant SG-152 to MRP.

References

1. Paule MR, Iida CT, Perna PJ, Harris GH, Knoll DA, D'Alessio JM: *In vitro* evidence that eukaryotic ribosomal RNA transcription is regulated by modification of RNA polymerase I. Nuc Acids Res 12: 8161–8180, 1984
2. Bateman E, Paule MR: Regulation of eukaryotic ribosomal RNA synthesis by RNA polymerase I modification. Cell 47: 445–450, 1986
3. Learned RM, Smale ST, Haltiner MM, Tjian R: Regulation of human ribosomal RNA transcription. Proc Natl Acad Sci USA 80: 3558–3562, 1983
4. Sollner-Webb B, Tower J: Transcription of cloned eukaryotic ribosomal RNA genes. Ann Rev Biochem 55: 801–830, 1986
5. Martindale DW: A conjugation-specific gene (cnjC) from *Tetrahymena* encodes a protein homologous to yeast RNA polymerase subunits (RPB3, RPC40) and similar to a portion of the prokaryotic RNA polymerase α subunit (rpoA). Nucleic Acids Res 18: 2953–2960, 1990
6. Learned RM, Learned TK, Haltiner MM, Tjian RT: Human rRNA transcription is modulated by the coordinate binding of two factors to an upstream control element. Cell 45: 847–857, 1986
7. Schnapp A, Clos J, Hädelt W, Schreck R, Cvekl A, Grummt I: Isolation and functional characterization of TIF-IB, a factor that confers promoter specificity to mouse RNA polymerase I. Nucleic Acids Res 18: 1385–1393, 1990
8. Pikaard CS, McStay B, Schultz MC, Bell SP, Reeder RH: The Xenopus ribosomal gene enhancers bind an essential polymerase I transcription factor, xUBF. Genes and Development 3: 1779–1788, 1989
9. Smith SD, Oriahi E, Lowe D, Yang-Yen H-F, O'Mahony D, Rose K, Chen K, Rothblum LI: Characterization of factors that direct transcription of rat ribosomal DNA. Mol Cell Biol 10: 3105–3116, 1990
10. Jantzen H-M, Admon A, Bell SP, Tjian R: Nucleolar transcription factor hUBF contains a DNA binding motif with homology to HMG proteins. Nature 1990
11. Bell SP, Pikaard CS, Reeder RH, Tjian R: Molecular mechanisms governing species-specific transcription of ribosomal RNA. Cell 59: 489–497, 1989
12. Pikaard CS, Smith SD, Reeder RH, Rothblum L: rUBF, an RNA polymerase I transcription factor from rats, produces DNase I footprints identical to those produced by xUBF, its homolog from frogs. Mol Cell Biol 10: 3810–3812, 1980
13. Iida CT, Kownin P, Paule MR: Ribosomal RNA transcription: Proteins and DNA sequences involved in preinitiation complex formation. Proc Natl Acad Sci USA 82: 1668–1672, 1985
14. Bateman E, Iida CT, Kownin P, Paule MR: Footprinting of ribosomal RNA genes by transcription initiation factor and RNA polymerase I. Proc Natl Acad Sci USA 82: 8004–8008, 1985
15. Galas D, Schmitz A: DNAse footprinting: a simple method for the detection of protein-DNA binding specificity. Nuc Acids Res 5: 3157–3170, 1978
16. Bateman E, Paule MR: Events during eukaryotic rRNA transcription initiation and elongation: Conversion from the closed to the open promoter complex requires nucleotide substrates. Mol Cell Biol 8: 1940–1946, 1988
17. Kownin P, Iida CT, Brown-Shimer S, Paule MR: The ribosomal RNA promoter of *Acanthamoeba castellanii* determined by transcription in a cell-free system. Nuc Acids Res 13: 6237–6247, 1985
18. Kownin P, Bateman E, Paule MR: Effects of single-base substitutions within the *Acanthamoeba castellanii* rRNA promoter on transcription and on binding of transcription initiation factor and RNA polymerase. Mol Cell Biol 8: 747–753, 1988
19. Kownin P, Bateman E, Paule MR: Eukaryotic RNA polymerase I promoter binding is directed by protein contacts with transcription initiation factor and is DNA sequence-independent. Cell 50: 693–699, 1987
20. Hertzberg RP, Dervan PB: Cleavage of double helical DNA by methidiumpropyl-EDTA iron(II). J Am Chem Soc 104: 313–315, 1982
21. Reeder RH: Regulatory elements of the genetic ribosomal gene. Curr Opinion Cell Biol 1: 466–474, 1989

21a. Yang-Yen H-F, Rothblum LI: Purification and characterization of an HMGlike, DNA-binding protein that stimulates ribosomal RNA synthesis *in vitro*. Mol Cell Biol 8: 3406–3414, 1988

22. Mendel D, Dervan PB: Hoogsteen base pairs proximal and distal to echinomycin binding sites on DNA. Proc Natl Acad Sci USA 84: 910–914, 1987
23. Bunick D, Zandomeni R, Ackerman S, Weinmann R: Mechanism of RNA polymerase II-specific initiation of transcription *in vitro:* ATP requirement and uncapped run-off transcripts. Cell 29: 877–886, 1982
24. Detke S, Paule MR: DNA-dependent RNA polymerases from *Acanthamoeba castellanii:* Multiple forms of the class III enzyme and levels of activity of the polymerase classes during encystment. Biochim Biophys Acta 520: 131–138, 1978

25. Spindler S, Duester GL, D'Alessio JM, Paule MR: A rapid and facile procedure for the preparation of RNA polymerase I from *Acanthamoeba castellanii:* Purification and subunit structure. J Biol Chem 253: 4669–4675, 1978
26. D'Alessio JM, Perna PJ, Paule MR: DNA-dependent RNA polymerases from *Acanthamoeba castellanii:* Comparative subunit structure of the homogeneous enzymes. J Biol Chem 254: 11282–11287, 1979

Address for offprints: M.R. Paule, Department of Biochemistry and the Cell and Molecular Biology Program, Colorado State University, Fort Collins, CO 80523, USA

Molecular and Cellular Biochemistry **104**: 127–135, 1991.

Complementary *in vivo* and *in vitro* analyses of the interactions between the cis-acting elements of the rat rDNA promoter

Wen Qin Xie, Daniel J. O'Mahony, S. David Smith and Lawrence Rothblum
Sigfried and Janet Weis Center for Research, Geisinger Clinic, North Academy Avenue, Danville, PA 17822, USA

Key words: in vivo expression, *in vitro* transcription, rDNA

Abstract

Two transcription factors, rat UBF (rUBF) and rat SL-1 are required for the efficient transcription of the rat promoter *in vitro. In vitro* studies have established that two broadly defined cis-acting domains, the core promoter element and the upstream promoter element, cooperate to direct correct transcription by RNA polymerase I. The ability of UBF to bind to two linker-scanning mutants of the upstream promoter element, which did not respond to the addition of UBF in *in vitro* transcription assays, was assessed by DNase footprinting. UBF protected the same region of the promoter in the linker-scanning mutant in BSM 129/124 as it did in the wild-type, but did not yield a typical footprint over the promoter in the linker-scanning mutant BSM 106/101.

Previously we reported that promoters with mutant core promoters elements, either the guanine at − 16 or − 7 substituted by an adenine, were inactive *in vitro* unless the assays were supplemented with UBF. Those results suggested that the binding of UBF upstream of the core was required for the promotion of transcription. The interactions between the core and upstream promoter elements were assessed by constructing double mutants of the promoter. In two constructs the conserved guanines at either − 16 or − 7 were altered in a deletion mutant (− 86) that did not respond to UBF. In a third construct the guanine at − 16 in BSM 129/124 was changed to an adenine. These bidomain mutant constructs did not respond to the addition of UBF in an *in vitro* transcription assay, confirming that the rescue of the core promoter mutants requires an intact and functional upstream promoter element. These *in vitro* results have been further examined *in vivo* by transfecting various constructs into CHO cells. The *in vivo* studies indicate that besides the nucleotides mutated in BSM 129/124 and BSM 106/101, a third element in the UPE, circa nucleotides − 117/− 112, may also play a role in transcription.

Introduction

Functional analysis of the promoters of the mammalian rRNA genes indicates that despite significant sequence differences, the promoters apparently consist of elements with similar functions. That region of the promoter (~ − 31 to ~ + 6) required for transcription *in vitro* [1–5] is referred to as the core promoter element (CPE). Under more stringent conditions a requirement for distal sequences becomes apparent [6], and *in vivo* the CPE is almost inactive [7]. Two distal elements, the upstream promoter (control) element (UPE) and an upstream terminator (referred to as either T_0 or as T_3 [8–10]) have been identified. The UPE has been shown to be required for transcription *in vivo* and for elevated levels of transcription *in vitro* under stringent conditions, and it appears to be required for the formation of the stable preinitiation complex [1, 11, 14]. The precise role of the terminator in transcription initiation, if any, is not yet understood.

Unlike many of the genes transcribed by RNA polymerase II and III [12], promoter recognition by vertebrate RNA polymerase I is, with one known exception [13] species-specific [19], e.g. competent extracts from primate cells will not transcribe rodent rDNA and vice versa. This property has been ascribed to a trans-acting factor, referred to as either PC-D [14, 15], or TIF-IB [16] or SL-1 [17] for rodents, or PC-D [14, 15] or SL-1 [18] for primates. Exonuclease III protection assays indicate that mouse TIF-IB protects a domain including the core promoter element of the mouse ribosomal DNA repeat. Although DNase 1 footprinting experiments failed to detect an interaction between human SL-1 and the core element of the human rDNA promoter, human SL-1 did direct the transcription of a hybrid promoter consisting of the mouse UPE and the human CPE. Human SL-1 protected part of the UPE, when the DNase I footprinting experiments were supplemented with human UBF (hUBF), a DNA-binding activity that has been demonstrated to [1] increase the efficiency of template utilization, [2] protect the region from approximately − 70 to − 120 of the human rDNA promoter, and [3] may also interact with the CPE of the human rDNA repeat [20, 21].

We have studied the factors required for the transcription of both the rat rDNA 45S and spacer promoters [17, 24]. Our findings indicate that one rat factor, referred to as rat SL-1, interacts with the core promoter element and may also interact with the upstream promoter element. The interaction of SL-1 with the core appears to be sufficient for accurate initiation by RNA polymerase I, either rat or human. The addition of a second factor, referred to as rat UBF (rUBF), increases the efficiency of template utilization and may act by increasing the stability of interaction or affinity of either SL-1 and/or RNA polymerase I with the core promoter. Both factors were shown to form complexes with the promoters [17, 24]. The results of *in vitro* transcription experiments using deletion and nucleotide substitution mutants of the 45S promoter in studies of the effect of rUBF on transcription demonstrated that the UPE consisted of two separate domains. One domain mapped to the nucleotides between − 129 and − 124. The second domain mapped to between − 106/− 101. DNase I footprinting experiments with highly purified rat UBF and SL-1 demonstrated that rUBF protected the region between − 130 and − 50 and that SL-1 protected several regions, including a region circa − 130. Further, our results indicated that the binding of rat SL-1 to the core promoter element was sufficient for transcription *in vitro*, and the interaction between the core SL-1 complex and the complex formed between rUBF and the upstream promoter element is responsible for the activation of the rat ribosomal DNA promoter.

Herein we report the further characterization of the interaction of rat UBF with the 45S promoter, as well as data that further demonstrate the requirement for a functional interaction between the UPE and the CPE. Furthermore, we have carried out *in vivo* expression assays that confirm and extend the *in vitro* results.

Materials and methods

Templates

The templates used in the study are derived from the plasmid p5.1 [35] and many of them were described previously [17]. Nucleotide substitution mutants were created using synthetic oligonucleotides to prime the incorporation of mutagenized bases using the polymerase chain reaction. After transformation into *E. coli*, minilysates were prepared and the presence of the substitution was confirmed by chain-termination DNA sequencing [22]. The constructs used in this study are described in Fig. 1.

In vitro *transcription*

The conditions for *in vitro* transcription were those described previously [5, 17]. A standard 50 μl reaction contained between 0.02 and 0.4 μg of truncated template and 1.0 μg of nonspecific DNA, either pBR322 or pUC18. In some cases upon termination of transcription a constant amount of an end-

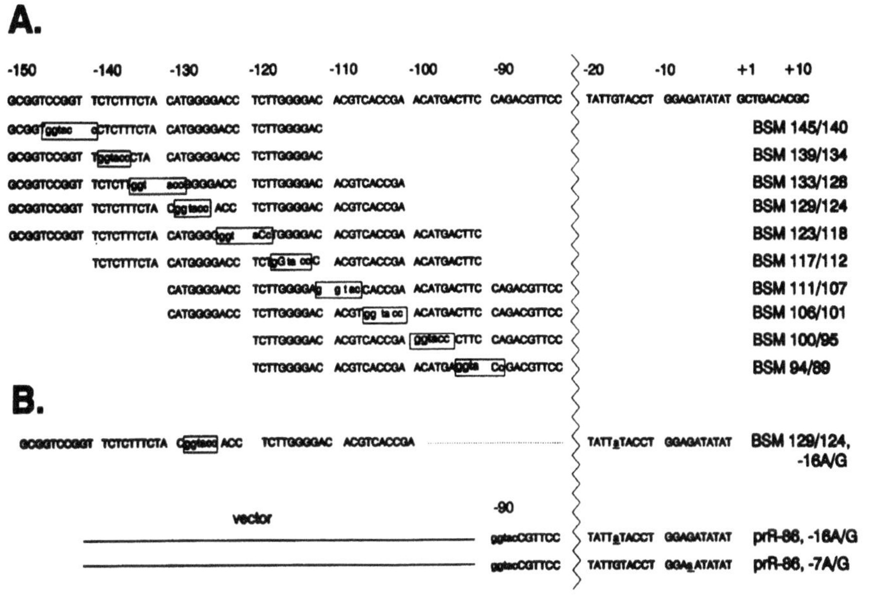

Fig. 1. Schematic depiction of the (A) linker scanning mutants and (B) bi-domain mutants used in these experiments. The nucleotides mutated are boxed and/or written in lower case.

labeled fragment of pBR322 was added to each reaction, to act as an internal standard for the efficiency of recovery of the nucleic acids. For routine assays demonstrating the effect of rUBF on transcription, the pooled RNA polymerase I containing fractions from the first DEAE-Sephadex column [17] were used both as the source of RNA polymerase I and as the source of SL-1. This fraction is referred to as 'UBF depleted extract'.

DNAase I footprinting

End-labeled promoter DNA probes were produced by PCR amplification following labeling one of the two primers with [γ^{32}P]-ATP and polynucleotide kinase. The labeled primer and a cold primer were then used in the polymerase chain reaction to generate end-labeled DNA for footprinting. Each footprint reaction contained between 3–10 ng of the labeled DNA fragment, one microgram of non-specific competitor, either pUC 18 or poly (dI · dC)- poly (dI · dC), or polydA/polydT, and rat UBF. The rat UBF used for footprinting was purified through the CM-Sephadex column step [17]. The binding reactions were carried under the same conditions as described previously [17, 24]. After DNAase I digestion, the reactions were extracted with phenol/chloroform/isoamyl alcohol, chloroform/isoamyl alcohol, precipitated with ethanol, washed in 70% ethanol, lyophilized, resuspended in 90% formamide, heat-denatured and analyzed on 6% acrylamide-urea sequencing gels.

Fractionation of nuclear extracts

Nuclear extracts were prepared from Novikoff hepatoma ascites cells or from logarithmic cultures of HeLa cells essentially as described [23], and fractionated as described previously [17].

In vivo *expression assay*

CHO cells grown in DMEM containing 10% fetal bovine serum were plated on 80 mm dishes at 40% of confluence. One day after plating the cells were transfected with the construct indicated essentially as described [25]. Whole cell RNA was isolated 20–30 hours post transfection, and the level of transcription measured using a primer elongation assay, using a primer specific for the transcript from the transfected genes [26, 27].

BrdUTP crosslinking
Photochemical crosslinking was carried out essentially as described [29]. The DNA was uniformly labelled with [α-^{32}P]dCTP, and where dTTP was substituted with bromodeoxyuridine, by PCR using primers based on the sequences of the rat rDNA promoter spanning the region from − 160 and + 40.

Results

We have previously demonstrated that promoters in which either the guanine at − 16 or the guanine at − 7 are replaced with adenines are not transcribed with UBF-depleted extract, but are rescued by the addition of UBF [17]. There were several explanations for those results.

Model 1. One interpretation of these experiments was that the mutants had altered the binding of either SL-1 and/or RNA polymerase I with the core promoter, and that upon the addition of UBF to the assay the interaction of UBF with the UPE had acted to stabilize the weakened interactions of SL-1 and/or RNA polymerase I with the CPE. In this model initiation occurs without a requirement for a productive interaction between rUBF and the core promoter.

Model 2. An alternative explanation for our finding was that the point mutants had weakened the binding of UBF to the CPE. Thus, extracts depleted of UBF would have insuffícent UBF to drive the formation of a UBF · CPE complex. This model requires that [1] rat UBF be capable of interacting productively with the CPE and [2] that DE-175 fraction (UBF-depleted) contains rUBF. One corollary of this model would suggest that the binding of rUBF to the CPE is required for transcription from the core promoter.

Model 3. In this model UBF binds to both the UPE and the CPE. UBF bound to the UPE might stabilize or cooperatively interact with or stabilize the binding of a molecule of UBF bound to the CPE. The binding of two molecules of UBF to the promoter would result in the efficient formation of initiation complexes.

We have attempted to differentiate between models one and two by further investigating the putative role of the UPE in stabilizing the interactions of the various transcriptional factors/pol I at the CPE. Also, as described below, we have attempted to confirm our earlier observations that DE-175 does not contain detectable levels of rUBF.

Productive interactions of rUBF with the rat 45S promoter requires an intact UPE

Model 1 requires that the UPE must be intact for UBF to rescue transcription, whereas Model 2 would not require an intact UPE. Therefore we constructed two types of mutants. The first series was based on a deletion mutant, prR-86, that did not respond to UBF in *in vitro* transcription analysis [16]. The guanines at − 16 and − 7 were then mutated to adenines, resulting in two clones referred to as prR-86, − 16 A/G and prR-86, − 7 A/G. The second class of mutant was based on the linker scanning mutant BSM 129/124 in which the guanine at − 16 was changed to an adenine, resulting in the construct BSM 129/124, − 16 A/G. These constructs were then assayed *in vitro* in the presence or absence of UBF using DE-175 as the source of RNA polymerase I and rat SL-1. As shown in Table 1, − 16 A/G, and − 7 A/G, prR-86, − 16 A/G and prR-86, − 7 A/G were not transcribed in the absence of UBF. When the reactions were supplemented with UBF, − 16 A/G and − 7 A/G were transcribed, but prR-86, − 16 A/G and prR-86, − 7 A/G were not. These results suggested that an intact UPE was required for UBF to rescue the core mutants.

When BSM 129/124, − 16 A/G was added to a transcription assay in the absence of UBF, it was inactive, as one might predict (Table 1). Further, this construct was not rescued by the addition of UBF to the transcription reaction (Table 1). This result confirms and extends the results obtained with the prR-86 based mutations. It confirms that the transcription rescue experiment required a functional UPE. Moreover, this result demonstrates that a functional UPE consists of more than just a UBF binding domain, as we have demonstrated using DNAase I footprinting [16] that UBF does protect the UPE in BSM 129/124. This conclusion led us to examine the binding of UBF to our other linker-scanning mutants through the region from −145 to −89.

UBF does not protect the linker-scanning mutant BSM 106/101

We have reported [17] that UBF protected BSM 129/124. However, in those experiments [1] the ratio of UBF to DNA was twenty-fold greater than that used in transcription and [2] we did not examine the interaction of UBF with BSM 106/101, a second mutant that did not respond to the addition of UBF. Thus, we carried out DNAse footprinting assays on the linker scanning mutants spanning the region from from −145 to −89. (starting with BSM 145/140 and ending with BSM 94/89). In these experiments the ratio of UBF to DNA was approximately 3 fold greater than that required for the maximum effect of UBF on transcription [17].

As summarized in Table 2 UBF protected all of these constructs except for BSM 106/101. These results [1] confirm that UBF does protect BSM 129/124, and [2] suggest that BSM 106/101 does not respond to the addition of UBF because it does not bind UBF, at least in a way that would be detected in these assays.

Thus our *in vitro* results, DNase footprinting and transcription assays, appeared to be consistent. However, this does not mean that those sites, e.g. −129/−124 and −106/−101, are significant *in vivo*. Therefore we examined the behavior of those mutants *in vivo*.

In vivo *expression assays*

In these experiments CHO cells were transfected with the indicated constructs (Fig. 2) and the level of transcription determined using a primer elongation assay. The sequence of the primer was chosen such that it would not hybridize to CHO 45S rRNA (Fig. 2, lane 1) but would prime authentic rat rRNA (Fig. 2, lane 2). We observed that both BSM 129/124 and BSM 106/101 were down mutants in this assay, although the effect of mutating nucleotides −129 to −124 was not as great as mutating the bases between −106 to −101. Furthermore, both BSM 123/118 and BSM 117/112 were also down mutants. In fact BSM 117/112 is a very powerful mutation in this assay.

Table 1. Effect of rUBF on the transcription of mutant promoters

Promoter	Transcription*	
	− UBF	+ UBF
prR-16A/G	n.d.**	Rescued
prR-7A/G	n.d.	Rescued
prR-86,− 16A/G	n.d.	n.d.
prR-86, − 7A/G	n.d.	n.d.
BSM129/124,− 16A/G	n.d.	n.d.

* Transcription was carried out using UBF depleted extract, as described previously [17].

** n.d. no transcript detected.

Table 2. DNase footprinting of rUBF on linker-scanning mutants of the rat rDNA promoter

Mutant	Footprint	Response to UBF[a]
BSM 145/140	+[b]	+[b]
BSM 139/134	+	+
BSM 133/128	+	+
BSM 129/124	+	−[c]
BSM 123/118	+	+
BSM 117/112	+	+
BSM 111/106	+	+
BSM 106/101	−	−
BSM 100/95	+	+
BSM 94/89	+	+

[a] Where transcription of the template was stimulated by the addition of UBF to an assay using UBF-depleted extract.

[b] UBF protected the template in a DNase footprinting assay, or stimulated transcription *in vitro*.

[c] UBF did not protect the template, or stimulate transcription *in vitro*.

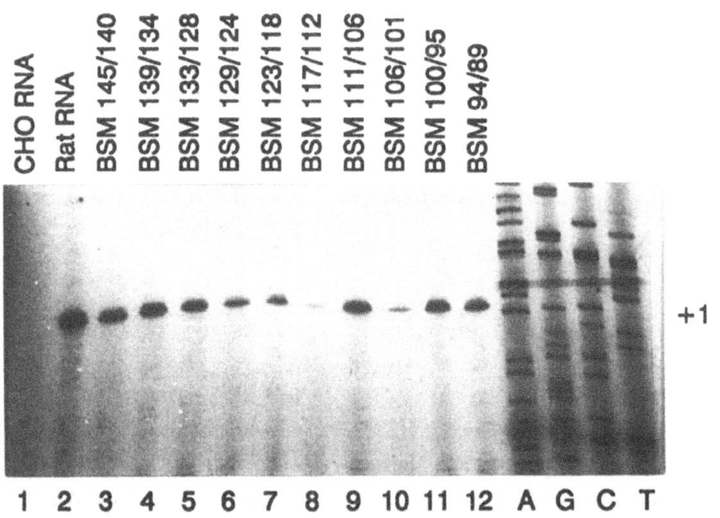

Fig. 2. In vivo analysis of the effects of the linker scanning mutants on transcription *in vivo*. Following transfection into CHO cells the ability of the mutant promoters was assessed by reverse transcriptase elongation as described in Materials and methods. CHO cell 45S pre-rRNA will not serve as a template in this reaction (Lane 1) whereas authentic rat pre-rRNA does (Lane 2). The lanes marked A, G, C, T contain sequencing reactions of the rat promoter using the same primer as that used for the reverse transcriptase reactions.

Purified rUBF consists of two DNA-binding proteins

We have previously reported [17, 28] that rat UBF purified as a dimer of approximately 97 and 94 kDa. However no evidence was presented that either one of those polypeptides was capable of recognizing the rDNA promoter. This left open the possibilities that the footprints of rat UBF were due to one or both of these polypeptides, or to the presence of an unseen component in the UBF preparations. To examine this possibility, we carried out UV crosslinking experiments using a fragment of the rat promoter and partially purified rat UBF (CM-Sephadex fraction). As shown in Fig. 3 two bands in the CM-Sephadex purified fraction were covalently coupled to the BrdU substituted promoter. The molecular weights of these bands were estimated to be 97 and 94 kd. Control lanes in which molecular weight markers or UBF-depleted extract (DE-175) were incubated with the same fragment under the same conditions did not demonstrate any DNA-binding proteins. The results confirmed that the DNA-binding activity present in the affinity purified forms of UBF could be ascribed to the proteins visible upon SDS/PAGE. Furthermore, these results indicate that DE-175, or UBF depleted extract, does not contain a significant amount of UBF, as the 97 and 94 kDa band were not detected in that fraction. In Southwestern blotting experiments we have detected DNA binding protein in DE-175, but we have not determined if these are the DNA binding components of RNA polymerase or SL-1.

These results also suggest that the factor required for the rescue of the point mutants was UBF, as that is the only DNA binding protein detected in the fractions used for those experiments.

Discussion

This manuscript extends our studies on the domains of the UPE of the rat rDNA promoter, and the interaction between the UPE and CPE of the rat rDNA promoter. We have demonstrated that the rescue of point mutants within the core promoter element of the rat rDNA promoter requires an intact and functional UPE. We have extended our

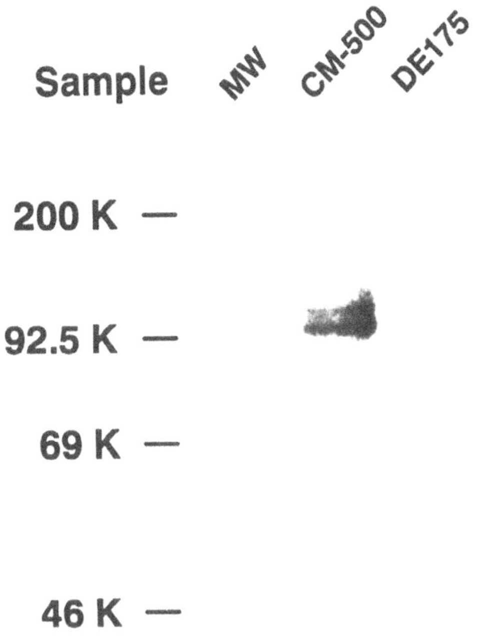

Fig. 3. Rat UBF consists of two polypeptides, both of which can bind DNA. Equal amounts of the indicated protein fractions were incubated with BrdU substituted promoter as described in Materials and methods and the resultant products analyzed by SDS-PAGE.

original observations on the domains of the UPE by demonstrating that two different mutants BSM 129/124 and BSM 106/101 inactivate the UPE by two different mechanisms. Furthermore, our *in vitro* observations have been complemented with results from *in vivo* expression assays.

In our studies of the interaction of rat UBF with the rat 45S and spacer promoter [17, 24] we observed that both promoters required upstream UBF binding sites if they were to respond to rUBF. This was in contrast to the observations reported for the effects of hUBF on transcription from the human rDNA promoter [20] where it was found that deletion of the UPE reduced but did not obviate the effect of hUBF on transcription. To extend our observations we created dual promoter mutants, i.e. promoter constructs with mutations in both the UPE and CPE of the rat rDNA promoter, and assayed them in the presence or absence of UBF. The results of those assays appear to eliminate the possibility that rUBF can functionally interact with the CPE of the rat promoter. These results, by themselves, do not eliminate the possibility that productive interactions of rUBF with the promoter require both intact UPE and CPE binding domains. However, our DNAase footprinting experiments demonstrate that: (1) rUBF can bind to the UPE in BSM 129/124 while (2) rUBF binding to the UPE is blocked by the mutation found in BSM 106/101. Thus, by itself, the binding of rUBF to the UPE (BSM 129/124) appears to be insufficient in these *in vitro* transcription rescue experiments (BSM 129/124, − 16 A/G). These results also suggest that the UPE, is a complex element, consisting of two or more domains.

The footprinting experiments suggest that the nucleotides mutated in BSM 106/101 are critical for the binding of rUBF to the rat promoter. As the rat and *Xenopus* forms of UBF bind to each others 45S promoter with essentially equivalent footprints [28], we searched the region of the *Xenopus* promoter protected by rUBF for the nucleotides cgtCACCGAa. The sequence was not found. Either these nucleotides are only significant in the context of the rat promoter, or they are only significant for the binding of rat UBF to the rat promoter. Experiments with xUBF and hUBF should clarify this point.

Finally, our *in vivo* expression experiments indicate that our *in vitro* mapping of the domains of the UPE are accurate, but only so far as concerns the functioning of the UPE with respect to UBF.

The fact that BSM 117/112 is a down mutant in the *in vivo* experiment suggests that UBF and SL-1 may not be the only proteins which interact with the UPE. Clearly this result needs to be examined in greater detail. An *in vitro* assay that duplicates this result will need to be established before a clear biochemical role for this interaction can be defined.

Acknowledgements

The authors would like to thank Thomas Smink for

his technical assistance and Brian Shoop for his aid in producing the figures. This work was supported by funds from the National Institutes of Health and the Geisinger Foundation. The authors are grateful to the Geisinger Foundation for sponsoring the meeting at which the results were presented.

References

1. Cassidy B, Haglund R, Rothblum L: Regions upstream from the core promoter of the rat ribosomal gene are required for the formation of a stable transcription initiation complex by RNA polymerase I *in vitro*. Biochim Biophys Acta 909: 113–144, 1987
2. Grummt I: Nucleotide sequence requirements for specific initiation of transcription by RNA polymerase I. Proc Natl Acad Sci USA 79: 6908–6911, 1982
3. Haltiner M, Smale S, Tjian R: Two distinct promoter elements in the human rRNA gene identified by linker scanning mutagenesis. Mol Cell Biol 6: 227–235, 1986
4. Learned R, Smale S, Haltiner M, Tjian R: Regulation of human ribosomal RNA transcription. Proc Natl Acad Sci USA 80: 3558–3562, 1983
5. Yamamoto O, Takakusa N, Mishima Y, Kominami R, Muramatsu M: Determination of the promoter region of the mouse ribosomal RNA gene by an *in vitro* transcription system. Proc Natl Acad Sci USA 81: 299–303, 1984
6. Miller K, Tower J, Sollner-Webb B: A complex control region of the mouse rRNA gene directs accurate initiation by RNA polymerase I. Mol Cell Biol 5: 554–562, 1985
7. Smale S, Tjian R: Transcription of Herpes Simplex Virus tk Sequences under the control of wild-type and mutant human RNA polymerase I promoters. Mol Cell Biol 5: 352–362, 1985
8. Grummt I, Kuhn A, Bartsch I, Rosenbauer H: A transcription terminator located upstream of the mouse rDNA initiation site affects rRNA synthesis. Cell 47: 901–911, 1986
9. Henderson S, Sollner-Webb B: A transcriptional terminator is a novel element of the promoter of the mouse ribosomal RNA gene. Cell 47: 891–900, 1986
10. McStay B, Reeder RH: A termination site for Xenopus RNA polymerase I also acts as an element of an adjacent promoter. Cell 47: 913–920, 1986
11. Tower J, Culotta V, Sollner-Webb B: Factors and nucleotide sequences that direct ribosomal DNA transcription and their relationship to the stable transcription complex. Mol Cell Biol 6: 3451–3462, 1986
12. Breathnach R, Chambon P: Organization and expression of eukaryotic split genes coding for proteins. Ann Rev Biochem 50: 349–383, 1981
13. Culotta V, Wilkinson J, Sollner-Webb B: Mouse and Frog violate the paradigm of species-specific transcription of ribosomal RNA genes. Proc Natl Acad Sci USA 84: 7498–7502, 1987
14. Miesfeld R, Arnheim N: Species-specific rDNA transcription is due to promoter-specific binding factor. Mol Cell Biol 4: 21–227, 1984
15. Mishima Y, Financsek L, Kominami R, Muramatsu M: Fractionation and reconstitution of factors required for accurate transcription of mammalian ribosomal RNA genes: identification of a species-specific factor. Nucl Acids Res 10: 6659–6670, 1982
16. Clos J, Buttgereit D, Grummt I: A purified transcription factor (TIF-1B) binds to essential sequences of the mouse rDNA promoter. Proc Natl Acad Sci USA 83: 604–608, 1986
17. Smith SD, Oriahi E, Lowe D, Yang-Yen H-F, O'Mahony D, Rose K, Chen K, Rothblum LI: Characterization of factors that direct transcription of rat ribosomal DNA. Mol Cell Biol 10: 3105–3116, 1990
18. Learned R, Cordes S, Tjian R: Purification and characterization of transcription factor that confers promoter specificity. Mol Cell Biol 5: 1358–1369, 1985
19. Grummt I, Roth E, Paule M: Ribosomal RNA transcription *in vitro* is species specific. Nature 296: 173–174, 1982
20. Bell S, Learned R, Jantzen H-M, Tjian R: Functional Cooperativity between transcription factors UBF1 and SL1 mediates human ribosomal RNA synthesis. Science 241: 1192–1197, 1988
21. Learned R, Learned T, Haltiner M, Tjian R: Human rRNA transcription is modulated by the coordinate binding of two factors to an upstream control element. Cell 45: 847–857, 1986
22. Sanger F, Nicklen S, Coulson A: DNA sequencing with chain-terminating inhibitors. Proc Natl Acad Sci USA 74: 5463–5467, 1977
23. Haglund RE, Rothblum LI: Isolation, fractionation and reconstitution of a nuclear extract capable of transcribing ribosomal DNA. Mol Cell Biochem 73: 11–20, 1987
24. Smith SD, Oriahi E, Yang-Yen H-F, Xie WQ, Chen C, Rothblum LI: Interaction of RNA polymerase I transcription factors with a promoter in the nontranscribed spacer of rat ribosomal DNA. Nucl Acids Res 18: 1677–1685, 1990
25. Wigler M, Pellicer A, Silvestein S, Axel R: Biochemical transfer of single copy genes using total cellular DNA as donor. Cell 22: 725–731, 1987
26. Favoloro J, Triesman R, Kamen R: Transcription maps of polyoma virus-specific RNA: Analysis by two-dimensional nuclease S1 gel mapping. Methods in Enzym 65: 718–749, 1980
27. Zinn K, Maio D, Maniatis T: Identification of Two Distinct Regulatory Regions Adjacent to the Human β-Interferon Gene. Cell 34: 865–879, 1987
28. Pikaard C, Smith D, Reeder R, Rothblum L: rUBF, an RNA Polymerase I Transcription Factor from Rats, Produces DNase I Footprints Identical to Those Produced by xUBF, Its Homolog from Frogs. Mol Cell Biol 10: 3810–3812, 1990

29. Linn S, Riggs A: Photochemical attachment of *lac* repressor to bromodeoxyuridine substituted *lac* operator by ultraviolet radiation. Proc Natl Acad Sci USA 71: 947–951, 1974

Address for offprints: L.Rothblum, Sigfried and Janet Weis Center for Research, Geisinger Clinic, North Academy Avenue, Danville, PA 17822, USA

Molecular and Cellular Biochemistry **104**: 137–147, 1991.

Trans-acting factors involved in species-specificity and control of mouse ribosomal gene transcription

Andreas Schnapp, Horst Rosenbauer and Ingrid Grummt
Institute of Cell and Tumor Biology, German Cancer Research Center, D-6900 Heidelberg, Germany

Key words: RNA polymerase I, transcription factors, signal transduction, protein phosphorylation, cdc2 protein kinase

Abstract

Faithful and efficient transcription initiation at the mouse ribosomal gene promoter requires besides RNA polymerase I (pol I) four polypeptide *trans*-acting factors, termed TIF-IA, TIF-IB, TIF-IC, and mUBF. We have partially purified these proteins from cultured Ehrlich ascites cells and show that in the presence of TIF-IA and TIF-IB, pol I directs very low amounts of specific transcripts. Neither TIF-IC nor mUBF on their own significantly stimulate the efficiency of template utilization. However, both factors together strongly activate transcription. Interestingly, factor TIF-IB – the murine homologue of human SL1 – fails to program a human extract to transcribe the murine template, but requires its homologous RNA polymerase I. This finding implicates that not only some rDNA transcription factors but also pol I exhibits species-specific differences. The growth-related factor TIF-IA, on the other hand, stimulates both mouse and human rDNA transcription. This regulatory factor whose amount or activity fluctuates according to the proliferation rate of the cells, is functionally inactivated by antibodies against cdc2 protein kinase. This result together with the observation that transcription is stimulated by ATP-γS, an ATP analogue which is a substrate for protein kinases but not for protein phosphatases, strongly suggests that post-translational protein modification is involved in rDNA transcription regulation.

Introduction

The availability of cell-free systems that faithfully transcribe cloned rDNA has facilitated the identification of *cis*-acting sequences and *trans*-acting factors involved in the transcription process. Contrary to the remarkable conservation of defined sequence motifs required for accurate transcription initiation by RNA polymerases II and III, sequence homologies in the intitiation region of pol I genes is found only between closely related species (for review, see 1). The *cis*-acting control signals for pol 1 consist of multiple, complex sequence elements. Initial experiments with deletion mutants located the region that is absolutely required for accurate initiation of mouserDNA transcription between residue ca. − 40 and a position approximately at or a few nucleotides downstream of the initiation site (reviewed in 2). This minimal or 'core' promoter is sufficient to direct accurate and species-specific transcription initiation. This core element contains the targetse-09quence for a specific DNA binding protein – the transcription initiation factor TIF-IB [3, 4] which confers promoter selectivity to mouse RNA polymerase I (pol I). Besides this species-specific factor a second protein termed UBF is required for specific transcription initiation [5]. UBF has been purified to homogeneity, and homologues of UBF have been shown to be present in human, mouse, frog and rat [6, 7]. In the human system it has been demonstrated that UBF plus the human-specific factor SL1 together form a complex which binds to the promoter and then directs pol I to initiate at a specific site.

In addition to UBF and TIF-IB which are sufficient to form the stable initiation complex, two other factors are required for efficient initiation at the mouse rDNA promoter. TIF-IC is a stimulatory factor, and TIF-IA is a regulatory protein whose amount or activity correlates with cell proliferation [8, 9]. Factor TIF-IA interacts with pol I and thus enables it to initiate specifically at the rDNA promoter in the presence of the other auxiliary factors. With regard to several criteria the growth-dependent factor TIF-IA behaves like a bacterial sigma factor. It is required for initiation of transcription, it is present in limiting amounts, and it physically associates with pol I thus converting a transcriptionally inactive 'core' polymerase into a holoenzyme which is able to initiate specifically at the rDNA promoter. Here we show that control of rRNA synthesis by factor TIF-IA is not restricted to mouse cells. Supplementation of HeLa cell extracts with TIF-IA strongly stimulates transcription at the human rDNA template. Furthermore, we present experimental evidence that TIF-IA may be associated or regulated by the cell cycle-dependent protein kinase cdc2. The results suggest that post-translational modification of pol I or other essential proteins play an essential role in the chain of events by which extracellular signals are transferred into the nucleolus and thus adapt ribosome synthesis to cell growth.

Materials and methods

rDNA templates

The mouse rDNA template pMr600 contains a 618 bp Pvu II fragment covering nucleotides from − 326 to + 292 with respect to the transcription initiation site cloned into the Sma I site of pUC9. The human rDNA template pHrP contains in a Sau3A fragment extending from position − 411 to + 379 which has been inserted into the BamHI site of pUC9.

Cultivation of cells and extract preparation

Ehrlich ascites cells were cultured in RPMI medium containing 5% new born calf serum for 20–40 hours. Transcriptionally active extracts were obtained from logarithmically growing cells (9×10^5 cells/ml). Transcriptionally inactive extracts were prepared from stationary phase cells, which were taken either directly from the cavity of mice, or were grown to maximal density ($1.5 - 2 \times 10^6$ cells/ml) in culture medium and incubated for an additional 24 hours before harvesting. For nutritional starvation logarithmically growing cells were transferred into histidine- and isoleucine-free RPMI medium containing 5% dialyzed calf serum and cultured overnight.

In vitro *transcription assays*

The soluble cell-free transcription system and the analysis of the RNA synthesized have been described before [3, 11]. Usually 50–100 ng of pMr600 template DNA truncated with EcoRI were incubated with 15 μl of S-100 or nuclear extracts, or with partially purified transcription factors, in a total volume of 25 μl containing 12 mM HEPES pH 7.9, 0.1 mM EDTA, 0.5 mM DTE, 5 mM $MgCl_2$, 75 mM KCl, 10 mM creatine phosphate, 12% glycerol (v/v), 0.66 mM each of ATP, CTP and UTP, 0.01 mM GTP and 1.5 μCi of (α-^{32}P)GTP (400 Ci/mmol). After incubation for 60 min at 30° C the nucleic acids were extracted, precipitated and analyzed on nondenaturing 5% polyacrylamide gels.

Total pol I activity was assayed in a 25 μl reaction with activated calf thymus DNA as template as described by Schnapp et al. [9].

Purification of transcription factors

A typical factor preparation was started from 100–200 ml of a mixture of nuclear and cytoplasmic extracts from cultured Ehrlich ascites cells. All buffers used contained 0.5 mM DTE and 0.5 mM PMSF added immediately prior use. Extract proteins were applied onto a DEAE-Sepharose CL-6B column, washed with buffer A (20 mM Tris- HCl pH 7.9, 0.1 mM EDTA, 20% glycerol) containing 100 mM KCl and step eluted at 280 mM KCl. This fraction which contains pol I and the transcription factors TIF-IA, TIF-IB, TIF-IC, and mUBF was loaded directly (at 200 mM KCl) onto a Heparin-Ultrogel A4-R column and bound proteins were eluted at 400, 600, and 1000 mM KCl, yielding fractions H-200, H-400, H-600, and H-1000 respectively. The fraction H-200 contained the growth-rate related activity TIF-IA and a stimulatory factor termed TIF-IC which were further purified on Q-Sepharose and separated from each other by re-chromatography on Heparin-Ultrogel. We have shown before that the chromatographic behaviour of TIF-IA on the heparin column depends on the salt conditions used for chromatography [9] and, therefore, when the second heparin column was run in the presence of 5 mM $MgCl_2$ TIF-IC was in the flowthrough, whereas TIF-IA bound to the column and was eluted at 400 mM KCl. Further purification of TIF-IA and of pol I (present in the H-400 fraction) was achieved by chromatography on a Mono-Q FPLC column. The H-600 fraction containing TIF-IB was purified on a CM-Sepharose column as described [4] and mUBF (present in the H-1000 fraction) was purified on a Mono-Q column. mUBF activity was eluted at 450 mM KCl (Rosenbauer, unpublished results).

Transcription in permeabilized cells

One ml of cultured cells (9×10^5 cells) were spun down, washed with 1 ml PBS, and resuspended in 100 μl of a buffer containing 50 mM Tris-HCl (pH 7.5), 5 mM $MgCl_2$, 75 mM $(NH_4)_2SO_4$, 0.2 M sucrose, 1 mM dithiothreithol, and 0.1% NP40. The activity of the permeabilized cells to incorporate (^{3}H)UMP into acid-precipitable RNA was determined by incubating 5 μl of the cell suspension for 15 min in a 50 μl reaction mixture containing 50 mM Tris-HCl (pH 7.9), 5 mM $MgCl_2$, 50 mM $(NH_4)_2SO_4$, 5 mM dithiothreithol, 0.6 mM ATP, GTP, CTP, and 0.012 mM (^{3}H)UTP (1.3 mCi/mmol) and 200 μg/ml α-amanitin.

Immunoprecipitation

The preparation of anti cdc2 serum G6 was described by Draetta and Beach [23]. All solutions used for immunoprecipitation reactions contained the following set of protease inhibitors: aprotinin (5 μg/ml), pepstatin (5 μg/ml), leupeptin (5 μg/ml), benzamidinehydrochloride (2 mM), antipain (2 μg/ml) and chymostatin (2 μg/ml).

For immunoprecipitation reactions, 40 μl of TIF-IA (10 μg of protein eluted from the MonoQ-column) were incubated with 1 μg of antipeptide serum, and in some cases, about 100 nmol of competing antigenic peptide were added. After a 1.5 h incubation at 4° C, the reactions were transferred to Eppendorf tubes containing 10 μl of protein A Sepharose and incubated for further 30 min. After a short centrifugation the supernatants were assayed for their activity to stimulate transcription of extract prepared from stationary cells. To control the immunoprecipitations, pellets were assayed for H1 kinase activity as described by Draetta *et al.* [24].

Results

Several factors in addition to RNA polymerase I are required for rDNA transcription initiation

In previous studies we have fractionated extracts from cultured Ehrlich ascites cells by chromatography on DEAE-Sepharose followed by Heparin Ultrogel and have shown that three factors in addition to pol I are required for specific transcription initiation at the mouse rDNA promoter. Step-elution at 200, 400, and 600 mM KCl, respectively, of proteins bound to the Heparin column yielded three protein pools (designated H-200, H-400, and H-600) each of which on its own is incapable to direct specific transcription, but which after combination with the two other fractions reconstitute transcriptional activity [4]. The transcription initiation factors present in the individual fractions have been functionally characterized and were termed TIF-IA, TIF-IB, and TIF-IC according to their order of discovery. TIF-IA, a regulatory factor which fluctuates with the growth rate of the cells, tends to associate with pol I and, therefore, is usually found in the H-400 fraction together with pol I. TIF-IB present in the H-600 fraction has been shown to be a 44 kD protein which specifically binds to the core region of the mouse rDNA promoter, and TIF-IC appears to be an essential stimulatory activity.

Further purification of the factors present in the H-400 and H-600 fraction by high resolution chromatography on FPLC columns resulted in a strong decrease of transcriptional activity indicating that one more factor may be present in these two fractions which has been purified away by subsequent chromatographic steps. Indeed, transcriptional activity could be recovered by addition of proteins eluting at 1 M KCl (H-1000) from Heparin Ultrogel (data not shown). Chromatography of this H-1000 fraction on a Mono-Q column yielded transcriptionally active fractions eluting at approximately 450 mM KCl. The activity correlates with two closely spaced proteins of apparent size 94 and 97 kD which specifically bind to both the murine 140 bp enhancer repeats and to the mouse promoter (Rosenbauer, unpublished results). These chromatographic, electrophoretic and binding properties suggest that the factor present in the H-1000 fraction is the murine homologue of hUBF [6, 10].

In the following experiment we studied whether mUBF would functionally substitute TIF-IC or whether both factors complement each other. For this, pol I, TIF-IA and TIF-IB (each purified after chromatography on Heparin Ultrogel on at least one additional column using gradient elutions) were complemented with either TIF-IC or mUBF or a combination of both factor fractions. In the absence of TIF-IC and mUBF, a hardly detectable basal level of transcription occurs (Fig. 1, lane 1). Addition of either factor TIF-IC (lane 2) or mUBF (lane 3) results in a slight increase in transcription. However, addition of both TIF-IC and mUBF together leads to strong activation of rDNA transcription (lane 4). This result demonstrates that TIF-IC and mUBF are different auxiliary factors. We conclude that faithful and specific initiation of mouse rDNA transcription requires besides pol I four additional protein factors, TIF-IA, TIF-IB, TIF-IC, and mUBF.

Transcriptional activity of cell extracts after nutritional shift-up and shift-down

We have previously established a cell-free system that faithfully reflects growth rate-dependent regulation of ribosomal gene transcription [11]. The activity of extracts prepared from cultured Ehrlich ascites cells closely mirrors the rRNA synthetic activity of the cells from which they have been prepared. Fig. 2 shows the rDNA transcriptional activity after nutritional shift-up and shift-down. Cultured cells were starved for the essential amino acids histidine and isoleucine for different time periods and transcriptional activity was measured both in permeabilized cells (Fig. 2A and C) and in extracts prepared from the cultured cells (Fig. 2B and D). Transcription in the permeabilized cells in the presence of high concentrations of α-amanitin is proportional to the number of actively transcribing RNA polymerase I molecules and thus yields a rough estimate of cellular rRNA synthetic activity. As shown in Fig. 2A there is a rapid and drastic decrease in cellular transcriptional activity after deprivation of the essential amino acids. The down-regulation of rRNA synthesis occurs very rapidly as indicated by half-maximal inhibition within 30 min. After adding back the essential amino acids to the culture medium transcriptional activity is recovered with a very similar time course (Fig. 2C).

Interestingly, the cell-free transcription system mirrors the alterations of cellular rDNA transcription. Extracts prepared from cells grown in complete medium show a strong signal in the run-off assay which is progressively reduced with prolonged starvation (Fig. 2B). Conversely, the extremely low activity of extracts derived from starved cells is markedly stimulated during the recovery period (Fig. 2D). The kinetic parameters of rDNA transcriptional regulation are very similar both in the *in vivo* and the *in vitro* assay system. However, there are quantitative differences in the residual transcription after amino acid depletion.

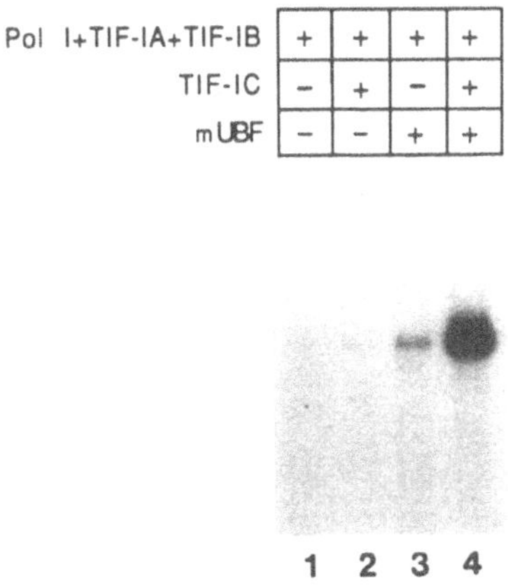

Fig. 1. Factors required for mouse rDNA transcription initiation
Individual factor fractions purified according to the fractionation procedure described in Materials and methods were tested under standard conditions for their ability to support specific transcription. Each assay contained 5 μl of pol I plus TIF-IA and 3 μl of TIF-IB. The reactions were complemented with 3 μl of TIF-IC (lane 2), 4 μl of mUBF (lane 3) or a mixture of both TIF-IC and mUBF (lane 4).

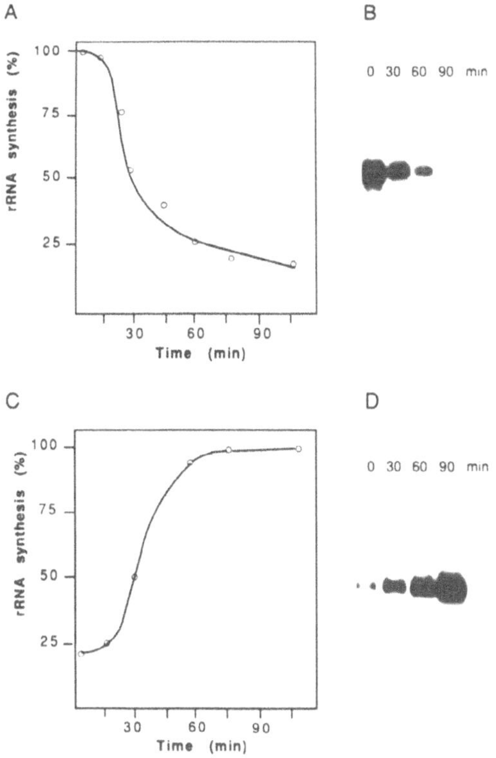

Fig. 2. rDNA transcriptional activity after nutritional shift-up and shift-down
A and B: Exponentially growing Ehrlich ascites cells (9×10^5 cells/ml) were spun down and resuspended in prewarmed RPMI medium lacking histidine and isoleucine. At the times indicated aliquots of the cell suspension were removed and the rRNA synthetic activity was measured both in permeabilized cells (A) or in soluble extracts using cloned rDNA as template (B). C and D: Cells (9×10^5/ml) were cultured for 4 hr in histidine- and isoleucine-free medium before the missing amino acids were added and rDNA transcription was measured at different times after nutritional shift-up in permeabilized cells (C) or in cell free extracts (D).

Ribosomal RNA synthesis *in vivo* is not completely switched-off but is reduced to a basal level which corresponds to 20–25% of maximal activity. In contrast, extracts from cells starved for about 2 hours show virtually no transcriptional activity. These quantitative differences may be attributed to the fact that *in vitro* a threshold concentration of some essential transcription factor(s) is required to yield measurable amounts of transcripts.

The growth-dependent murine factor TIF-IA also stimulates human rDNA transcription

The down-regulaton of rDNA transcription observed after nutritional shift-down and after inhibition of protein synthesis by cycloheximide has been shown to be due to low levels of a growth-dependent essential transcription initiation factor termed TIF-IA [8, 9] or factor C [12]. Addition of this partially purified TIF-IA to transcriptional inactive extracts from starved, stationary or cycloheximide-treated cells stimulates transcription to control levels, indicating that TIF-IA activity is markedly decreased in growth-inhibited cells. As shown in Fig. 3A, the factor-mediated transcriptional stimulation is proportional to the amounts of TIF-IA added to the inactive extract (lanes 5–8). Even the activity of an active extract from exponentially growing cells is increased in the presence of exogenous TIF-IA (lanes 1–4). This finding suggests that the adaptation of rRNA synthesis to the physiological state of the cells may be solely

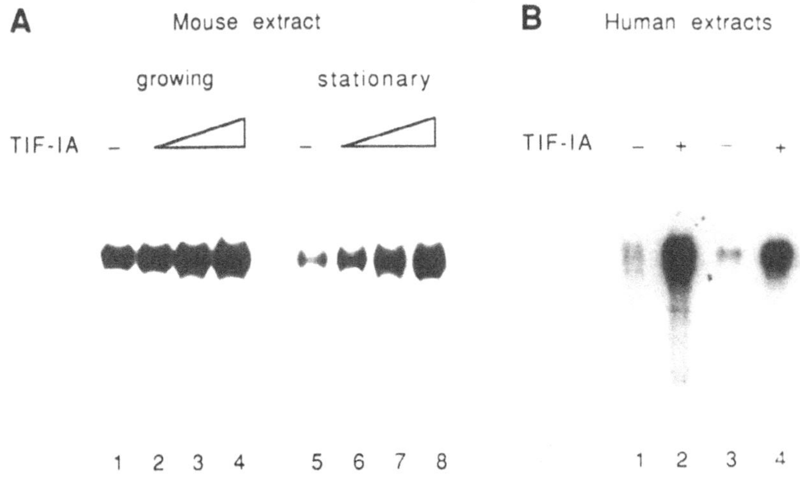

Fig. 3. Transcriptional stimulation of mouse and human cell extracts with factor TIF-IA
A) Transcriptional activity of extracts prepared from exponentially growing cells (lanes 1–4) and stationary phase cells (lanes 5–8) in the presence of increasing amounts of factor TIF-IA. Each assay contained 7 μl of the respective extract and either no additional factor (lanes 1 and 5) or 2 μl (lanes 2 and 6), 4 μl (lanes 3 and 7), or 6 μl (lanes 4 and 8) of TIF-IA. B) Stimulation of human extract by factor TIF-IA. 7 μl of two different preparations of HeLa cell extracts were assayed in the absence (lanes 1 and 3) and presence (lanes 2 and 4) of partially purified TIF-IA.

brought about by alterations in the level or activity of factor of TIF-IA.

Next we tested whether this regulatory factor would also stimulate transcription in a heterologous system. It is a well known fact that there are species-specific borders of rRNA transcription. Extracts from mouse cells fail to transcribe human ribosomal gene templates and, on the other hand, HeLa extracts are not capable to transcribe mouse rDNA [13]. Apparently one (or more) of the proteins required for rDNA transcription initiation is specific for a given organism. To find out whether the activity of the growth-dependent factor TIF-IA is restricted to mouse cells or whether this mechanism of transcription regulation is conserved among different species, partially purified TIF-IA was added to a human rDNA transcription system containing HeLa extract proteins and cloned human rDNA (Fig. 3B). In several extracts tested there is a significant stimulation of human rDNA transcription brought about by TIF-IA. This activation of transcription in the heterologous system demonstrates that TIF-IA function is not restricted to mouse cells and, therefore, its mode of action may be conserved during evolution. Whether or not more distantly related species will also respond to TIF-IA action remains to be investigated.

Species-specificity of mouse rDNA transcription is mediated by factor TIF-IB and RNA polymerase I

To study which of the factors required for pol I transcription initiation are shared by mouse and human cells and which of them are species-specific, we complemented HeLa extracts with partially purified mouse factors and assayed for transcription of mouse rDNA (Fig. 4). Lanes 1 and 2 are control reactions showing the activity of nuclear extract from HeLa cells to transcribe its homologous rDNA template and its inability to initiate specific transcription at the heterologous mouse template. In the presence of the mouse factors TIF-IA, TIF-IB, TIF-IC, and pol I, the murine template is specifically transcribed (lane 3). Addition of increasing amounts of individual mouse fractions, i.e. TIF-IB (lanes 4–6) or pol I (lanes 7, 8), respectively, was not sufficient to enable the human extract to transcribe the mouse template. Similarly, addition of factor mUBF and TIF-IA either alone or in combination with another factor fraction did not reprogram the human extract (data not shown). However, if a combination of both TIF-IB and pol I was used to complement the human extract specific transcription of mouse rDNA was observed. This finding indicates that the pol I present in the human cell extract cannot functionally replace the mouse enzyme.

This result is in apparent contrast to the human system where a single factor, SL1, has been shown to be sufficient to program mouse extract to transcribe human rDNA [14]. It seems that the formation of a productive initiation complex on the mouse rDNA promoter requires both TIF-IB, the mouse analog of human SL1, and mouse RNA polymerase I. This result suggests that pol I itself exhibits species-specific differences. Apparently human pol I requires its own factor SL1 for initiation, whereas the mouse enzyme is more promiscuous and interacts with preinitiation complexes which contain either human SL1 or mouse TIF-IB.

Protein phosphorylation plays a role in rDNA transcription initiation

The activity of a number of pol II transcription factors has been shown to be affected by posttranslational modifications, such as glycosylation and phosphorylation [15–19]. Phosphorylation of the respective proteins affects their ability to bind to their target sequences or their ability to activate transcription and, therefore, in many cases protein phosphorylation is the final target in the chain of events by which extracellular signals are transferred into the cell nucleus. In view of the rapid and reversible alterations of TIF-IA activity in response to extracellular signals, we investigated whether pol I transcription intiation requires TIF-IA to be specifically phosphorylated. As a first step to study this question we used ATP analogues which inhibit the activity of protein kinases or protein phosphatases, respectively. Two different ATP analogues were applied, AMP-PNP which contains an imido group between the β and γ phosphate of ATP and, therefore, is a non-hydrolyzable substrate for protein kinases, and ATP-γS containing in γ-position a thiophosphate group which can be transferred to acceptor proteins. The thiophosphorylated proteins, on the other hand, cannot be dephosphorylated by specific protein phosphatases (see 20 and references therein). Fig. 5 shows the effect of both substrates on specific transcription. The assay system contained partially purified pol I, TIF-IB (H-400 and H-600 fractions both of which contain UBF and do not need to be complemented with H-1000), TIF-IC as well as two different amounts of factor TIF-IA. Lanes 1 and 4 are controls performed in the presence of unmodified ATP. The transcriptional activity of both reactions varies by a factor of about 2–3. This difference in transcription is due to the different amounts of TIF-IA added. In the presence of AMP-PNP specific transcription is inhibited by approximately 40% (lanes 2 and 5). As a similar degree of inhibition (ca. 30%) was also observed when the catalytic activity of pol I was measured on nonspecific calf thymus DNA (data not shown), the reduced transcription may be caused by a negative effect on the elongation reaction rather than by an inhibition of protein phosphorylation. On the other hand, a strong stimulation of rDNA transcription was observed in the presence of ATP-γS. Interestingly, the degree of transcriptional stimulation correlates with the amount of TIF-IA present in the assay. Whereas the reaction containing the higher concentration of TIF-IA is stimulated about 5-fold (lane 3), transcription in the assay with less TIF-IA is increased only 2–3 fold. This intriguing correspondence between the amount or activity of the growth-dependent regulatory factor TIF-IA suggests that the stimulation of rDNA transcription by ATP-γS is mediated by TIF-IA. Whether TIF-IA itself has kinase activity or is the substrate for a protein kinase, or whether its activity is modified indirectly by ATP-dependent processes which in turn require an unmodified γ-phosphate group is unknown.

TIF-IA activity is precipitated by antibodies against cdc2 protein kinase

The previous observations suggest that protein

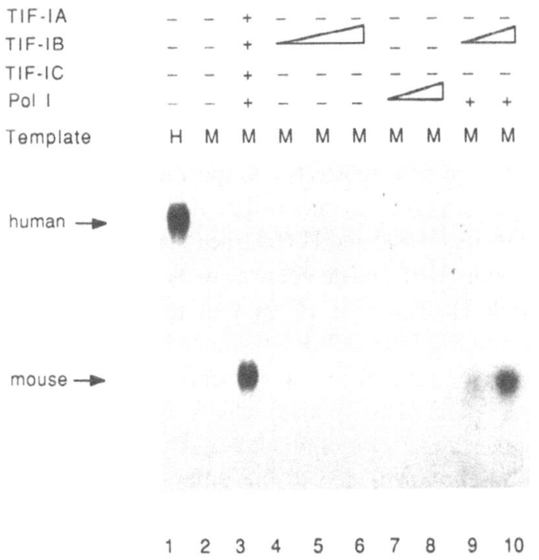

Fig. 4. Species-specificity of murine pol I transcription factors
The individual reactions contained 6 μl of HeLa nuclear extracts and 70 ng of either a human rDNA template (pHrP/Pvu II, lane 1) or mouse rDNA (pMr600/Eco RI, lanes 2–10). Lane 3 is a control showing the activity of a mixture of mouse factors (equal amounts of TIF-IA, TIF-IB, TIF-IC and pol I) in the presence of 6 μl of HeLa cell extract. Increasing amounts of either TIF-IB (1 μl, 3 μl and 6 μl, lanes 4–6), pol I (4 μl or 8 μl, lanes 7 and 8), or a mixture of 4 μl of pol I, and 2 μl or 4 μl of TIF-IB, respectively, were added (lanes 9 and 10). The transcripts derived from the human and mouse rDNA template are indicated.

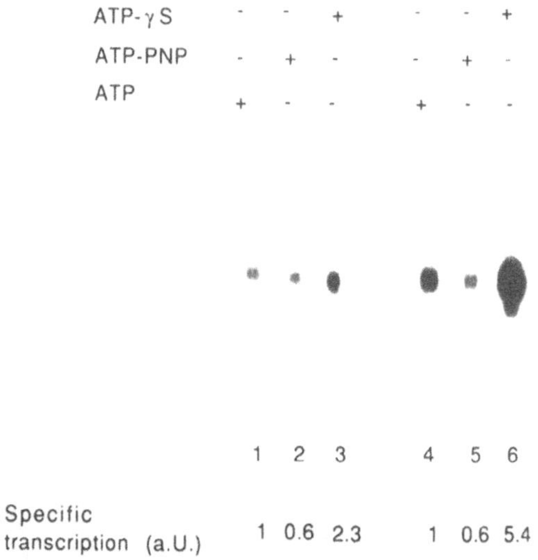

Fig. 5. Effect of ATP-analogues on specific rDNA transcription
In vitro transcription assays were performed as described in Materials and methods with the exception that ATP (lanes 1 and 4) was replaced by either AMP-PNP (lanes 2 and 5) or by ATP-γS (lanes 3 and 6). Each reaction contained 9 μl of a mixture of partially purified pol I, TIF-IB, TIF-IC and mUBF, and either 2 μl (lanes 1–3) or 5 μl (lanes 4–6) of TIF-IA. The values below each lane indicate the relative efficiencies of transcription (in arbitrary units) as determined by densitometric scanning of appropriately exposed autoradiograms.

phosphorylation is involved in the control of rDNA transcription, and that most likely TIF-IA is the protein which mediates this regulation. Up to now we have failed to functionally replace TIF-IA activity by available protein kinases, such as cAMP and cGMP-dependent protein kinase or by HeLa cell p34, the homologue of the yeast cdc2/CDC28 protein kinase. p 34 is involved in regulation of the cell cycle in a wide range of eukaryotes [21–23] and has been shown to interact with other cellular proteins, such as a 13 kd suc1-related polypeptide [24] and a 62 kd protein, cyclin, which acts as the major endogenous substrate of the protein kinase complex [23, 25, 26]. Other cellular polypeptides may also associate with p34 [27]. Because of the marked fluctuations of TIF-IA activity according to the proliferation rate of the cells, we suspected that this regulatory factor might be regulated by cdc2 which plays such a central role in the cell cycle. In the experiment shown in Fig. 6, TIF-IA was incubated with an anti-cdc2 serum (G6, prepared against a carboxy-terminal peptide of human cdc2; 23) in the presence or absence of a competing antigenic peptide. The supernatant of the immunoprecipitates was assayed for TIF-IA activity by complementation of an extract from growth-arrested cells. This extract on its own exhibits a hardly detectable transcriptional activity (lane 1). In the presence of TIF-IA a strong signal is obtained (lane 2). Interestingly, the intensity of this signal was significantly reduced when the TIF-IA fraction was incubated with the anti-cdc2 serum (lane 3). This decrease in transcription is specific for the antibodies against p34, since in the presence of the competing synthetic peptide (lane 4) or in the presence of a control serum (lane 5) no effect on TIF-IA activity was observed. This result is compatible with any of the following conclusions: (I) TIF-IA is antigenically related to cdc2, (II) TIF-IA is associated with cdc2, (III) TIF-IA interacts with cdc2 kinase indirectly via a cdc2-associated polypeptide, or (IV) the activity of TIF-IA is regulated by cdc2 kinase.

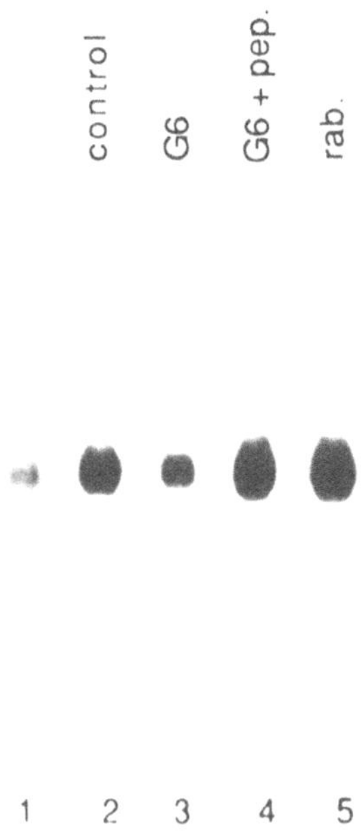

Fig. 6. Immunoprecipitation of TIF-IA by antibodies against cdc2

Partially purified TIF-IA was incubated with anti-cdc2 (G6) or control serum (rab) in the absence or presence of 100 nmol of competing antigenic peptide (pep) as described in Materials and methods. The supernatants of the immunoprecipitation reactions were assayed for TIF-IA activity by complementation of an inactive extract from stationary cells. Lane 1: 7 μl of inactive extract plus buffer; lane 2: extract plus 6 μl TIF-IA; lanes 3–5: extract plus 8 μl of the supernatant of the individual immunoprecipitation reactions.

Discussion

We have demonstrated that accurate and efficient transcription initiation at the mouse rDNA promoter requires four factors in addition to RNA polymerase I, termed TIF-IA, TIF-IB, TIF-IC, and mUBF. It is still difficult to compare the different pol I transcription factors studied in different laboratories since different species are investigated and, in addition, each group has its own factor nomenclature. With the recent purification to homogeneity of UBF from HeLa cells and cloning of the gene coding for hUBF, the situation begins to clarify. UBF homologues are present in all four vertebrates so far examined, i.e. in human, frog, mouse, and rat. Surprisingly, UBF from each of these species has similar DNA-binding properties and produces not only nearly identical footprints on both their own and on heterologous rDNA promoters, but also on the frog 60/81 bp and the mouse 140 bp enhancer elements despite the lack of any obvious conservation in their recognition sequence [6, 7]. There is a general consensus that in addition to UBF another factor is required for promoter recognition by pol I. This selectivity factor has been termed in human and rat SL1 [14, 28], in mouse TIF-IB [3], TFID [29], or factor D [30]. There is experimental evidence demonstrating that SL1 and UBF together form a complex at the rDNA promoter which in turn is recognized by the transcription machinery.

In contrast to UBF, the SL1-type factor is specific for a given class of vertebrates, and most likely is responsible for the species-specific nature of rDNA transcription [13]. Both in the human and rat system this factor has been shown to program heterologous, i.e. mouse or human cell extracts, respectively, to transcribe human or rat rDNA templates [14, 28]. We have performed analogous experiments using highly purified TIF-IB and always failed to make HeLa extracts to transcribe mouse rDNA. Transcription of the murine template required in addition to TIF-IB the homologous mouse RNA polymerase. This finding implicates that the pol I present in the human cell extract cannot functionally replace the mouse enzyme and suggests that pol I itself exhibits species-specific differences. Apparently human pol I requires its own factor SL1 for initiation, whereas the mouse enzyme is less stringent and interacts with preinitiation complexes containing either human SL1 or mouse TIF-IB.

These interesting similarities and differences between mouse and human pol I-specific transcription factors are not restricted to proteins which direct transcription initiation but also extend to proteins mediating transcription termination. We have shown that pol I termination at the mouse rDNA transcription unit requires the binding of a nuclear factor (TTF I) to an 18 bp terminator sequence (the Sal boxes) which are located in several copies downstream of the 45S precursor coding region [31, 32]. In human rDNA the Sal box terminator element is shorter encompassing only 11 bp of the proximal part of the murine sequence [33]. Because of the different size of its target sequence, the murine factor does not bind to the human terminator and, therefore, does not stop human pol I transcription. The human factor, on the other hand, recognizes the murine signal and directs termination of transcripts that have been started from the mouse rDNA promoter [34]. Thus, although the structural organization of eukaryotic ribosomal genes is strongly conserved, even in related species like human and mouse both at the 5′ and 3′ end of the rDNA transcription unit changes in *cis*-acting regulatory DNA sequences occurred which in turn have been accompanied by compensatory changes in the DNA binding domain of the proteins interacting with the respective signal sequence.

The other two auxiliary pol I transcription initiation factors TIF-IA and TIF-IC appear to be shared between mouse and human. Firstly, we could show that human rDNA transcription is stimulated by TIF-IA. Secondly, mouse rDNA transcription could be reconstituted in HeLa extracts by TIF-IB and pol I, indicating that both UBF and TIF-IC are present in the human extract. Apparently UBF and TIF-IC are common factors which are not directly involved in promoter recognition but rather play a role in facilitating the assembly or stabilizing the preinitiation complex.

We also begin to get some insight into the mode of action of the growth-dependent regulatory factor TIF-IA. In this communication we present three lines of evidence that regulation of rDNA transcription is brought about by specific protein phosphorylation, and that TIF-IA is causally involved in this process. (I) ATP-γS which is used by protein kinases to transfer thiophosphate residues to proteins which in turn cannot be removed by protein phosphatases, stimulates rDNA transcription. (II) The degree of stimulation correlates with the amount of factor TIF-IA present in the assay. (III) TIF-IA is inactivated by antibodies against protein kinase cdc2. These results suggest that TIF-IA may be a specific protein kinase whose activity is regulated by cdc2. Alternatively, TIF-IA may be a substrate for this growth-regulated kinase. In this scenario, factor TIF-IA is the final target in the sequence of events by which extracellular signals are transferred into the nucleolus and cdc2 plays a role in regulating this signal transduction pathway.

References

1. Moss T, Michelson K, De Winter R: The promotion of ribosomal transcription in eukaryotes. Oxford Surveys on Eukaryotic Genes 2: 207–250, 1985
2. Grummt I: Mammalian ribosomal gene transcription. In: Nucleic Acids & Molecular Biology, ed. by F. Eckstein and D.M.J. Lilley, Springer-Verlag Berlin Heidelberg, Vol 3, p. 148–163, 1989
3. Clos J, Buttgereit D, Grummt I: A purified transcription factor (TIF-IB) binds to essential sequences of the mouse rDNA promoter. Proc Natl Acad Sci USA 83: 604–608, 1986
4. Schnapp A, Hädelt W, Clos J, Schreck R, Cvekl A, Grummt I: Isolation and functional characterization of TIF-IB, a factor that confers promoter specificity to mouse RNA polymerase I. Nucl Acids Res 18: 1385–1393, 1990a
5. Bell SP, Learned M, Jantzen HM, Tjian RT: Functional cooperativity between transcription factors UBF1 and SL1 mediates human ribosomal RNA synthesis. Science 241: 1192–1197, 1988
6. Bell SP, Pikaard CS, Reeder RH, Tjian RT: Molecular mechanisms governing species-specific transcription of ribosomal RNA. Cell 59: 489–497, 1989
7. Pikaard CS, Mc. Stay B, Schultz MC, Bell SP, Reeder RH: The Xenopus ribosomal gen enhancers bind an essential polymerase I transcription factor, xUBF. Genes Dev 3: 1779–1788, 1989
8. Buttgereit D, Pflugfelder G, Grummt I: Growth-dependent regulation of rRNA synthesis is mediated by a transcription initiation factor (TIF-IA). Nucl Acids Res 13: 8165–8170, 1985
9. Schnapp A, Pfleiderer C, Rosenbauer H, Grummt I: A growth dependent transcription initiation factor (TIF-IA) interacting with RNA polymerase I regulates mouse ribosomal RNA synthesis. EMBO J 9: 2857–2863, 1990b
10. Jantzen H-M, Admon A, Bell SP, Tjian R: Nucleolar transcription factor hUBF contains a DNA-binding motif with homology to HMG proteins. Nature 344: 830–836, 1990
11. Grummt I: Specific transcription of mouse ribosomal DNA in a cell-free system that mimics control *in vivo*. Proc Natl Acad Sci USA 78: 727–731, 1981

12. Tower J, Sollner-Webb B: Transcription of mouse rDNA is regulated by an activated subform of RNA polymerase I. Cell 50: 873–883, 1987
13. Grummt I, Roth E, Paule M: Ribosomal transcription *in vitro* is species-specific. Nature 296: 173–174, 1982
14. Learned RM, Cordes S, Tjian RT: Purification and characterization of a transcription factor that confers promoter specificity to human RNA polymerase I. Mol Cell Biol 5: 1358–1369, 1985
15. Sorger PK, Pelham HRB: Yeast heat shock factor is an essential DNA-binding protein that exhibits temperature-dependent phosphorylation. Cell 54: 855–864, 1988
16. Prywes R, Dutta A, Cromlish JA, Roeder RG: Phosphorylation of serum response factor, a factor that binds to the serum response element of the c-Fos enhancer. Proc Natl Acad Sci 85: 7206–7210, 1988
17. Bagchi S, Raychaudhuri P, Nevins JR: Phosphorylation-dependent activation of the adenovirus inducible E2F transcription factor in a cell-free system. Proc Natl Acad Sci 12: 4352–4356, 1989
18. Jackson SP, Tjian RT: O-glycosylation of eukaryotic transcription factors: implications for mechanisms of transcription regulation. Cell 55: 125–133, 1988
19. Lichtensteiner S, Schibler U: A glycosylated liver-specific transcription factor stimulates transcription of the albumin gene. Cell 57: 1179–1187, 1989
20. Edelman A: Protein serine/threonine kinases. Ann Rev Biochem 57: 567–613, 1987
21. Reed SI, Hadwinger JA, Lörincz AT: Protein kinase activity associated with the product of the yeast cell-cycle gene CDC28. Proc Natl Acad Sci USA 82: 4055–4059, 1985
22. Gautier J, Norbury C, Lohka M, Nurse P, Maller J: Purified maturation-promoting factor contains the product of a Xenopus homolog of the fission yeast cell cycle control gene cdc2 +. Cell 54: 433–439, 1988
23. Draetta G, Beach D: Activation of cdc2 protein kinase during mitosis in human cells: cell-cycle dependent phosphorylation and subunit rearrangement. Cell 54: 17–26, 1988
24. Draetta G, Brizuela L, Patashkin J, Beach D: Identification of p34 and p13, human homologs of the cell cycle regulators of fission yeast encoded by cdc2 + and suc1 +. Cell 50: 319–325, 1987
25. Draetta G, Luca F, Westendorf J, Brizuela L, Ruderman J, Beach D: cdc2 protein kinase is complexed with both Cyclin A and B: evidence for proteolytic inactivation of MPF. Cell 56: 829–838, 1989
26. Brizuela L, Draetta G, Beach D: Activation of human CDC2 protein as a histone H1 kinase is associated with complex formation with the p62 subunit. Proc Natl Acad Sci USA 86: 4362–4366, 1989
27. Giordano A, Whyte P, Harlow E, Franza BR, Beach D, Draetta G: A 60 kd cdc2 associated polypeptide complexes with the E1A proteins in adenovirus-infected cells. Cell 58: 981–990, 1989
28. Smith SD, Oriahi E, Lowe D, Yang-Yen H-F, O'Mahony D, Rose K, Chen K, Rothblum LI: Characterisation of factors that direct transcription of rat ribosomal DNA. Mol Cell Biol 6: 3105–3116, 1990
29. Mishima Y, Financsek I, Kominami R, Marumatsu R: Fractionation and reconstitution of factors required for accurate transcription of mammalian ribosomal RNA genes: identification of a species-dependent initiation factor. Nucl Acids Res 10: 6659–6669, 1982
30. Tower J, Cizewski-Culotta V, Sollner-Webb B: Factors and nucleotide sequences that direct ribosomal DNA transcription and their relationship to the stable transcription complex. Mol Cell Biol 6: 3451–3462, 1986
31. Grummt I, Maier U, Öhrlein A, Hassouna N, Bachellerie J.P: Transcription of mouse rDNA terminates downstream of the 3'end of 28S RNA and involves the interaction of factors with repeated sequences in the 3' spacer. Cell 43, 801–810, 1985
32. Grummt I, Rosenbauer H, Niedermeyer I, Maier U, Öhrlein A: A repeated 18 bp sequence motif in the mouse rDNA spacer mediates binding of a nuclear factor and transcription termination. Cell 45, 837–846, 1986
33. Bartsch I, Schoneberg C, Grummt I: Evolutionary changes of sequences and factors that direct transcription termination of human and mouse ribosomal genes. Mol Cell Biol 7: 2521–2529, 1987

Address for offprints: A. Schnapp, Institute of Cell and Tumor Biology, German Cancer Research Center, im Neuenheimer Feld 280, D-6900 Heidelberg, Germany

Molecular and Cellular Biochemistry **104**: 149–154, 1991.

Expression of mouse and frog rRNA genes: transcription and processing

Barbara Sollner-Webb, Louise Pape[1], Kenneth Ryan, Edward B. Mougey, Regina Poretta, Emil Nikolov, Mark H. Paalman, Inara Lazdins and Cathy Martin
The Johns Hopkins University, School of Medicine, 725 North Wolfe Street, Baltimore, Maryland 21205, USA; [1] Current address: Department of Chemistry, New York University, New York, NY 10003, USA

Key words: rRNA genes, species selectivity for gene expression, promoter, enhances, rRNA processing

Abstract

This article summarizes a number of lines of investigation of rRNA gene expression that are ongoing in the laboratory. These studies focus on mouse and frog, two distant vertebrate species. One major conclusion is that the basic properties of rRNA gene expression appear remarkably well conserved in evolution, with only relatively minor perturbations between frog and mouse, contrary to the common interpretation of the species-selectivety between mouse and human rDNA transcription (e.g., 1). This is true both for the process of rDNA transcription and for the subsequent rRNA processing event.

The rDNA promoter

We have shown that the rDNA promoter of both Xenopus and mouse consists of a core and an upstream domain. The ~40 bp core domain surrounds the initiation site and is sufficient to promote efficient transcription under conditions chosen to support maximal rRNA production by the complete promoter [2–4]. The upstream domain, surrounding residue ~ − 140, has a stimulatory effect that becomes important for expression under more stringent *in vitro* conditions [2, 4] and is virtually essential under most *in vivo* expression conditions [3, 5] in frog as well as in mouse. The sequence of the 100 bp segment between the two important domains of the Xenopus rDNA promoter can be greatly altered without a major effect on transcription, but the relative position of these two domains is rigidly constrained, for transcription is abolished when the spacing is altered by more than 2 bp [6]. This suggests that the two domains must be appropriately aligned and that they function in concert. Presumably, it is the binding of the rDNA transcription factor D to the upstream promoter domain and to the upstream half of the core promoter domain (see below; 7) that dictates this helical alignement.

Since the primary sequences of the essential mouse and frog promoter domains are not obviously similar, we were surprised to find that the mouse transcription machinery could transcribe frog rDNA [8]. This heterologous transcription requires the same core and upstream promoter domains of the frog rDNA as does transcription in the homologous frog system [9], suggesting that the natural promoter interactions are utilized. However, the heterologous rDNA expression was less efficient than that of the homologous mouse system and intiation occurred at residue − 4 (not + 1) suggesting that there might be appreciable differences between the heterologous and homologous expression systems. We have recently found that this initiation at residue − 4 is due to the relative helical face of the two promoter domains; when a half helical turn is introduced anywhere between the core and upstream domains of the frog promoter (altering spacing by − 1/2, + 1/2, or + 1 1/2 turns of DNA) the heterologous transcription now initiates at residue + 1 [10]. This correctly initiated heterologous transcription is extremely efficient; indeed, the frog half helical turn spacing change mutants can even out-compete mouse rDNA for the mouse transcription factors [10]. Furthermore, this efficient heterologous transcription is not restricted to *in vitro* systems, but it also occurs when frog templates with such spacing changes are introduced into intact rodent cells (Lazdins and Sollner-Webb, in preparation). From these results, we conclude that frog and mouse rDNA promoters share a very similar organization and that the nucleoprotein interactions that direct transcription are also well conserved across this evolutionary distance.

Upstream transcriptional elements

The similarity of functional domains between Xenopus and mouse rDNA extends to all identified transcriptional stimulatory elements. Terminator-like elements are found not only at the 3′ end of the rDNA transcription unit, where they would be expected to occur, but also in a novel position just upstream of the promoter. In both mouse and frog, these promoter-proximal terminators greatly affect the rDNA transcriptional level [11, 12]. This effect occurs, at least in part, by their preventing upstream polymerase from transcribing across the promoter region, an event that dislodges the transcription complex from the promoter and thereby prevents all the subsequent rounds of transcription which that promoter complex would have directed [13].

Just upstream from the promoter-proximal terminator are repetitive elements that have been shown to be polymerase I transcriptional enhancers in mouse [14, 15] as well as in frog [16, 17]. These enhancers can stimulate rDNA promoters in both orientations and over considerable distances [16, 17]. However, we have recently shown that their stimulatory effect is much more position-dependent than initially believed, falling off markedly when the repeats are moved 1.5 kb away from the promoter [17]. Although enhancement has traditionally been assessed under conditions of template competition (e.g., 16), it can also readily be demonstrated using a direct (polymerase II-type) enhancer assay with a single kind of input template [15, 17]. In X. laevis, the promoter-proximal half of the intergenic spacer is largely composed of 60 and 81 bp repetitive enhancer elements [16], and the analogous 140 bp repetitive enhancers of mouse are also in a promoter-proximal location [15]. Recently, we have found that the 100 bp and the 35 bp repetitive elements that comprise the bulk of the promoter-distal half of the X. laevis intergenic spacer are also polymerase I enhancers, with similar but not identical properties to the 60/81 bp enhancers (Mougey, Pape, and Sollner-Webb, in preparation). Thus, virtually the entire (~4 kb) intergenic spacer of X. laevis rDNA serves to stimulate transcription. The intergenic spacer of mouse rDNA is much larger than that of frog, and most of its ~30 kb has not been examined for additional transcriptional effectors. It is noteworthy that, despite the lack of apparent sequence similarity between the mouse 140 bp and frog 60/81 bp enhancer repeats, the mouse enhancer is fully functional in a Xenopus expression system [15], again demonstrating a great conservation in rDNA transcriptional machinery.

There is yet another conserved feature of Xenopus and mouse rDNA. Upstream of the promoter-proximal enhancer repeats is an additional polymerase I promoter called the 'spacer promoter'. First recognized in Xenopus [14, 19] and shown to stimulate transcription from the downstream rRNA gene promoter in Xenopus [20], spacer promoters have since been found in rodent rDNA [21–23]. We have recently shown that rodent spacer promoters can similarly augment transcription from the downstream gene promoter (Paalman and Sollner-Webb, unpublished observations). While the mechanism of this stimulation remains a matter of speculation, it seems likely to be similar in Xenopus and mouse.

Factors

Pursuing our earlier finding that two fractions, called C and D, were sufficient to direct rDNA transcription [7], we have recently shown that both of these fractions can be separated into two activities: fraction C into C* and polymerase I, and fraction D into factor D and UBF (Ryan and Sollner-Webb, unpublished observation).

The C component, the 'active polymerase I', has been purified extensively and co-purifies with bulk polymerase I through virtually complete chromatographic purification of the polymerae I polypeptides [24]. This C/pol could, however, be separated from most of the bulk polymerase I by two methods. (Bulk polymerase I is classically defined by its ability to catalyze non-specific elongation, independent of its specific initiation capacity.) The first method consists of a functional fractionation involving pelleting of the stable transcription complex (including the C component) away from unbound polymerase; in the second, the C/pol species can be enriched by velocity centrifugation since it forms the rapidly sedimenting portion of the total polymerase I peak [24]. The C/pol, but not the separated bulk polymerase I, is able to initiate specific transcription on an rDNA promoter that is complexed with the factors of the D fraction.

Factor C is especially intriguing because of its potential role in rDNA transcriptional regulation. Cells whose growth has been inhibited by any of a number of treatments are known to down-regulate their rDNA transcription (see 14). This modulated transcriptional activity is mimicked in extracts prepared from such growth-inhibited cells and this is due to a diminution in the activity/amount of factor C [14, 25]. We have recently found that modulations in C activity according to growth rate also occur in the intact animal. Livers of control rats, rats in which rDNA transcription is down-regulated (by cycloheximide administration) and rats in which rDNA transcription is up-regulated (by partial hepatectomy) have correspondingly regulated amounts of C activity (Nikolov, Ryan, and Sollner-Webb, in preparation). These results suggest that alterations in the level of C activity may also be responsible for the regulation of rDNA transcription *in vivo*.

Recently, we have made two other critical observations concerning the C activity. The first is that a C* activity can be separated from all detectable polymerase I activity by mono Q chromatography (but not by any other matrices examined) (Ryan and Sollner-Webb, unpublished observations), indicating that C/pol and bulk polymerase I are distinguished by the presence of additional polypeptide(s) and not solely by covalent modification. This C* activity is the growth-regulated component of the rDNA transcriptional machinery. [Here C* is assayed by its ability to restore rDNA transcriptional activity to a down-regulated cell extract and/or by its ability to allow specific initiation in combination with bulk polymerase I plus the factors of the D fraction, D and UBF.] Similar separations of the regulated activity from polymerase I have also been obtained by the Thompson [26] and Grummt [27] laboratories. The fact that C* and polymerase I co-fractionated on gradient elution from seven different chromatographic matrices makes it very improbable that C* is a separate, fortuitously co-purifying protein; rather, it is likely to be a polymerase-associated activity which can be removed by certain columns, possibly akin to the sigma subunit of E. coli RNA polymerase.

The second intriguing finding concerning factor C is that the C* activity is stoichiometrically consumed at each round of transcription (Porretta, Ryan and Sollner-Webb, in preparation). Pursuing earlier observations that C appeared to be short-lived (e.g., 24), we observed that its rapid turnover *in vivo* requires ongoing rDNA transcription, for C was stable *in vivo* in the absence of transcription. Analogous results were then obtained in the *in vitro* expression system. This phenomenon of C consumption has now been found to occur at a post-initiation stage that is still early in the transcription process, when the nascent RNA is ~ 15–25 nucleotides in length (Porretta, Ryan and Sollner-Webb, in preparation). [In these experiments, C* activity was assayed following its liberation from transcription complexes whose progress had been arrested at defined positions using chain terminating nucleotide analogs.] Such an alteration in the elongating polymerase is reminiscent of the conversion of E. coli RNA polymerase from the holoenzyme to the core enzyme via release of the sigma factor (at nucleotide ~ + 10), thereby converting the polymerase from an initiation form into an elongation form [28]. Eukaryotic RNA polymerase II has also been found to be modified at the start of transcription, in this case by phosphorylation, converting the II_A to the II_O form [29]. The identity of the change that causes the loss of C activity after polymerization of ~20 nucleotides remains to be determined.

The original fraction D can be chromatographically separated into a factor D and UBF (Ryan, Pape, and Sollner-Webb, in preparation). Mouse UBF, now purified to homogeneity, has rDNA binding properties virtually identical to frog and human UBF [15]. In combination with UBF, factor D binds selectively to the rDNA promoter, utilizing sequences in the upstream half of the core promoter domain [7] and the upstream promoter domain [9]. Factor C binds to this D/UBF/promoter complex without requiring additional sequence interactions; the residues in the downstream half of the core promoter domain are, however, required for transcriptional initiation [7]. The interaction of UBF plus D with the promoter sequences is considerably stronger than that of either UBF or D alone (Pape, Ryan, and Sollner-Webb, in preparation). Evidence for this conclusion has been obtained by exonuclease III footprinting studies (where the specific nuclease protection that arises from D plus UBF was not similarly obtained even with twice the amount of D or UBF individually). In addition, analysis of enhancer-bearing linker scanning mutants has shown the D binding regions of the promoter to be necessary for enhancer function and the primary UBF binding site to be involved as well. While the UBF protein has been identified, renatured from an SDS gel band [15 and references therein] and now cloned [30], D activity has been found to co-purify with a small number of polypeptides (Ryan and Sollner-Webb, unpublished observations), whose identity, potential relevance, and function remain to be defined. D is almost assuredly the SL1 activity of the Tjian laboratory and is contained within the B activity of the Grummt laboratory.

Enhancer function appears to involve the same fundamental factors that function in promoter activation. UBF binds to the Xenopus 60/81 bp enhancer and to the mouse 140 bp enhancer as well as to these species' promoters [15, 31], prompting speculation that the enhancer repeats act as reiterated initial docking sites for UBF which could facilitate more efficient UBF binding in the adjacent promoter [31]. However, we have recently obtained evidence by exonuclease III footprinting analysis and preferential co-pelleting studies that factor D binds to the UBF/enhancer complex (Pape, Ryan, and Sollner-Webb, in preparation). These data suggest that the relatively abundant UBF could bind to both the enhancer and the promoter, and enhancement may result from factor D rapidly interacting with one of the many tandem UBF/enhancer complexes and then preferentially transferring to its tighter binding site on the UBF-bound promoter region.

This model can readily be extended to include the other conserved polymerase I transcriptional stimulatory elements. Specifically, transcription from the spacer promoters could also be stimulated by the enhancer repeats, although presumably less efficiently than is transcription from the gene promoter (since the enhancer appears to act preferentially on downstream promoters; 17). Thus, following activation of the gene promoter by the enhancer, the spacer promoter could be activated to transcribe across the enhancer region, releasing the D and UBF bound to the enhancer, just as transcription through a promoter region causes release of its bound D/UBF [13]. The liberated D/UBF would then be available to activate additional rDNA promoters, thereby enhancing total transcription. The promoter-proximal terminator would have the crucial role of preventing the spacer polymerases from progressing through the gene promoter and dissociating its transcription complex. Experiments are in progress to access the validity of this model of rDNA transcriptional activation.

Processing of the primary transcript

Once the pre-rRNA is transcribed, it is processed to give rise to the 18S, 5.8S, and 28S rRNAs of the ribosome. Already by the mid 1970's, it had been shown that there was a preferential, but not obligatory, order of the cleavages at the ends of the mature rRNA regions (reviewed in 14). However, we found that the first processing that mouse pre-rRNA undergoes is at residue + 650, within the 5′ external transcribed spacer, nearly 4 kb upstream from the 18S region (32; see also 33). This cleavage within the mouse external transcribed spacer (ETS) was the first protein dependent eukaryotic rRNA processing event to be reproduced *in vitro* [32]. It involves a large ribonucleoprotein complex [34] that assembles on the rRNA sequences just downstream of the cleavage site which direct the processing [35]. This processing has recently been shown to require the U3 snRNP [36], thus demonstrating that an rRNA processing event, like mRNA splicing, requires an RNA-containing cofactor. Experiments to assess the mechanism of this processing and the secondary interactions that allow snRNP binding to the processing region are currently underway (Martin and Sollner-Webb, in progress).

The 5′ primary ETS processing event is not unique to mouse. Indeed, in human rRNA there is a virtually identical 5′ ETS processing [38], and Xenopus also has an analogous processing in its 5′ spacer (Mougey and Sollner-Webb, in preparation). Like the mammalian processing, the Xenopus processing also requires the U3 snRNP. In contrast to the rest of the ETS sequence which is highly divergent, the mouse and human processing signals are conserved, and there is even an 11 nucleotide homology between the Xenopus processing region and the minimal mouse processing signal. In fact, the Xenopus processing signal is appropriately recognized by the mouse processing machinery (Mougey and Sollner-Webb, in preparation), suggesting great evolutionary conservation in this processing event. [Processing within the 5′ ETS has also been reported in numerous lower eukaryotic species (reviewed in 36), but their mechanistic relationship to the vertebrate processing remains to be established.]

Although the biological role of the U3-requiring ETS processing is not yet understood, the fact that it is specifically conserved, at least in vertebrate evolution, suggests that it provides a selective advantage. One hypothesis is that this processing site (near the 5′ end of the very long pre-rRNA transcript) serves as a rapid initial assembly site for processing factors which will function at downstream processing sites; these sites might otherwise be shielded by the large nascent rRNA transcript and not be as accessible for direct association with the processing components. In any event, it is likely that the ETS processing complexes are the 'terminal knobs' which were first visualized in electron micrographs of nucleolar chromatin 20 years ago [37] and are characteristic of ribosomal transcription units, decorating the 5′ ends of nascent pre-rRNA fibrils.

Acknowledgements

We thank the Geisinger foundation for making possible the meeting at which these results were presented. This work was supported by funds from the NIH and the ACS.

References

1. Sommerville J: RNA polynmerase I promoters and transcription factors. Nature 310: 189–190, 1984
2. Sollner-Webb B, Wilkinson J, Roan J, Reeder R: Nested control regions promote Xenopus ribosomal RNA synthesis by RNA polymerase I. Cell 35: 199–206, 1983
3. Windle J, Sollner-Webb B: Upstream domains of the frog rDNA promoter are revealed in microinjected oocytes. Mol Cell Biol 6: 1228–1234, 1986
4. Miller K, Tower J, Sollner-Webb B: A complex control region for the mouse rRNA gene directs accurate transcription by RNA polymerase I. Mol Cell Biol 5: 554–562, 1985
5. Henderson S, Sollner-Webb B: The mouse rDNA promoter has more stringent requirements *in vivo* than *in vitro*. Mol Cell Biol 10: 4970–4973, 1990
6. Windle J, Sollner-Webb B: Two distinct and precisely positioned domains promote transcription of Xenopus rRNA genes: Analysis with linker scanning mutants. Mol Cell Biol 6: 4585–4593, 1986
7. Tower J, Culotta V, Sollner-Webb B: The factors and nucleotide sequences that direct rDNA transcription and their relationship to the stable transcription complex. Mol Cell Biol 6: 3451–3462, 1986
8. Wilkinson J, Sollner-Webb B: Transcription of Xenopus ribosomal RNA genes by RNA polymerase I *in vitro*. J Biol Chem 257: 14375–14383, 1982
9. Culotta V, Wilkinson J, Sollner-Webb B: Mouse and frog

violate the paradigm of species-specific ribosomal RNA transcription. Proc Natl Acad Sci USA 84: 7498–7502, 1987
10. Pape L, Windle J, Sollner-Webb B: Half helix turn spacing changes convert a 'frog' into a 'mouse' rDNA promoter: a distant upstream domain determining the helix face of the initiation site. Genes and Dev 4: 52–62, 1990
11. Henderson S, Sollner-Webb B: A transcriptional terminator is a novel element of the promoter of the mouse ribosomal RNA gene. Cell 47: 891–900, 1986
12. McStay B, Reeder R: A termination site for Xenopus RNA polymerase I also acts as an element of an adjacent promoter. Cell 47: 913–920, 1986
13. Henderson S, Ryan K, Sollner-Webb B: The promoter-proximal rDNA terminator augments initiation by preventing disruption of the stable transcription complex caused by polymerase read-in. Genes and Dev 3: 212–223, 1989
14. Sollner-Webb B, Tower J: Transcription of cloned eukaryotic ribosomal RNA genes. Ann Rev Biochem 55: 801–830, 1986
15. Pikaard C, Pape L, Henderson S, Ryan K, Paalman M, Lopata M, Reeder R, Sollner-Webb B: Enhancers for RNA polymerase I in mouse ribosomal DNA. Mol Cell Biol 10: 4816–4825, 1990
16. Reeder R: Enhancers and ribosomal gene spacers. Cell 47: 349–351, 1984
17. Pape L, Windle J, Mougey E, Sollner-Webb B: The Xenopus rDNA 60/81 bp repeats are true 5′ enhancers and function at the establishment of the preinitiation complex: analysis *in vivo* and in an enhancer-responsive *in vitro* system. Mol Cell Biol 9: 5093–5104, 1989
18. Moss T, Birnstiel ML: The putative promoter of a Xenopus laevis ribosomal gene is reduplicated. Nucleic Acids Res 6: 3733–3743, 1979
19. Sollner-Webb B, Reeder R: The nucleotide sequence of the initiation and termination sites for rRNA transcription in Xenopus laevis. Cell 18: 485–499, 1979
20. DeWinter RFJ, Moss T: Spacer promoters are essential for efficient enhancement of X. laevis ribosomal transcription. Cell 44: 313–318, 1986
21. Cassidy B, Yang-Yen H-F, Rothblum L: Additional RNA polymerase I initiation site within the nontranscribed spacer region of the rat rRNA gene. Mol Cell Biol 7: 2388–2396, 1987
22. Kuhn A, Grummt I: A novel promoter in the mouse rDNA spacer is active *in vivo* and *in vitro*. EMBO J 6: 3487–3492, 1987
23. Tower J, Henderson S, Sollner-Webb B: An RNA polymerase I promoter located in the CHO and mouse rDNA spacers: functional analysis, factor, and sequence requirements. Mol Cell Biol 9: 1513–1525, 1989
24. Tower J, Sollner-Webb B: Transcription of mouse rDNA is regulated by an activated subform of RNA polymerase I. Cell 50: 873–883, 1987
25. Sollner-Webb B, Mougey E: News from the nucleolus: rRNA gene expression. T.I.B.S. Invited review, in press, 1990
26. Mahajan PB, Thompson EA: Hormonal regulation of transcription of rDNA. J Biol Chem 265: 16225–16233, 1990
27. Schnapp A, Pfleiderer C, Rosenbauer H, Grummt I: A growth-dependent transcription initiation factor (TIF-IA) interacting with RNA polymerase I regulates mouse ribosomal RNA synthesis. EMBO J 9: 2857–2863, 1990
28. Hansen UM, McClure WR: Role of the sigma subunit of Escherichia coli RNA polymerase in initiation. J Biol Chem 255: 9569–9570, 1980
29. Payne JM, Laybourn PJ, Dahmus ME: The transition of RNA polymerase II from initiation to elongation associated with phosphorylation of the carboxyl-terminal domain of subunit II_A. J Biol Chem 264: 19621–19629, 1989
30. Jantzen H-M, Admon A, Bell SP, Tjian R: Nucleolar transcription factor hUBF contains a DNA-binding motif with homology to HMG proteins. Nature 344: 830–836, 1990
31. Pikaard CS, McStay B, Schultz MC, Bell SP, Reeder RH: The Xenopus ribosomal gene enhancers bind an essential polymerase I transcription factor, xUBF. Genes and Dev 3: 1779–1788, 1989
32. Miller K, Sollner-Webb B: Transcription of mouse rRNA genes by RNA polymerase I: *in vitro* and *in vivo* initiation and processing sites. Cell 27: 165–174, 1981
33. Gurney T: Characterization of mouse 45S ribosomal RNA subspecies suggests that the first processing cleavage occurs 600 ± 100 nucleotides from the 5′ end and the second 500 ± 100 nucleotides from the 3′ end of a 13.9 kd precursor. Nucleic Acids Res 13: 4905–4919, 1985
34. Kass S, Sollner-Webb B: The first pre-RNA processing event occurs in a large complex: analysis by gel retardation, sedimentation, and UV crosslinking. Mol Cell Biol 10: 4920–4931, 1990
35. Craig N, Kass S, Sollner-Webb B: Nucleotide sequence determining the first cleavage site in the processing of mouse precursor ribosomal RNA. Proc Nat Acad Sci USA 84: 629–633, 1987
36. Kass S, Tyc K, Steitz J, Sollner-Webb B: The U3 small nucleolar ribonucleoprotein functions in the first step of pre-ribosomal RNA processing. Cell 60: 897–908, 1990
37. Miller O, Beatty B: Visualization of nucleolar genes. Science 164: 955–957, 1969
38. Kass S, Craig N, Sollner-Webb B: The primary processing of mammalian rRNA involves two adjacent cleavages and is not species specific. Mol Cell Biol 7: 2891–2898, 1987

Address for offprints: B. Sollner-Webb, The John Hopkins University, School of Medicine, 725 North Wolfe Street, Baltimore, MA 21205, USA

Molecular and Cellular Biochemistry **104**: 155–162, 1991.

Multiple functional enhancer motifs of rat ribosomal gene

Samson T. Jacob, Ji Zhang, Lalit C, Garg[1] and Carol-Beth Book[2]
Department of Pharmacology and Molecular Biology, The Chicago Medical School, 3333 Green Bay Road, North Chicago, IL 60064, USA; [1] Present address: National Institute of Immunology, Shahid Jeel-Sing Marg, New Delhi 110067, India; [2] Present address: Department of Pharmacology, The Pennsylvania State University College of Medicine, Hershey, Pennsylvania 17033, USA

Key words: multiple functional motifs, enhancer, rat ribosomal gene

Abstract

Previous studies from this laboratory have characterized a 174 bp enhancer element which is located 2 kb upstream of the initiation site. Half of the enhancer action is controlled by a 37 bp element at the 3′ end of the 174 bp region. We now report that a 43 bp adjacent domain which is located upstream of the 37 bp element constitutes an additional motif of the rDNA enhancer. When the plasmid consisting of the 43 bp DNA upstream of the rDNA core promoter was transcribed in a fractionated rat tumor cell extract (fraction DE-B), transcription of rDNA was augmented 4 fold. Electrophoretic mobility shift and DNAase I footprinting analyses showed that the purified 37 bp enhancer (E_1)-binding protein, (E_1BF) not only interacted with the enhancer motif E_1 but also interacted with the neighbouring 43 bp enhancer domain E_2. The specificity of the binding was demonstrated by competition with unlabeled 37 bp and 43 bp fragment and lack of competition with nonspecific DNAs in the mobility shift assay. These studies have shown that a single pol I transcription factor can bind to multiple enhancer domains with no significant sequence homologies and such multiple interactions may result in maximal transcription of ribosomal gene from the core promoter.

Introduction

Regulation of eukaryotic gene expression is accomplished by two basic types of components, namely cis-acting elements and trans-acting factors. In addition to the promoter, the enhancer constitutes another major cis-acting element. The enhancer, first identified in Simian virus 40, could stimulate transcription from homologous or heterologous promoters irrespective of their orientation or distance from the initiation site [1, 2]. Several viral and cellular enhancers have since been identified (for reviews, see references 3–6). The mechanism of enhancer action has not been completely elucidated. The enhancers appear to stimulate transcription by forming a tight complex with their putative binding factor(s) that causes stable activation of the promoter. A similar enhancer element has also been reported for ribosomal genes from Xenopus [7] and [8]. In X·laevis, enhancer elements share stretches of nucleotides with the promoter [9], and thus may act as a loading site for the essential transcription factors. However, the rat rDNA enhancer (174 bp) characterized earlier in our laboratory [8] lies far upstream (between positions − 2.357 and − 2.183 kb) of the promoter and does not share any detectable sequence homology with the promoter. The rat rDNA enhancer exhibits all the characteristics of the RNA polymerase II enhancer [8]. Thus, it can function in either orientation or when inserted downstream of the start site or independent of the distance from nucleotide + 1. Further analysis of this 174 bp enhancer region revealed that one element lies within the 37 bp from the 3′ end (− 2.219 kb to − 2.183 kb) [10] and binds to a trans-acting factor E_1BF [11]. However, the stimulation of transcription by he 37 bp element was only half of that obtained with the 174 bp spacer [10]. This finding suggested that other enhancer motif(s) must be present upstream of the 37 bp region. The present study was undertaken to delineate the additional motifs of the rat rDNA enhancer element.

Materials and methods

Reagents
Restriction endonuclease, DNA modification enzymes, and DNA size markers (ØX 174 RF DNA digested with Hae III) were purchased from Bethesda Research Laboratories. All enzymes were used as directed by the manufacturers. E. Coli HB101 (rec A) frozen competent cells were also obtained from Bethesda Research Laboratory and were used as a host for subcloning.

Plasmids
Plasmids pB7-2.0 [12] and pEH-5.0 [13] were provided by Dr. Lawrence Rothblum (Geisinger Clinic, Sigfried and Janet Weis Center for Research) and Dr. Christina Harrington (Tufts University School of Medicine) respectively. These plasmids were used to construct plasmid pDJ4-3′ ΔSK which contained 174 bp upstream region of rDNA spacer (− 2.357 kb to − 2.183 kb) ligated to − 167 bp to + 2.0 kb region of the rat rDNA [8].

A plasmid containing 43 bp of rDNA spacer (spanning from − 2.262 kb to − 2.220 kb), rDNA core promoter and part of the coding region (− 167 bp to + 2.0 kb) was constructed. (Figs. 1A, B, C). For this purpose a 43-mer oligonucleotide was synthesized. A 43-mer ladder was made by standard ligation method. The ends of multimers were repaired by using DNA polymerase I (Klenow), ligated to BamHI linkers at both ends and cloned into pB7-2.0 linearized with Bam HI. E Coli HB101 cells were transformed with this DNA and the positive clones were identified by standard colony lift and hybridization with nick translated ^{32}P-labeled mixture of 43-mer multimers. Restriction digestion of the positive clones with Bam HI showed the size of insert. A clone containing the shortest insert (a − 86 bp dimer of 43-mer) was selected in the present study and is designated plasmid pR43X2 (Fig. 1C).

In vitro *transcription assay*
In vitro transcription was carried out with plasmids linearized with Xho I, which should yield 635 nucleotides-long transcripts following a run-off transcription assay [14]. The transcription reaction was carried out for 30 minutes at 30° C and the reaction products were analyzed as described [14]. The amount of the transcript was determined by densitometric analysis of the autoradiograms [14].

Purification of E_1BF
E_1BF (E_1 binding factor which interacts with the enhancer motif E_1) was purified as described previously [11]. Briefly, the purification procedure consisted of fractionation of whole cell extract derived from rat mammary adenocarcinoma. Ascites cells on DEAE-sephadex, heparin-sepharose, CM-sepharose and 37 bp oligo columns. The final preparation was subjected to polyacrylamide gel electrophoresis under denaturing conditions [11].

Electrophoretic mobility shift assay
The synthetic oligonucleotides containing 43 bp spacer region (− 2.220 kb to − 2.262 kb) and 37 bp spacer region (− 2.183 kb to − 2.219 kb) of the rat ribosomal gene were labelled with ^{32}P-αdATP using the Klenow fragment and were gel purified. The electrophoretic mobility shift assays were performed as described [11]. The assays were performed with highly purified E_1BF in absence of Poly (dIdC). For competition experiments, E_1BF was incubated with unlabeled DNA fragment for 10 minutes at room temperature prior to the addition of labelled 37 bp probe. The samples were then incubated for an additional 30 minutes at 30° C.

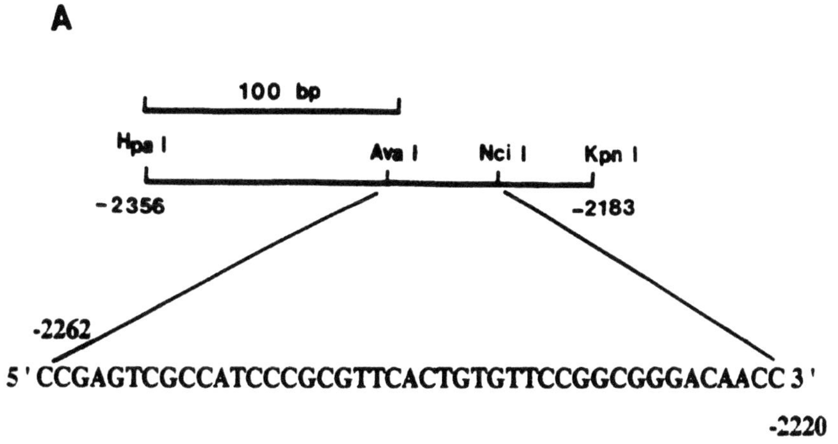

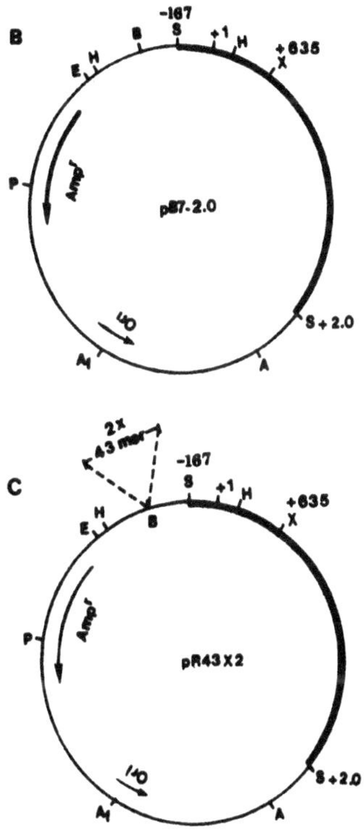

Fig. 1. A. Partial restriction map of rat rDNA 174 bp enhancer and sequence of 43-mer deoxy oligonucleotide encompassing Nci I – Ava I (– 2.220 to – 2.262 kb) region of 174 bp enhancer. B. Partial restriction map of pB7-2.0 consisting of – 167 bp to + 2.0 kb of rat rDNA coding spacer region (shown by thick line) inserted at Sal I site of pBR322. C. Partial restriction map of recombinant plasmid consisting of – 167 bp to + 2.0 kb of rat rDNA coding spacer region inserted at Sal I site of pBR322 and a dimer of 43-mer oligo deoxynucleotide inserted at Bam HI site of pBR322. Hp = Hpa I; A = Ava I; N = NciI; K = Kpn I; S = Sal I; H = Hind III; X = Xho I; B = Bam HI; P = Pvu II; Amp^r – Ampicillin resistant region of plasmid pBR322.

Results

Evidence for a direct role of the 43 bp cis-acting sequence in rat rDNA transcription

Since the 37 bp sequence at the 3′ end of the 174 bp spacer region (previously identified to contain enhancer element, see ref. 10) exhibited only half as much enhancer activity as the entire 174 bp region, we investigated the potential enhancer role of 43 bp sequence just upstream of the 37 bp sequence. To explore this possibility, a dimer of the 43-deoxy oligonucleotide (Fig. 1C) was cloned into the Bam HI site of the plasmid pB7 – 2.0 as described under Material and methods. This plasmid has been designated pR43X2. When plasmids pB7 – 2.0 and pR43X2 linearized with XhoI were transcribed (Fig. 2) in equimolar amounts, the latter plasmid was transcribed 4-fold more efficiently (lane 1) than the control plasmid pB7 – 2.0 (lane 2), as determined by densitometric analysis of the autoradiograms (see Experimental Procedures). This

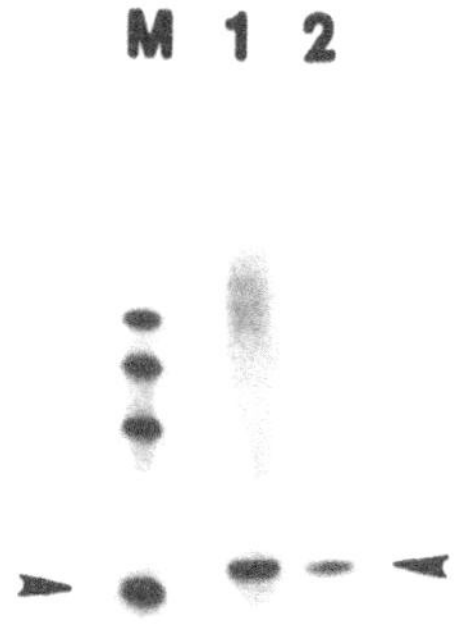

Fig. 2. In vitro transcription of rat rDNA plasmids. 50 ng of XhoI linearized pB7 – 2.0 (lane 2) and pR43X2 (lane 1) were transcribed with 12 μg of fraction DE-B. M Indicates DNA size markers. Arrow points to the 635 nucleotides-long transcript.

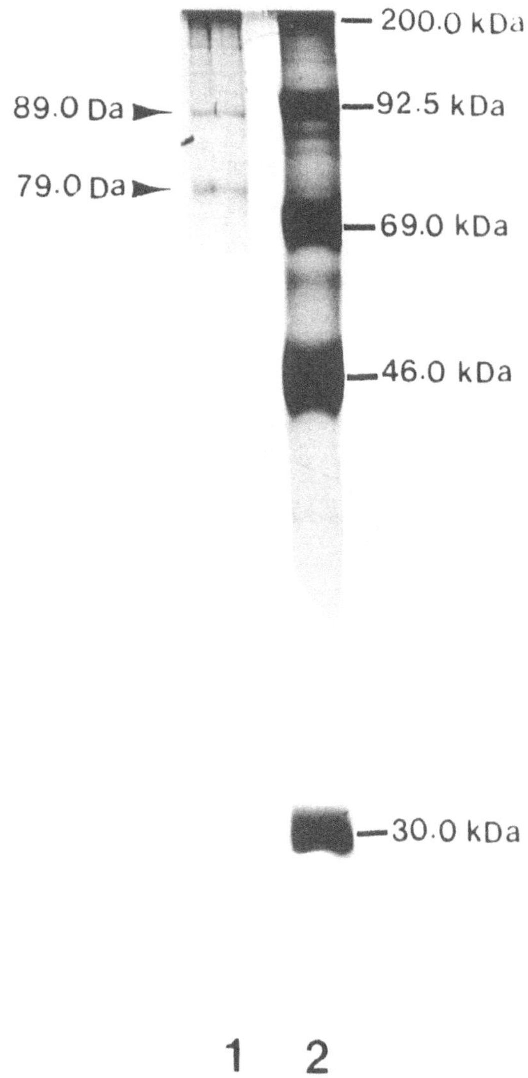

Fig. 3. SDS-polyacrylamide gel electrophoresis and visualization of purified E_1BF. Sample of purified E_1BF was precipitated with 15% TCA, subjected to electrophoresis on SDS-polyacrylamide gel (10%), and visualized by silver staining. Lane 1, purified E_1BF. Lane 2, marker proteins and their molecular size (in thousands) indicated on the right.

data demonstrated that the 43 bp spacer element within the 174 bp enhancer region is also a functional enhancer motif which stimulates rDNA transcription.

The protein which interacts with the 37 bp (E_1) enhancer motif also binds to the 43 bp (E_2) enhancer motif

We have recently purified and characterized a factor (E_1BF) which binds to the 37 bp enhancer (E_1) motif as well as the core promoter and stimulates rDNA transcription [11]. The purification protocol consisted of chromatography of fraction DE-B which contains pol I and all pol I trans-acting factors [14] on heparin Sepharose, CM-Sepharose and 37 bp oligo affinity columns [11]. The final preparation consisted of two polypeptides of molecular masses 89, 100 and 79, 400 following silver staining (see Fig. 3) and is collectively called E_1BF. Both polypeptides interacted with the E_1 element and the core promoter region, particularly the downstream promoter sequence spanning from +2 to +50 [11]. Since a single factor is known to interact with multiple enhancer domains for maximal stimulatory effects [15], we investigated the potential

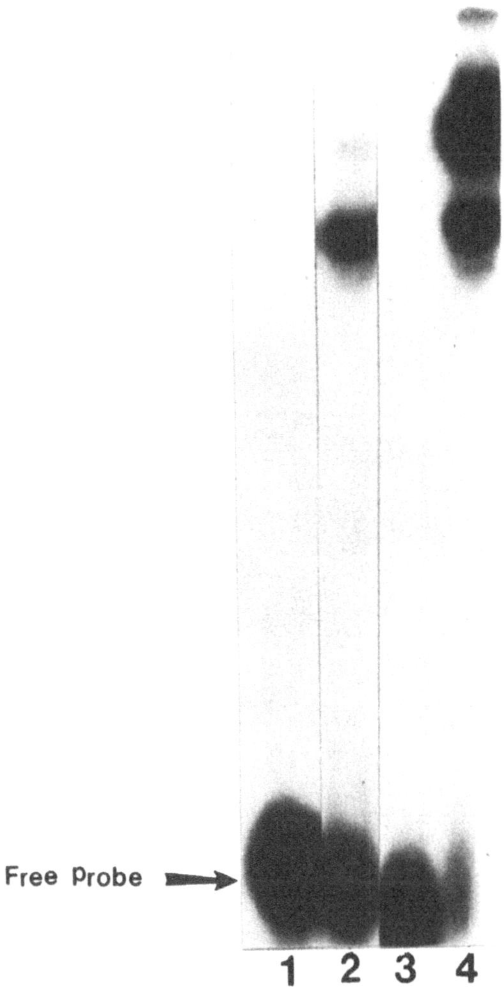

Fig. 4. Electrophoretic mobility shift assay using purified E_1BF with labelled 37 bp and 43 bp enhancer probes. The purified E_1BF was incubated with 2 ng of the 3′-end labelled probes at 30° C for 30 minutes. Lane 1, labelled 37 bp enhancer fragment alone. Lane 2, labelled 37 bp enhancer plus 2 μl purified E_1BF. Lane 3, labelled 43 bp enhancer fragment alone. Lane 4, labelled 43 bp enhancer plus 2 μl purified E_1BF.

interaction of E_1BF with the sequence next to the 37 bp domain within the original 174 bp enhancer region identified in our laboratory [8]. In fact, DNAase I footprinting analysis had shown interaction of E_1BF with the 43 bp sequence as well as the 37 bp domain. To confirm further interaction of E_1BF with the 43 bp motif and to get preliminary information on the nature and specificity of the complexes formed between 37 bp and 43 bp sequences, electrophoretic mobility assays were done using labeled enhancer motifs (Fig. 4). A distinct complex was observed with the 37 bp fragment whereas two complexes were formed with the 43 bp sequence. An analysis of the DNAase I footprinting profile [11] showed two distinct protection regions in the 43 bp domain whereas only one protection region was observed in the 37 bp sequence. Thus, the results obtained with the gel shift analysis are in perfect agreement with the footprinting assay.

We then investigated whether the 43 bp and 37 bp enhancer motifs share the same DNA binding protein. This question was addressed by competitive electrophoretic mobility shift assays which were performed using purified E_1BF, labelled 37 bp enhancer probe and various unlabeled DNA fragments. The experiment was done by preincubation of E_1BF with different amounts of unlabeled competitor DNA fragments prior to addition of [^{32}P] labelled 37 bp enhancer probe. Only one complex was observed, as labeled 37 bp fragment and not the 43 bp fragment was used in this study. The results (Fig. 5) showed that both unlabeled 37 bp and 43 bp enhancer fragments could abolish complex formation (see lanes 2 through 9 in Fig. 5), but unlabeled control DNAs such as pBR322 fragment (lane 10, Fig. 5) or other DNAs (data not shown) had no effect on complex formation. This study demonstrated that the 43 bp and 37 bp enhancer motifs bind to the same protein involving either the same domain or a very closely related domain of E_1BF. This result further confirms the specificity of the interaciton between E_1BF and the 43 bp enhancer motif as well as the 37 bp enhancer domain.

Discussion

This study was undertaken to delineate the enhancer motifs associated with the previously characterized 174 bp rcgion in thc rat ribosomal gcnc. We have previously shown that the 37 bp sequence (− 2.183 kb to − 2.219 kb) located at the 3′ end of the 174 bp spacer contains one of the enhancer elements. The present study has identified an addi-

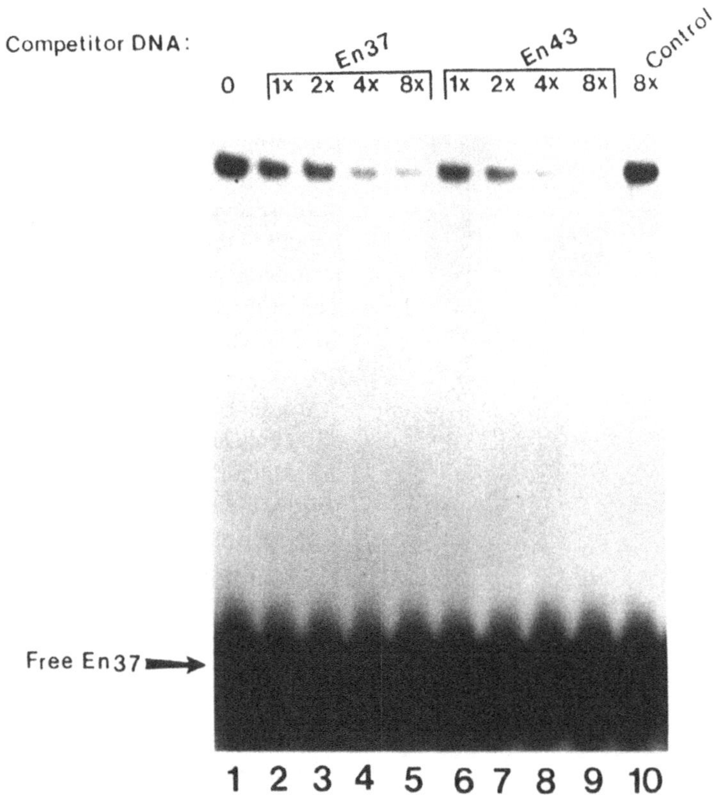

Fig. 5. Electrophoretic mobility shift assay in the presence of competitor DNAs. Purified E_1BF ($2 \mu l$) was incubated with various unlabeled competitor DNAs under the optimal binding condition for 10 minutes at room temperature prior to addition of 2 ng of 3′-end labelled 37 bp enhancer. The reaction mixtures were further incubated at 30° C for 30 minutes. lane 1, labelled 37 bp enhancer without competitor DNA. Lanes 2–5, unlabelled 37 bp enhancer DNA as competitor: 2 ng (lane 2), 4 ng (lane 3), 8 ng (lane 4) and 16 ng (lane 5). Lanes 6–9, unlabeled 43 bp core promoter as competitor: 2 ng (lane 6), 4 ng (lane 7), 8 ng (lane 8) and 16 ng (lane 9). Lane 10, 16 ng 154 bp HinfI fragment of PBR322 as nonspecific control DNA. The folds of amounts of unlabeled competitors were relative to labelled 37 bp enhancer.

tional enhancer element upstream of the 37 bp enhancer element. The degree of enhancement of rDNA transcription obtained *in vitro* with the 37 bp and 43 bp spacer motifs is approximately equal to that achieved with the 174 bp element. The majority of the enhancer domain within the 174 bp spacer element should, therefore, reside in the two shorter motifs.

The rat rDNA enhancers differs from the only other well characterized RNA polymerase I (pol I) enhancer of Xenopus laevis [7, 9] in several respects. First, Xenopus rDNA enhancers are imperfect duplicated copies of the core promoter associated with the 60/81 bp repeat elements, whereas the rat enhancer sequence does not appear to be homologous to the core promoter. Second, the 174 bp rat rDNA enhancer is not part of any repeating region in the rat gene located further downstream of this element. Third, the rat enhancer is located far upstream of the promoter, which is analogous to the observation [16] that an enhancer located as much as 10 kb upstream is essential for efficient, liver-specific expression of the albumin gene. Fourth, the 37 bp or 43 bp regions do not contain binding sites which binds Xenopus enhancer-binding protein UBF-[17]. Interestingly, despite the lack of any detectable sequence homology between the rat rDNA 37 bp and 43 bp enhancer motifs, the two cis-acting elements (37 bp and 43 bp) interact with a common trans-acting factor,

E_1BF. The binding of a trans-acting factor to different cis-acting sequences is not an isolated case. For example, the yeast HAPI activator can bind to two different upstream elements [18] and a mitochondrial trans-acting factor can form complexes with two non homologous promoters [19]. Similarly, the other well-characterized pol I trans-acting factor, UBF-I, can interact with different cis-acting elements showing very limited homology [20]. Evidently, different domains of the factor should be interacting with different sequences (e.g. enhancer and core promoter of rDNA) or a protein-protein interaction could direct the factor to different sites on the gene [21]. More specifically, it has been shown [22] that the HeLa cell protein TEF-1 binds specifically and co-operatively to two SV40 enhancer motifs of unrelated sequence. Mutagenesis studies in the enhancer motifs followed by analysis of DNA-protein interaction using purified E_1BF are needed to explore the mechanism of the enhancer action at this level.

Recent studies in our laboratory (A. Ghosh and S. Jacob, manuscript in preparation) have shown that the 174 bp spacer sequence can function as an enhancer *in vivo*. This conclusion was reached by transfecting two plasmids, one containing rDNA core promoter, the rDNA enhancer and the CAT gene and the other one consisting of the core promoter and the CAT gene as the control. Following transfection of the above plasmids into CHO cells, the CAT activity was assayed. In the presence of the rDNA enhancer, the CAT activity increased 16 fold over that in the absence of the enhancer. That the initiation of rDNA transcription under the influence of the enhancer occurred at + 1 site *in vivo* was confirmed by primer extension studies.

In conclusion, the present studies indicate that 43 bp spacer region just upstream of 37 bp enhancer region is another functional motif in 174 bp enhancer region and that E_1BF, the E_1 binding protein, also interacts with the 43 bp motif (E_2). In fact, DNAase I footprint analysis using labelled 174 bp enhancer as probe indicated that E_1BF produced three protected regions which include 37 bp and 43 bp enhancer regions as well as a further upstream motif [11]. The maximal augmentation of rDNA transcription under the direction of 174 bp enhancer region should, therefore, be due to cumulative effects of these multiple enhancer motifs that interact with E_1BF. Multiple cis-acting elements within a relatively short segment of the 5′ flanking region have also been observed in RNA polymerase II genes. For instance, multiple enhancer motifs include the three enhancer elements found in the α-fetoprotein gene [23] and the three short segements of the SV40 enhancer each with a different cell type specificity [24]. Similarly, the immunoglobulin heavy and light chain enhancer contain multiple copies of an enhancer motif as well as recognition sites for the $NF\text{-}_{k}B$ and cell specific octamer binding proteins [25]. A similar mechanism appears to exist for pol I-directed transcription of ribosomal gene as well.

Acknowledgements

We thank Dr. P. Sutrave for synthesizing the oligonucleotide and many helpful discussions. We also thank Katrina Piotrowski for preparation of the manuscript. This work was supported by U.S. Public Health Service Grants CA31894 and CA25078 to STJ.

References

1. Moreau P, Hen R, Wasylyk B, Everett R, Gaub MP, Chambon P: The SV40 72 base pair repeat has a striking effect on gene expression both in SV40 and other chimeric recombinants. Nucleic Acids Res 9: 6047–6068, 1981
2. Banerji R, Rusconi S, Schaffner W: Expression of a β-globin gene is enhanced by remote SV40DNA sequence. Cell 27: 299–308, 1981
3. Chambon P, Dierich A, Gaub MP, Jakowlev SB, Jongstra J, Krust A, LePennec JP, Oudet P, Reudelhuber T: Promoter elements of genes coding for proteins and modulation of transcription by estrogens and progesterone. Recent Prog Horm Res 40: 1–42, 1984
4. Gruss P: Magic enhancers. DNA 3: 1–5, 1984
5. Yaniv M: Two distinct enhancers with different cell specificities coexist in the regulatory region of polyoma. Cell 39: 653–662, 1984
6. Serfling E, Jasin M, Schaffner W: Enhancers and eukaryotic gene transcription. Trands Genet 1: 224–230, 1985
7. Reeder RH, Roan JG, Dunaway M: Spacer regulation of

Xenopus ribosomal gene transcription. Competition in oocytes. Cell 35: 449–456, 1983
8. Dixit A, Garg LC, Chao W, Jacob ST: An enhancer element in the far upstream spacer region of rat ribosomal RNA gene. J Biol Chem 262: 11616–11622, 1987
9. DeWinter RFJ, Moss T: A complex array of sequences enhances ribosomal transcription in Xenopus laevis. J Mol Biol 196: 813–827, 1987
10. Garg LC, Dixit A, Jacob ST: A 37-base pair element in the far upstream spacer region can enhance transcription of rat rDNA *in vitro* and can bind to the core promoter-binding factor(s). J Biol Chem 264: 220–224, 1989
11. Zhang J, Jacob ST: Purification and characterization of a novel factor which stimulates rat ribosomal gene transcription *in vitro* by interaction with an enhancer element and core promoter. Mol Cell Biol 10: 5177–5186, 1990
12. Rothblum LI, Reddy R, Cassidy B: Transcription initiation site of rat ribosomal DNA. Nucleic Acids Res 10: 7345–7362, 1982
13. Harrington CA, Chikaraishi DM: Identification and sequence of the initiation site for rat 45S ribosomal RNA synthesis. Nucleic Acids Res 11: 3317–3332, 1983
14. Kurl RN, Rothblum LI, Jacob ST: A purified fraction containing RNA polymerase I that can accurately transcribe rat ribosomal RNA gene. Proc Natl Acad Sci USA 81: 6672–6675, 1984
15. Schatt MD, Rusconi S, Schaffner W: A single DNA-binding transcription factor is sufficient for activation from a distant enhancer and/or from a promoter position. EMBO J 9: 481–487, 1990
16. Pinkert CA, Ornits DM, Brinster RL, Palmiter RD: An albumin enhancer located 10 kb upstream functions along with its promoter to direct, liver-specific expression in transgenic mice. Genes and Dev 1: 268–276, 1987
17. Pikaard CS, McStay B, Schultz MC, Bell SP, Reeder RH: The Xenopus ribosomal gene enhancers bind an essential polymerase I transcription factor, xUBF. Genes Dev 3: 1779–1788, 1989
18. Pfeifer K, Prezant T, Guarante L: Yeast HAP1 activator binds to two upstream activation sites of different sequence. Cell 49: 19–27, 1987
19. Fisher RP, Clayton D: Purification and characterization of human mitochondrial transcription factor 1. Mol Cell Biol 8: 3496–3509, 1988
20. Bell SP, Pikaard CS, Reeder RH, Tjian R: Molecular mechanisms governing species-specific transcription of ribosomal RNA. Cell 59: 489–497, 1989
21. Ptashne M: How eukaryotic transcriptional activators work. Nature 335: 683–689, 1988
22. Davidson I, Xiao JH, Rosales R, Staub A, Chambon P: The HeLa cell protein TEF-1 binds specifically and cooperatively to two SV40 enhancer motifs of unrelated seqence. Cell 54: 931–942, 1988
23. Hammer RE, Krumlauf R, Camper SA, Brinster RL, Tilghman SM: Diversity of alpha-fetoprotein gene expression in mice is generated by a combination of separate enhancer elements. Science 235: 53–58, 1987
24. Shirm S, Jiricny J, Schaffner W: The SV40 enhancer can be dissected into multiple segments, each with a different cell type specificity. Genes and Dev 1: 65–74, 1987
25. Lenardo M, Pierce JW, Baltimore D: Protein-binding motifs in immunoglobin enhancers determine transcriptional activity and inducibility. Science 236: 1573–1577, 1987

Address for offprints: S.T. Jacob, Department of Pharmacology and Molecular Biology, The Chicago Medical School, 3333 Green Bay Road, North Chicago, IL 60064, USA

Molecular and Cellular Biochemistry **104**: 163–168, 1991.

Termination of transcription of ribosomal RNA in *Saccharomyces cerevisiae*

Stewart P. Johnson[1] and Jonathan R. Warner
Department of Cell Biology, Albert Einstein College of Medicine, Bronx, NY 10461, USA; [1] Present address: Dept. of Pathology, Duke University Medical Center, Durham, NC 27710, USA

Key words: ribosomal RNA transcription, DNA binding protein sites

Abstract

We have attempted to determine the site of termination of transcription of ribosomal RNA in the yeast, *Saccharomyces cerevisiae.* While a quantitative description of the termination sites of RNA polymerase I is not possible using presently available methods, we conclude that transcription of most molecules continues through a large portion of the adjacent enhancer region. There are two potential termination sites within the enhancer, one of which is near the binding site of the DNA binding protein REBI. In addition there is an apparently fail-safe termination site approximately 950 nucleotides beyond the 3′ end of 35S ribosomal precursor RNA. Processing at the end of 35S RNA influences the choice of downstream termination site. Conversely downstream sequences also influence the site of termination.

Introduction

The transcription of ribosomal RNA is of particular interest both because it makes up a major portion of the transcription of the cell and because the ribosomal RNA genes are present in the genome as a tandem array. It has become clear that a processing event is responsible for the 3′ end of the rRNA precursor molecules found in the nucleolus in yeast [1] as well as in higher organisms [2, 3]. Therefore, an important problem has been the identification of the actual site of the termination of transcription, in particular, because the tandem nature of the genes suggests the possibility that the termination of transcription of one gene may be coupled to the initiation of transcription at the next one. Experimentally the question is a difficult one, both for POL I transcripts and for POL II transcripts (reviewed in 4), because of the very rapid degradation of the 3′ products of such a processing reaction.

Transcription of ribosomal RNA genes by RNA polymerase I, including the question of termination, has been reviewed recently [5]. Briefly, in mouse, termination appears to occur 565 nucleotides downstream of the end of the 45S rRNA precursor in front of a group of repeated sequences termed 'Sal boxes' [6]. An additional termination site is found not far upstream of the initiation site. Whether its function is as a 'fail-safe' terminator to protect the initiation complex from errant polymerases, or whether it is related to the presence of cryptic promoters in the non-transcribed spacer (NTS) is still unclear. In *Xenopus laevis,* there has been some evidence that transcription reads through most of the NTS, terminating just upstream of the promoter [2], but in other *Xenopus* species, transcription terminates about 235 nucleotides from the end of the precursor RNA [5]. In the yeast *Saccharomyces cerevisiae* there is also some evidence that transcription continues beyond the end of the 35S precursor [1]. This is particularly interesting because the major enhancer of rRNA transcription lies just there [7, 8] (see Fig. 1). In this paper, we have used three independent methods to assess the site(s) of termination, and find that there

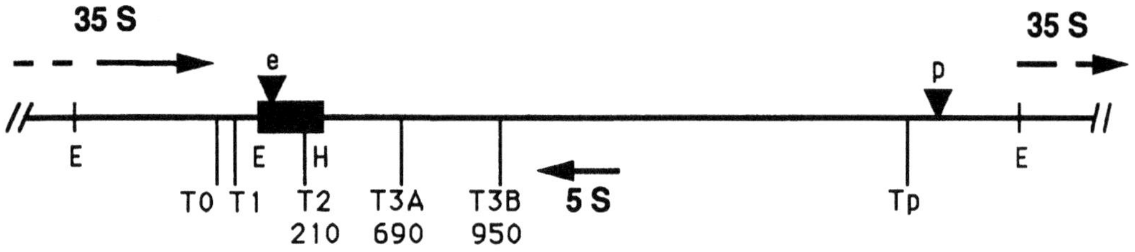

Fig. 1. Schematic representation of a portion of the rDNA repeat. Sequences are numbered from the 3′ end of the 25S RNA. The 5S transcript and the beginning and the end of the 35S transcript are shown. The enhancer element is indicated, as are the potential termination sites, T2, T3A, T3B, and Tp, described by Van der Sande, *et al.* [9]. T0 and T1 represent processing sites responsible for generating the 3′ end of 35S precursor RNA. E and H indicate EcoRI and HindIII restriction sites. The inverted triangles represent the REB1 binding sites at the enhancer(e) and promoter(p).

are likely to be several between the end of the 35S RNA gene and the end of the 5S RNA gene. After these experiments had been completed, Van der Sande *et al.* reported results in general agreement with ours [9].

Materials and methods

Saccharomyces cerevisiae, strains W303 and J400, was cultured, and its RNA was prepared as described previously [7, 8]. Run-on transcription was carried out as described [8]. Construction and analysis of the products of transcription of the artificial rDNA gene have been described in detail [7].

Results

Measurement of the rate of transcription of a particular piece of DNA can be difficult if the lifetime of the product is very short. As a way around that problem, run-on transcription assays have been developed for mammalian cells [10, 11], based on the observation that, in isolated nuclei, RNA polymerase molecules will continue a transcription in which they are involved for a few hundred nucleotides, but will not initiate new chains. Furthermore, processing and degradation of this RNA is minimal. We have developed a run-on transcription protocol for *Saccharomyces,* in which the membrane is permeabilized with detergent, leaving the cell wall intact [8]. Radiolabeled nucleoside triphosphates added to such cells are incorporated into RNA for a brief period. The radiolabeled RNA can be hybridized to complementary sequences; the amount of radioactivity in double-stranded form is then a rough measure of the transcriptional activity of that sequence. We have utilized this protocol to measure transcription across

Table 1. Cells of *Saccharomyces cerevisiae,* strain W303a, were permeabilized and labeled with α^{32}P UTP as described previously [8]. RNA was prepared and hybridized to slotblot filters, where each slot contained a different species of complementary RNA, transcribed from a pGEM plasmid carrying the indicated sequences of the rDNA repeat. After hybridization, the filter was washed, treated with pancreatic RNAse to remove single-stranded tails, and subjected to autoradiography. Then, each slot was excised and its radioactivity determined in a scintillation counter.

Complementary rDNA sequences[a]	Relative transcription[b]
−496 to +101	12.3[c]
+96 to +289	52.2
+288 to +400	1.9
+401 to +671	3.1
+774 to +969	2.2
+992 to +1179	1.9
+1484 to +1644	0.95
+1797 to +1910	0

[a] Position is stated relative to the 3′ end of 25S rRNA. 5S RNA is transcribed from +1255 to +1137.

[b] Calculated as cpm relative to the U content of the hybridized RNA.

[c] This value is artifically low probably reflecting the saturation of the filter with mature rRNA molecules. In other experiments, the relative transcription of this region was 2 to 3 fold greater than +96 to +289.

the rDNA spacer region using as hybridization probes anti-sense RNAs synthesized *in vitro* from pGEM plasmids that contained small fragments of the spacer DNA.

As shown in Table 1, the region of the spacer from 96 to 289 bp downstream of the 3′ end of 25S rRNA is as transcriptionally active as sequences coding for 25S rRNA. Beyond this region transcription drops off markedly, although a low level of transcription is still measurable 1 kb beyond the end of 25S RNA. These results suggest that 80–90% of the RNA polymerase I molecules are terminating within or, perhaps slightly beyond, the EcoRI-HindIII enhancer element, and the remainder are terminating much further downstream. This is in relative agreement with the results of similar experiments carried out by Van der Sande *et al.* [9]. These results suggest that termination actually occurs within or shortly downstream of the rRNA enhancer.

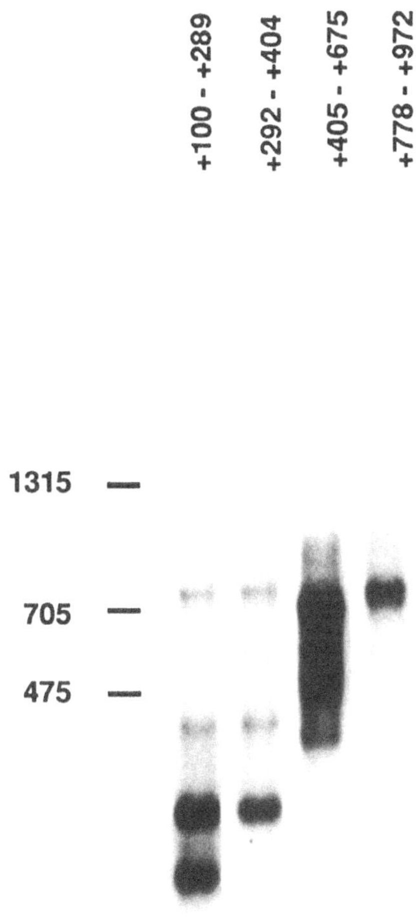

Fig. 2. Transcripts of the NTS sequences. Total RNA prepared from log-phase cells was subjected to Northern analysis, and the resulting filter was hybridized sequentially to a series of radioactive anti-sense RNA probes from the NTS, as indicated. The relative intensity between the lanes is arbitrary, depending on the specific activity of the probe and the length of exposure of the autoradiograph. The relative concentration of the different RNA species is apparent from comparing the intensity of the bands in the first lane.

Transcripts from the NTS1 region of spacer rDNA

As another approach, we have searched for transcripts of the spacer region, using anti-sense RNAs radiolabeled *in vitro*. RNA was isolated from an exponentially growing culture of cells, separated by denaturing gel electrophoresis, and transferred to a nylon filter. The RNA was cross-linked to the filter by UV irradiation and then probed sequentially with complementary RNAs synthesized from pGEM plasmids containing discrete segments of the rDNA spacer. As shown in Fig. 2, a probe from + 96 to + 289 hybridizes to four RNAs. These RNAs are approximately 240, 320, 470, and 850 bases in length. When the filter was stripped and reprobed with RNA complementary to + 292 to + 404, the smallest of these was no longer observed. Similarly, a probe of + 405 to + 675 hybridizes to the two largest RNAs, and a probe of + 778 to + 972 hybridizes to only the largest of the RNAs. An antisense probe that spans + 915 to + 1060 hybridizes to none of these RNAs (not shown).

The simplest explanation for these results is that these RNAs possess a common 5′ end, perhaps that generated by the processing event that generates the end of 25S rRNA. If that were the case, the 3′ end of the smallest of the RNAs would be near the T2 site [9] and that of the largest RNA ncar the T3B site [9]. The two other RNAs do not correlate to any of the 3′ end generating sites previously described and might represent intermediates in the degradation of the longest transcript.

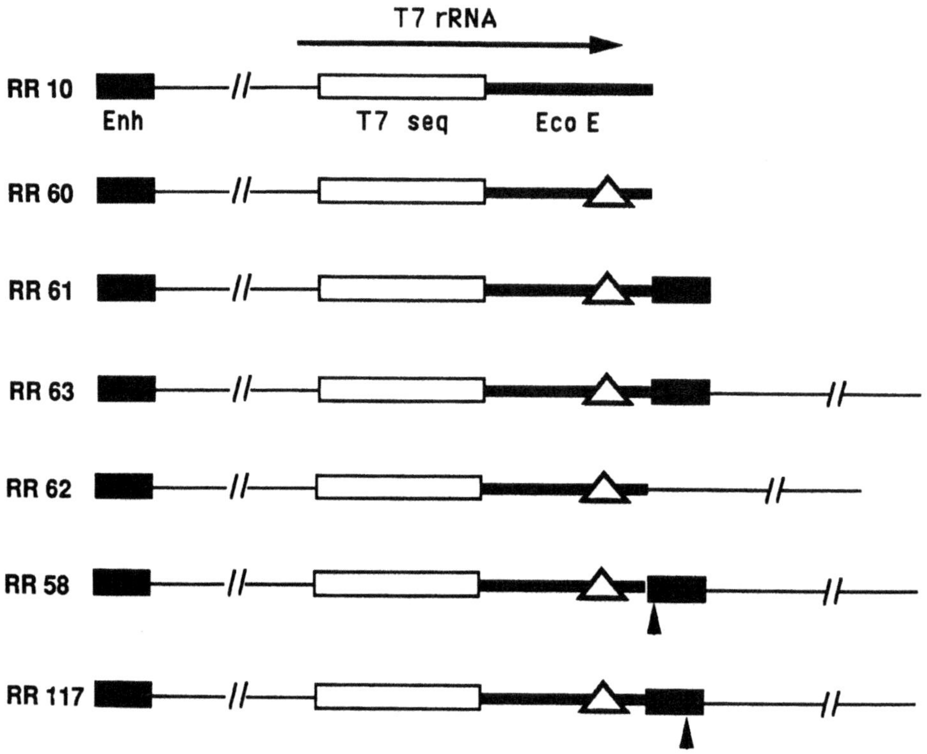

Fig. 3. Artificial rRNA transcription unit containing bacteriophage T7 sequences. The original construct, RR10, has been described in detail [8]. Briefly, the sequences shown are carried on a centromere plasmid, YCp50, that is maintained at about one copy per cell. *Enh* is the 190 bp enhancer sequence; *T7* is the 600 bp fragment of bacteriophage T7 that is used to identify the transcript of this gene, T7 rRNA. *EcoE* is the 591 bp EcoRi 'E' fragment that contains the T0 and T1 processing sites. The T0 and perhaps the T1 sites were removed by deleting the sequence from − 149 to + 37 (large triangle), to generate plasmid RR60, in which there are no NTS sequences beyond + 100. The other constructs are as follows: RR61: Includes the EcoRI-HindIII enhancer fragment, i.e. sequences from + 100 to + 290. RR63: Includes the entire NTS. RR62: Includes the entire NTS except the enhancer fragment. RR58: Includes the entire NTS except nucleotides + 108 to + 112 (arrowhead), that comprise the REB1 binding site. RR117: Includes the entire NTS except nucleotides + 218 to + 246 (arrowhead), that include the major run of T residues within the enhancer.

Expression of an rDNA mini-gene that lacks a 3′ processing site

We have previously described the use of an rDNA mini-gene to study the promoter and enhancer functions of the spacer rDNA [7, 8; see Fig. 3]. The transcript can be identified because it carries sequences of bacteriophage T7. Its 3′ end is generated by the same processing event that produces the 35S precursor RNA. In order to ask how far transcription continues, we have deleted the sequence from − 149 to + 37, removing the T0 site and perhaps the T1 sites (Fig. 1). This was then substituted for the fragment present in RR10 to generate RR60 (Fig. 3). As shown in Fig. 4, transcription from this minigene terminates at multiple sites within the plasmid sequences. When the minigene sequence was extended by adding the EcoRI-HindIII enhancer fragment (RR61), most of the transcripts still read through into the plasmid sequences, but a novel RNA species, R2, is produced whose 3′ end is near the HindIII site. When the complete spacer rDNA is present downstream of the mini-gene (RR63), there are three prominent RNA species, as well as a number of minor ones. In addition to an RNA that comigrates with R2, there is R1, whose 3′ end is near the EcoRI site, and R3, whose 3′ end is approximately 650 bases downstream of R2 (Fig. 1). When the enhancer fragment was deleted from this construct (RR62), neither R1 nor R2 were present. Almost all the transcript was in the form of R3, which is shorter by 190 nucleotides in this case, since it lacks sequences from the enhancer.

These results show that specific sequences are necessary for termination to occur. In particular the presence of sequences determining termination at the end of R3 appear to be 100% effective. When they are present, no longer transcripts are observed. The 3′ end of R3 appears to coincide with termination site T3B described by Van der Sande *et al.* [9] (Fig. 1). Precise identification of the 3′ ends of R1 and R2 is more problematical. It appears that R2 terminates at the site T2 described by Van der Sande *et al.* [9], but it may be beyond rather than before the stretch of T residues at +217 to +246. The 3′ end of R1 is slightly downstream of the EcoRI site that limits the 5′ end of the enhancer fragment. Close analysis of these results yields some curious conclusions. For instance, comparison of RNAs from YCpRR61 and YCpRR63 suggests that the production of R1 depends on sequences downstream of the HindIII site, well over 150 bases away.

We have identified a protein, termed REB1, that binds both to the enhancer region and to the promoter region of the NTS [12]. When a 6 bp deletion of the REB1 binding site within the enhancer is introduced into the test plasmid (yielding RR58), R1 is no longer produced (Fig. 4). Similarly, Van der Sande *et al.* [9] have identified an additional termination site Tp, near the promoter (−300 from the initiation site.) When they disrupted the REB1 binding site (at −200 from the initiation site) termination activity of Tp was abolished. Thus, in two instances, disruption of a REB1 binding site abolished a termination site.

The T2 site has been mapped to a point immediately upstream of a stretch of T residues that is present at +217 to +246, within the enhancer fragment. It should be noted that nuclease S1 analyses of this site show considerable protection into this T-rich region [9]. Deletion of this run of T residues (yielding RR117) leads to a decrease in the relative amount of R2, although one now sees a somewhat larger RNA that is a light band in RR63 but is normally obscured by the presence of R2

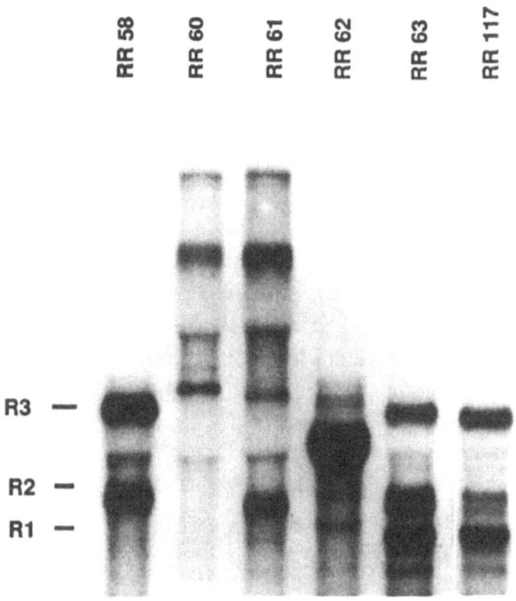

Fig. 4. Products of transcription units lacking T0. RNA prepared from strain J400 carrying the plasmids described in Fig. 3 (except RR10, see ref. 8) was subjected to Northern analysis and probed with labeled RNA complementary to the T7A sequence present in the construct. Bands R1, R2, and R3 are indicated. Their sizes were estimated from known marker bands.

(Fig. 4). Thus, the T-rich region may influence the efficiency of T2.

Discussion

One almost feels impelled to invoke the Heisenberg Uncertainty Principle for the study of the termination of transcription. Each experimental approach introduces its own perturbation into the system, thus casting into doubt the conclusions. In our case, although we have used three separate approaches, each has its potential drawbacks.

a) Run-on transcription is said to proceed only a few hundred nucleotides, not to intiate new transcripts at either correct or incorrect sites, and not to be subject to post-transcriptional modifications such aas rapid degradation [10, 11]. The data rigorously supporting each of those claims is sparse, especially for the *Saccharomyces cerevisiae* system.

b) Northern analysis of transcripts assumes that the products observed are the normal products of transcription, and that those containing NTS sequences are rare simply because they are degraded rapidly. An alternative possibility is that they are aberrant, rare transcripts unrepresentative of normal transcription events.

c) Finally, the deletion of the 3′ processing signal, as we have done in the RR plasmids shown in Fig. 3, may itself lead to aberrant termination. Indeed, in studies of POL II transcription it has been shown that deletion of the usual poly adenylation signal can alter termination at downstream sequences [4].

Nevertheless a number of conclusions can be safely drawn from the data:

a) In the presence of T0 and all the downstream sequences, substantial transcription continues into the enhancer, but probably less than 10% of the molecules continue beyond the HindIII site marking the end of the enhancer at + 292.

b) The presence of sequences within the enhancer element lead to termination. Sequences both at the REB1 binding site and the T-stretch are seen to lead to inefficient but specific termination sites, at least in the absence of T0 (Fig. 4).

c) Sequences both upstream and downstream can affect termination events. Thus, in the presence of T0 most transcripts terminate within the enhancer (Table 1), but in its absence, many read through the enhancer (Fig. 4). Furthermore termination leading to the transcript R1 (Fig. 4) occurs only in the presence of sequences relatively far downstream.

d) Little if any transcription is seen beyond T3B. Either T3B represents a fail-safe termination site, or transcripts beyond it are degraded immediately.

Acknowledgements

The authors acknowledge fruitful discussions with Bernice Morrow and the invaluable technical assistance of Mary Studeny. This work has been supported by grants from the NIH # GM25532 and CA13330.

References

1. Kempers-Veenstra AE, Oliemans J, Offenberg H, Dekker AF, Piper PW, Planta RJ, Klootwijk J: 3′-End formation of transcripts from the yeast rRNA operon. EMBO J 5: 2703–2710, 1986
2. Labhart P, Reeder RH: Characterization of three sites of RNA 3′ end formation in the *Xenopus* ribosomal gene spacer. Cell 45: 431–443, 1986
3. Bartsch I, Schoneberg C, Grummt I: Purification and characterization of TTF I, a factor that mediates termination of mouse ribosomal DNA transcription. Mol Cell Biol 8: 3891–3897, 1988
4. Proudfoot NJ: How RNA polymerase II terminates transcription in higher eukaryotes. Trends Bio Sci 14: 105–110, 1989
5. Reeder RH: Regulatory elements of the generic ribosomal gene. Curr Op in Cell Biol 1: 466–474, 1989
6. Grummt I, Maier U, Öhrlein A, Hassouna N, Bachellerie JP: Transcription of mouse rDNA terminates downstream of the 3′ end of 28S RNA and involves the interaction of factors with repeated sequences in the 3′ spacer. Cell 43: 801–810, 1985
7. Johnson SP, Warner JR: Unusual enhancer function in yeast rRNA transcription. Molec Cell Biol 9: 4986–4993, 1989
8. Elion EA, Warner JR: An RNA polymerase I enhancer in *Saccharomyces cerevisiae*. Molec and Cell Biol 6: 2089–2097, 1986
9. van der Sande CAFM, Kulkens T, Kramer AB, de Wijs IJ, van Heerikhuizen H, Klootwijk J, Planta RJ: Termination of transcription by yeast RNA polymerase I. Nucl Acids Res 17: 9127–9146, 1989
10. McKnight GS, Palmiter RD: Transcriptional regulation of the ovalbumin and conalbumin genes by steroid hormones in chick oviduct. J Biol Chem 254: 9050–9058, 1979
11. Weber J, Jelinek W, Darnell Jr. JE: The definition of a large viral transcription unit late in Ad2 infection of HeLa cells: mapping of nascent RNA molecules labeled in isolated nuclei. Cell 10: 611–616, 1977
12. Morrow BE, Johnson SP, Warner JR: Proteins that bind to the yeast rDNA enhancer. J Biol Chem 264: 9061–9064, 1989

Address for offprints: J.R. Warner, Department of Cell Biology, Albert Einstein College of Medicine, Bronx, NY 10461, USA

Molecular and Cellular Biochemistry **104**: 169–177, 1991.

Stimulation of ribosomal RNA synthesis during hypertrophic growth of cultured heart cells by phorbol ester

Takashi Haneda[1] and Paul J. McDermott[2]
Sigfried and Janet Weis Center for Research, Geisinger Clinic, Danville, PA 17822, USA; Present address: [1] 1st Department of Internal Medicine, Asahikawa Medical College, Nishikagura 4–5, Asahikawa, Hokkaido, Japan 078; [2] Department of Cardiology, Medical University of South Carolina, Veterans Administration Medical Center, 109 Bee Street, Charleston, SC 29403, USA;

Key words: myocyte, hypertrophy, rRNA, phorbol ester

Abstract

Primary cultures of neonatal cardiac myocytes were used to determine the effects of tumor-promoting phorbol esters on ribosomal RNA (rRNA) synthesis during myocyte growth. Treatment of myocytes with phorbol-12, 13-dibutyrate (PDBu) increased protein accumulation by 25% and RNA content by 20%. Rates of rRNA synthesis were measured to assess the mechanism by which rRNA accumulated during myocyte growth. Rates of rRNA synthesis were determined from the incorporation of [^{3}H]uridine into UMP of purified rRNA and the specific radioactivity of the cellular UTP pool. After 24 h of PDBu treatment, cellular rates of 18S and 28S rRNA synthesis were accelerated by 67% and 64%, respectively. The increased rate of rRNA synthesis accounted for the net increase in myocyte rRNA content after PDBu treatment.

During development of cardiac hypertrophy, a consistently observed change is an increase in myocardial RNA content [1]. An increase in RNA content per cell is indicative of an increased translational capacity for protein synthesis because approximately 85% of cytoplasmic RNA is ribosomal RNA (rRNA) [2]. Furthermore, an increase in ribosome content is the primary mechanism by which the rate of protein synthesis is accelerated during hypertrophic growth *in vivo* and in cardiac myocytes maintained in tissue culture [3, 4]. Thus, regulation of cardiac protein synthesis and ultimately growth may occur at one or many of the reactions involved in ribosome formation. These reactions include transcription of rDNA, processing of preribosomal RNA into 18S, 28S and 5.8S rRNA and assembly into mature ribosomes with 70–80 distinct ribosomal proteins [5]. The rate-limiting reactions for ribosome formation, such as transcription of rDNA, provide a potential mechanism through which an intracellular signal generated by a growth stimulus can accelerate the rate of protein synthesis.

Protein kinase C (PK-C) has been an extensively studied system for signal transduction in many diverse cell types, including cardiac myocytes [6]. PK-C is activated by diacylglycerol, formed by the hydrolysis of membrane phosphatidylinositides. The activation of phospholipase C is coupled to agonist-receptor binding and may occur through a guanine-nucleotide binding protein (G-protein). PK-C is also activated directly by the tumor-promoting agents referred to as phorbol esters, and there is substantial evidence that PK-C mediates the cellular response to phorbol esters [7]. It has been shown that an increased activity of PK-C may be responsible for stimulating growth of cardiac myocytes [8]. However, specific substrates for protein phosphorylation catalyzed by PK-C and the subsequent effects on the biochemical reactions involved in protein synthesis are not known.

In the present study, the relationship between rRNA accumulation and phorbol ester stimulated growth of myocytes was examined. We have previously demonstrated that rRNA accumulated during growth of contracting myocytes as a result of accelerated rates of rRNA synthesis [9]. Accordingly, rRNA synthesis rates were measured after phorbol ester treatment to determine whether a common mechanism exists in the myocyte for regulation of protein synthesis upon application of a growth stimulus.

Materials and methods

Myocyte culture

Primary cell cultures were prepared from minced ventricular myocardium of 2–4 day old neonatal rats with modifications as described previously [4]. The myocytes were plated at a concentration of 1×10^6 cells/60 mm dish as determined with a Coulter Counter (Coulter Electronics, Inc.). After an overnight incubation in minimal Eagles medium containing 10% newborn calf serum and 0.1 mM 5-bromo-2′deoxyuridine (BrdU, Sigma) at 37° C in an atmosphere of 5% CO_2/95% room air, the attached cells were rinsed and maintained in serum-free medium [4]. The media were changed every 48 h over the time course of the experiments. Phorbol-12,13-dibutyrate (PDBu, Sigma) stock solutions were prepared in dimethyl sulfoxide (DMSO) at a concentration of 10^{-2} M, stored at − 20° C, and diluted in culture medium at the time of use.

Myocyte protein, DNA and RNA content

Each individual culture dish was rinsed 3 times with ice-cold phosphate-buffered saline (PBS). The cell layer was scraped from each dish with two 0.5 ml volumes of 1 × standard sodium citrate (SSC) containing 0.25% (wt/vol) sodium dodecyl sulfate (SDS) and frozen at − 20° C. Prior to assay, the extracts were thawed and vortexed extensively. Protein content was assayed directly in aliquots of each extract by the method of Lowry using crystalline bovine serum albumin as a standard [10]. DNA content was measured fluorometrically in aliquots of each extract using calf thymus DNA as a standard [11]. For RNA determination, a 750 μl aliquot from each extract was precipitated with an equal volume of ice cold 1 N $HClO_4$. The material was centrifuged and the pellet was washed 3 × with 0.5 N $HClO_4$. The pellet was hydrolyzed in 0.3 N NaOH for 24 h at 37° C and the protein was reprecipitated by adding 4 N $HClO_4$ to a final concentration of 1 N. The supernatant was used to measure the absorbance at 260 and 232 nm and RNA concentration was calculated as described [9].

RNA synthesis

The cellular rate of RNA synthesis (ng RNA/μg DNA) · h^{-1} was calculated by multiplying the fractional rate of RNA synthesis (K_s) by the corresponding RNA/DNA ratio in each preparation. K_s was measured by pulse labeling using the following formula:

$$K_s = \frac{[\text{3'-UMP specific radioactivity (dpm} \cdot \text{pmol}^{-1})]}{[\text{5'-UTP specific radioactivity (dpm} \cdot \text{pmol}^{-1})]} \cdot h^{-1}$$

The specific radioactivities of the total cellular UTP pool and UMP in either total RNA or rRNA were measured after pulse labeling with [5,6-^{3}H]uridine (ICN, 42–50 Ci/mmol, 25 μCi/ml) in medium to which uridine was added to a final concentration of 50 μM uridine. After the labeling period (6 h) was completed, duplicate dishes were rinsed 3 times with ice-cold PBS, and the cells from each dish were scraped into 1 ml of 0.5 N $HClO_4$ with several rinses. The acid extract was centrifuged, and the $HClO_4$ soluble fraction was neutralized by addition of 2 M $KHCO_3$ to a final concentration of 0.5 M. The perchlorate was pelleted by centrifugation, and the supernatant was filtered through 0.2 μm filters (Millipore) and taken to dryness in a vacuum centrifuge. The specific radioactivity of 5′-UTP was measured after purification by high pressure liquid chromatography (HPLC, Beckman Instruments) using a 4 mm × 25 cm Bio-Sil ODS-5S column (Bio-Rad Laboratories). 5′-UTP was purified by the method of Folley *et al.* [12] with a buffer of 30 mM H_3PO_4-triethylamine, pH 6.5, and a linear gradient of 80 mM $MgSO_4$/min. Ultraviolet absorbance was monitored at 260 nm. The area under the peak was integrated and compared with known standards. Fractions corresponding to the retention time of UTP were used to determine specific radioactivity.

The specific radioactivity of 3′-UMP in total RNA was measured in the $HClO_4$ pellet after centrifugation of the whole cell acid extract from each dish. The pellet was washed 3 times with 0.5 N $HClO_4$ prior to incubation in 0.3 N NaOH for 24 hr at 37° C. The protein was reprecipitated by the addition of 4 N $HClO_4$ to a concentration of 1 N and centrifuged. Prior to HPLC analysis, $HClO_4$ was removed by extraction with a 22% (v/v) Alamine 336/Freon mixture by the method of Khym [13]. 3′-UMP was purified by HPLC as described above except that the buffer contained 3.2 mM $MgSO_4$, and a gradient of 160 μM $MgSO_4$/min was utilized to elute UMP.

To isolate 18S and 28S rRNA, 6–8 dishes from each preparation were rinsed 3 times with ice-cold PBS, and the myocytes from each dish were scraped in a buffer containing 20 mM Tris, pH 7.2, 250 mM KCl, 3 mM $MgCl_2$, 1 mM EDTA (ethylenediaminetetroacetic acid) and 0.13 mM puromycin. The cells were homogenized in a Dounce homogenizer, and the homogenate was treated with 0.3% (v/v) Triton X-100 and centrifuged at 10,000 g for 10 minutes. An aliquot of the postmitochondrial supernatant was layered over 3 ml of 0.5 M sucrose, 20 mM Tris, pH 7.6, 500 mM KCL and 3 mM $MgCl_2$ in the tubes of Ti 70.1 rotor (Beckman). The tubes were centrifuged for 2 hr at 100,000 g, and the pellet was used to extract RNA by the phenol/chloroform method. The RNA was further purified by fractionation on a 15–68% exponential sucrose gradient which contained 20 mM Tris, pH 7.6, 10 mM EDTA and 50 mM KCl during centrifugation at 100,000 g for 16 hr. The peak fractions of radiolabeled 18S and 28S rRNA were precipitated with ethanol, resuspended in 0.3 N NaOH and incubated at 37° C for 24 h. The samples were dried in a vacuum centrifuge and 3′-UMP from 18S and 28S rRNA was purified by HPLC as described above.

Results

Primary cultures of neonatal, cardiac myocytes were employed to determine the effects of phorbol esters, direct activators of the enzyme PK-C, on myocyte growth. Myocytes were plated at a relatively 'sparse' density of 1×10^6 cells dish. At this plating density, intercellular contacts were infrequent and synchronous contraction at regular rates was obviated. In agreement with previously reported findings, growth of sparsely plated myocytes as measured by protein/DNA increased during the first few days in culture, and reached a steady state with respect to cell protein [4]. The effects of PDBu were observed more easily because contraction as a stimulus for growth was eliminated. In fact, it was observed that PDBu increased the protein/DNA ratio by 25% and the RNA/DNA ratio by 15% at 2 days after the addition to 1×10^6 cells/dish, but only 5% and 7%, respectively, in contracting myocytes plated at 4×10^6 cells/dish (data not shown).

To determine whether treatment with phorbol ester stimulated RNA content and myocyte growth, PDBu was added to a final concentration of 10^{-7} M on culture day 3, and protein/DNA and RNA/DNA ratios were measured at 12 and 24 h intervals over the next 5 days in culture. The protein/DNA ratio increased over the first 2 days after the addition of PDBu (Fig. 1A). The protein/DNA ratios in PDBu-treated myocytes were significantly increased 18 to 25% as compared to controls and remained higher over the duration of the experiments. Myocyte growth occurred by hypertrophy because absolute values of DNA content in control and PDBu-treated cultures did not change significantly over the course of the experiments (data not shown). The RNA/DNA ratios increased significantly by 1 day after the addition of PDBu (Fig. 1B). The maximal increases in the RNA/DNA ratios by PDBu treatment coincided with the increase in protein/DNA ratios attained after 2 days. How-

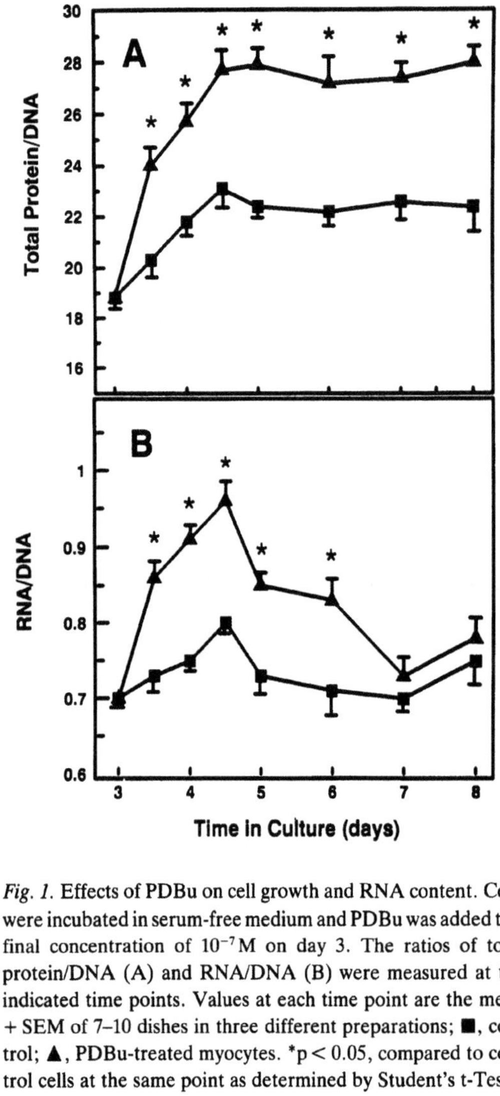

Fig. 1. Effects of PDBu on cell growth and RNA content. Cells were incubated in serum-free medium and PDBu was added to a final concentration of 10^{-7} M on day 3. The ratios of total protein/DNA (A) and RNA/DNA (B) were measured at the indicated time points. Values at each time point are the mean + SEM of 7–10 dishes in three different preparations; ■, control; ▲, PDBu-treated myocytes. $^{*}p < 0.05$, compared to control cells at the same point as determined by Student's t-Test

ever, a decline of the RNA/DNA ratios occurred after 2 days of PDBu treatment (Fig. 1B). These findings indicated that PDBu rapidly increased the capacity for protein synthesis, as measured by RNA content during growth. However, the increase in RNA content was not maintained in myocytes treated with PDBu for more than 2 days.

The dose-response relationships for the effects of PDBu on protein/DNA and RNA/DNA ratios were measured at 1 and 2 days after the addition of PDBu, the time when the maximal increase in these ratios were observed (Fig. 1A and 1B). As compared to control myocytes, PDBu at concentrations of 10^{-7} M and 10^{-6} M significantly increased protein/DNA ratios by 20 and 25% and RNA/DNA ratios by 20 and 26%, respectively. (Table 1). PDBu did not change absolute values of DNA significantly (Table 1). Since DMSO was used as the vehicle to prepare PDBu, the maximum concentration of DMSO (0.01%) in the culture medium was tested in two preparations to determine if it had any effect on myocyte growth and RNA content. DMSO was added to a final concentration of 0.01% on day 3 in culture and was compared after 1 and 2 days to untreated myocytes. No effects of DMSO were observed (data not shown). In the remainder of these studies, myocytes were treated with 10^{-7} M PDBu in order to achieve a maximal response.

To determine the mechanism(s) by which PDBu-treatment stimulated myocyte growth, rates of total RNA and rRNA synthesis were measured. Total RNA and rRNA synthesis rates were calculated by measuring the incorporation of [^{3}H]uridine into UMP of total RNA and 18S and 28S rRNA and dividing by the specific radioactivity of the UTP pool, the precursor for RNA synthesis [9]. A concentration of 50 μM [^{3}H]uridine in the medium was used to expand the pool of UTP during labeling in order to minimize isotopic dilution by an unlabeled pool of uridine nucleotide derived from RNA degradation via the salvage pathway [14]. The specific radioactivities of the UTP pool in both control and PDBu-treated myocytes rose rapidly during the first 1 h of labeling with 50 μM [^{3}H]uridine, the earliest time point measured (Fig. 2B). However, the specific radioactivity of UTP in PDBu-treated myocytes continued to increase and was significantly greater than control after 6 h (301 ± 20 dpm $\cdot$ pmol^{-1}, n = 8 and 214 ± 10 dpm $\cdot$ pmol^{-1}, n = 8, respectively). Because the UTP pool was expanded, the increased specific radioactivity was probably the result of direct stimulation of the enzyme uridine kinase by PDBu. The incorporation of [^{3}H]UMP into RNA was linear over 6 h of labeling (Fig. 2A). In order to calculate rates of RNA synthesis, the expanded pool of cellular [^{3}H]UTP was assumed to be well-mixed and served as the immediate precursor for RNA synthesis. UTP specific radioactivities were determined routinely at 6 h to correct for any differences between control and PDBu-treated myocytes and were used to calculate rates of RNA synthesis. If incorporation of [^{3}H]uridine into RNA alone was taken to represent rates of RNA synthesis, reported increases in PDBu treated myocytes would be overestimated because of the difference between UTP specific radioactivities.

Rates of RNA synthesis in myocytes treated with PDBu for 24 h were compared to control myocytes, the time when the increase in RNA content was maximal. The values for K_s of total RNA and 18S and 28S rRNA were calculated after 6 h of labeling with [^{3}H]uridine as described in Material and methods. K_s for total RNA was accelerated by 17% in PDBu-treated myocytes (Table 2). Cellular rates of total RNA synthesis were calculated by multiplying K_s by the corresponding RNA/DNA ratio (Fig. 3). The increase in RNA/DNA in PDBu-treated myocytes over control (0.91 ± 0.01 vs. 0.72 ± 0.01), resulted in a 49% increase in the corresponding cellular rate of RNA synthesis (Fig. 3).

Measurements of total RNA synthesis by pulse labeling with [^{3}H]uridine reflect the incorporation of radioactivity into pre-ribosomal RNA, mRNA and tRNA species. Because of the contribution of relatively high specific radioactivity UMP from rapidly synthesized RNA species obtained by pulse

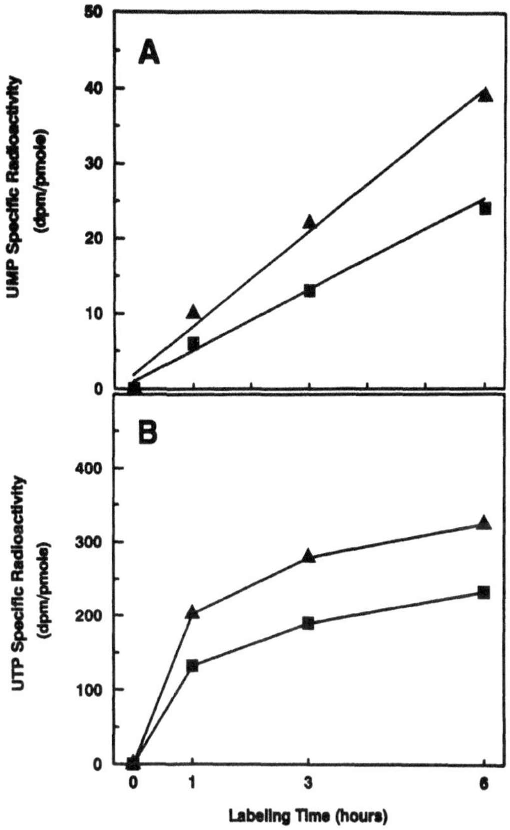

Fig. 2. Uridine nucleotide pools during labeling with [^{3}H]uridine. Control, ■ and PDBu-treated, ▲ myocytes were labeled in medium containing 50 μM [^{3}H]uridine (20 μCi/ml) on day 4. A. The specific radioactivity of UMP incorporated into RNA. B. The specific radioactivity of the cellular UTP pool. Values at each time point are the average of 2 experiments.

Table 1. Dose-response relationships for the hypertrophic effect of PDBu

	Agent	DNA (μg/dish)		Total protein/DNA		RNA/DNA	
		day 4	day 5	day 4	day 5	day 4	day 5
Control		17.6 ± 0.7	18.7 ± 1.1	17.9 ± 0.5	19.8 ± 0.9	0.70 ± 0.01	0.70 ± 0.02
PDBu	10^{-10} M	18.8 ± 1.5	18.3 ± 0.8	17.8 ± 0.6	20.3 ± 0.8	0.70 ± 0.02	0.71 ± 0.01
	10^{-9} M	18.0 ± 1.1	18.3 ± 1.0	18.7 ± 0.8	20.6 ± 1.1	0.74 ± 0.01	0.72 ± 0.02
	10^{-8} M	18.0 ± 1.0	18.9 ± 1.1	19.0 ± 0.6	21.3 ± 1.2	0.76 ± 0.02	0.74 ± 0.02
	10^{-7} M	19.2 ± 1.2	19.3 ± 0.6	22.0 ± 0.8*	24.6 ± 1.4*	0.87 ± 0.02*	0.84 ± 0.01*
	10^{-6} M	18.6 ± 1.2	20.2 ± 0.9	22.2 ± 0.5*	23.7 ± 1.0*	0.88 ± 0.02*	0.84 ± 0.02*

Cells were incubated in serum-free medium and PDBu was added on day 3 to the final concentrations indicated. The absolute values of DNA and the ratios of total protein/DNA and RNA/DNA were measured at days 4 and 5. Values are the mean ± SEM of 6–7 dishes in two different preparations.

*$p < 0.05$, compared to control cells as determined by Student's t-Test.

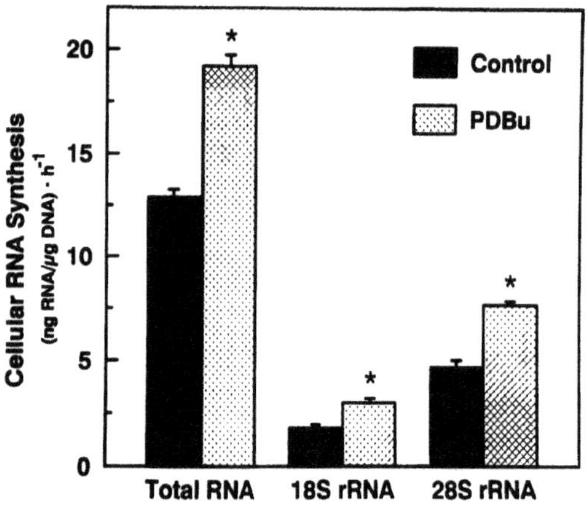

Fig. 3. Effects of PDBu on RNA synthesis. K_s for total RNA and rRNA were measured on day 4 as described in Table 2. The corresponding cellular rates of RNA synthesis were calculated by multiplying the K_s values by the RNA/DNA ratios measured in each preparation. Values are the mean + SEM of 4 different culture preparations. *$p < 0.05$, compared to control cells as determined by Student's t-Test.

labeling, the values of K_s for total RNA were not indicative of rRNA synthesis [9]. Therefore, K_s for 18S and 28S rRNA were measured using the specific radioactivity of [^{3}H]UMP from sucrose-gradient purified rRNA. As expected, the values for K_s of 18S and 28S rRNA were 43% and 35% lower as compared to the K_s value for total RNA, respectively (Table 2). However, PDBu treatment significantly accelerated K_s for both 18S and 28S rRNA by 33% and 30%, respectively (Table 2). Cellular rates of rRNA synthesis were calculated by multiplying K_s by the corresponding RNA/DNA ratio, assuming that 18S and 28S rRNA represent 24% and 61% of total RNA, respectively [15, 16]. PDBu increased cellular rates of 18S and 28S rRNA synthesis 67% and 64%, respectively (Fig. 3).

Discussion

Previous work has demonstrated that an increased ribosome content is the primary mechanism by which the rate of protein synthesis is accelerated during hypertrophic growth of cardiac myocytes [3, 4]. The accumulation of rRNA was the result of accelerated rates of rRNA synthesis [9]. Primary cultures of cardiac myocytes were used to determine if direct activation of PK-C by phorbol ester stimulated rRNA synthesis during hypertrophic growth. The major findings of the present experiments were 1) PDBu increased the capacity for protein synthesis, as measured by RNA content; and 2) The increased RNA content after PDBu treatment was due to accelerated rates of 18S and 28S rRNA synthesis. The rates of rRNA synthesis were accelerated approximately 65%, an increase that was sufficient for the accumulation of rRNA observed after 1 day of PDBu treatment because the pool of rRNA in the myocyte is large and has a half-life of approximately 4.5 days in the ribosome [9, 17].

PK-C is activated by diacylglycerol, a short-lived molecule generated during turnover of membrane phosphatidylinositol. This mechanism serves to transduce extracellular signals across the membrane into many Ca^{2+}-dependent reactions [6]. Tumor-promoting phorbol esters have a structure similar to diacylglycerol and activate PK-C directly both *in vitro* and *in vivo* [6, 7, 18]. Phorbol esters, like diacylglycerol, bind to the PK-C-phospholipid complex to cause stimulation of the enzyme [7]. It has been demonstrated that phorbol esters activate PK-C in perfused rat heart and neonatal cardiac myocytes in culture [8, 19]. The presence of specific

Table 2. Effects of PDBu on fractional rates of RNA synthesis

	Fractional rate, $K_s \cdot h^{-1}$	
	Control	PDBu
Total RNA	0.0181 ± 0.0013 (7)	0.0212 ± 0.0006* (7)
18S rRNA	0.0102 ± 0.0006 (4)	0.0136 ± 0.0006* (4)
28S rRNA	0.0108 ± 0.0006 (4)	0.0140 ± 0.0004* (4)

Cells were incubated in serum-free medium and PDBu was added to a final concentration of 10^{-7} M on day 3. Values for K_s were calculated as described in Methods on day 4. Values are the mean ± SEM, and the number of experiments are in parenthesis. *$p < 0.05$, compared to control cells as determined by Student's t-Test.

and saturable binding sites for phorbol esters has been reported in isolated cardiac myocytes [18]. Although phorbol ester is known to stimulate myocyte hypertrophic growth [8], the mechanism(s) by which phorbol ester accelerate(s) growth has not been elucidated. Therefore, both RNA content and rates of rRNA synthesis were measured during hypertrophic growth in PDBu-treated cells.

PDBu significantly increased myocyte RNA content and accelerated growth; the maximum response was reached at a concentration of 10^{-7} M. The extent and time course of PK-C activation were not measured in the present studies, but it has been demonstrated that treatment of neonatal myocytes with 10^{-7} M phorbol myristate acetate (PMA) resulted in only a transient increase in PK-C activity [8]. In agreement with these findings, we have observed a transient increase in membrane associated PK-C activity after phorbol ester treatment (S. Allo and P. McDermott, unpublished observations). Chronic exposure to phorbol ester resulted in both down-regulation of PK-C activity and dimunition of PK-C content [8]. Similar to our findings, a significant increase in total cell protein was observed 1 day after PMA treatment. If PK-C was similarly down-regulated by PDBu, this may be the reason for the inability to maintain the increased RNA content after exposure to PDBu for longer than 2 days. Further studies will be required to examine the temporal relationship between PK-C activation and the rates of rRNA synthesis and degradation during phorbol ester treatment.

The rates of 18S and 28S rRNA synthesis were calculated by measuring the incorporation of [^{3}H]uridine into 3′-UMP of mature cytoplasmic rRNA and dividing by the specific radioactivity of the UTP pool. The UTP pool is also the primary source of CTP utilized for RNA synthesis. Approximately 45% of the radioactivity incorporated into rRNA during pulse labeling with [^{3}H]uridine was found in CMP (data not shown). This problem was overcome by measuring the specific radioactivities of 3′-UMP after purification by HPLC. UMP specific radioactivities were considered to be valid measures of the corresponding rRNA pools for the following reasons. First, the cells were homogenized in a buffer containing 0.13 mM puromycin and the post-mitochondrial supernatant was centrifuged through a buffer containing 500 mM KCl to pellet the ribosomes. Second, the ribosome pellets were phenol extracted, ethanol precipitated and the 18S and 28S rRNA purified on sucrose gradients. Third, any residual UMP contamination derived from mRNA would be minor because the entire mRNA pool was approximately two orders of magnitude smaller than the rRNA pool.

The possibility that the specific radioactivity of UTP pools had changed was examined before concluding that an increase in rRNA synthesis had occurred. Thus, the importance of using specific radioactivity of cellular [^{3}H]UTP as the precursor pool for calculations of rRNA synthesis was demonstrated. The specific radioactivity of UTP in both control and PDBu-treated cells rose rapidly during the first 1 h of labeling and the UTP pool was expanded. This was followed by a second phase during which the specific radioactivity of the UTP pool increased slowly (Fig. 2). The specific radioactivities of UTP at 6 hr were significantly greater in PDBu-treated myocytes than in control myocytes. The potential increase in rRNA synthesis rates in PDBu treated myocytes was therefore not as great as that indicated by [^{3}H]uridine incorporation alone. The increased specific radioactivity of UTP in PDBu treated myocytes, even though the UTP pool was expanded, was probably due to direct stimulation of the enzyme uridine kinase by PDBu. Stimulation of uridine kinase by thyroid hormone has been demonstrated in heart [20]. We have observed that triiodothyronine increased UTP specific radioactivities by approximately 50% after pulse labeling with 50 μM [^{3}H]uridine as compared to control myocytes (unpublished observations). These observations underscore the importance of measuring UTP specific radioactivities for calculations of RNA synthesis rates using [^{3}H]uridine labeling. If the specific radioactivity of the precursor pool is not considered, rates of [^{3}H]uridine incorporation into RNA do not reflect rates of RNA synthesis accurately.

A regulatory link between the synthesis of rRNA and the regulation of ribosome formation has been studied in skeletal muscle cells. The synthesis of rRNA and r-proteins occur coordinately but accumulation of newly formed ribosomes during differentiation into myotubes is regulated by transcription of rDNA [21, 22]. In most instances, rRNA synthesis is regulated at the level of rDNA transcription [5, 23]. The rate of rRNA synthesis measured in the present experiments represents the product of the reaction rates for rDNA transcription, processing of 45S pre-ribosomal RNA and assembly with r-proteins into ribosomes. While phorbol esters have been reported to induce changes in transcriptional activity of certain genes [24, 25], further studies are needed to determine the precise reactions involved in ribosome formation that are affected by PDBu-treatment.

Acknowledgements

We thank Lois Carl for her excellent technical assistance and Theresa Nestico and Brian Shoop for preparation of the manuscript.

References

1. Morgan HE, Gordon EE, Kira Y, Chua BHL, Russo LA, Peterson CJ, McDermott PJ, Watson PA: Biochemical mechanisms of cardiac hypertrophy. Ann Rev Physiol 49: 533–543, 1987
2. Young VR: The role of skeletal and cardiac muscle in the regulation of protein metabolism. In: HN Munro and JB Allison (eds) Mammalian Protein Metabolism, Academic Press, New York, 1970, pp 585–662
3. Siehl D, Shua BHL, Lautensach-Belser N, Morgan HE: Faster protein and ribosome synthesis in thyroxine-induced hypertrophy of rat heart. Am J Physiol 248: C309–C319, 1985
4. McDermott PJ, Morgan HE: Contraction modulates the capacity for protein synthesis during growth of neonatal heart cells in culture. Circ Res 64: 542–553, 1989
5. Sollner-Webb B, Tower J: Transcription of cloned eukaryotic ribosomal RNA genes. Ann Rev Biochem 55: 801–830, 1986
6. Nishizuka Y: The role of protein kinase C in cell surface signal transduction and tumor production. Nature Lond 308: 693–698, 1984
7. Kikkawa V, Takai Y, Tanaka Y, Miyake R, Nishizuka Y: Protein kinase C as a possible receptor protein of tumor-promoting phorbol esters. J Biol Chem 258: 11442–11445, 1983
8. Henrich CJ, Simpson PC: Differential acute and chronic response of protein kinase C in cultured neonatal rat heart myocytes to α_1-adrenergic and phorbol ester stimulation. J Mol Cell Cardiol 20: 1081–1085, 1988
9. McDermott PJ, Rothblum LI, Smith SD, Morgan HE: Accelerated rates of ribosomal RNA synthesis during growth of contracting heart cells in culture. J Biol Chem 264: 18220–18227, 1989
10. Lowry DH, Rosenbrough NJ, Farr AL, Randall RJ: Protein measurement with Folin phenol reagent. J Biol Chem 193: 265–275, 1951
11. Cesarone CF, Bolognesi C, Santi L: Improved microfluorometric DNA determination in biological material using 33258 Hoechst. Anal Biochem 100: 188–197, 1979
12. Folley LA, Power SC, Poyton RO: Separation of nucleotides by ion-pair, reversed-phase high performance liquid chromatography. J Chromatogr 281: 199–207, 1983
13. Khym JX: An analytical system for rapid separation of tissue nucleotides at low pressures on conventional anion exchangers. Clin Chem 21: 1245–1252, 1975
14. Matsushita S, Fanburg BL: Pyrimidine nucleotide synthesis in the normal and hypertrophying rat heart. Circ Res 27: 415–428, 1970
15. Chan Y, Gutell R, Noller HF, Wool IG: The nucleotide sequence of a rat 18S ribosomal ribonucleic acid gene and a proposal for the secondary structure of 18S ribosomal ribonucleic acid. J Biol Chem 259: 224–230, 1984
16. Chan Y, Olvera J, Wool IG: The structure of rat 28S ribosomal ribonucleic acid inferred from the sequence of nucleotides in a gene. Nucleic Acids Res 11: 7819–7831, 1983
17. Ashford AJ, Pain VM: Effect of diabetes on the rates of synthesis and degradation of ribosomes in rat muscle and liver *in vivo*. J Biol Chem 261: 4059–4065, 1986
18. Limas CJ: Characterization of phorbol ester binding to isolated cardiac myocytes. Arch Biochem Biophys 238: 300–304, 1985
19. Yuan S, Sunahara FA, Sen AK: Tumor-promoting phorbol esters inhibit cardiac functions and induce redistribution of protein kinase C in perfused beating rat heart. Circ Res 61: 372–378, 1987
20. Gertz BJ, Haugaard ES, Haugaard N: Effects of thyroid hormone on UTP content and uridine kinase activity of rat heart and skeletal muscle. Am J Physiol 238: E443–E449, 1980
21. Jacobs FA, Bird RC, Sells BH: Differentiation of rat myoblasts. Regulation of turnover of ribosomal proteins and their mRNAs. Eur J Biochem 150: 255–263, 1985
22. Hammond ML, Bowman LH: Insulin stimulates the translation of ribosomal proteins and the transcription of rDNA in mouse myoblasts. J Biol Chem 263: 17785–17791, 1988
23. Wolf S, Sameshima M, Liebhaber SA, Schlessinger D: Regulation of ribosomal ribonucleic acid levels in growing, ^{3}H-arrested, and crisis-phase WI-38 human diploid fibroblasts. Biochemistry 19: 3484–3490, 1980
24. Angel P, Imagawa M, Chiu R, Stein B, Imbra RJ, Rahmsdorf HJ, Jonat C, Herrlich P, Karin M: Phobol ester-in-

ducible genes contain a common cis element recognized by a TPA-modulated trans-acting factor. Cell 49: 729–739, 1987
25. Lee W, Mitchell P, Tjian R: Purified transcription factor AP-1 interacts with TPA-inducible enhancer elements. Cell 49: 741–752, 1987

Address for offprints: P.J. McDermott, Department of Cardiology, Medical University of South Carolina, Veterans Administration Medical Center, 109 Bee Street, Charleston, SC 29403, USA

RIBOSOME BIOGENESIS SYNTHESIS OF RIBOSOMAL PROTEINS AND RIBOSOME FORMATION

Molecular and Cellular Biochemistry **104**: 181–187, 1991.

Coordinate expression of ribosomal protein genes in yeast as a function of cellular growth rate

Willem H. Mager and Rudi J. Planta
Biochemisch Laboratorium, Vrije Universiteit, de Boelelaan 1083, 1081 HV Amsterdam, The Netherlands

Key words: yeast, ribosomal protein genes, growth rate, multifunctional proteins

Abstract

The rate of ribosome formation in yeast is precisely adjusted to the physiological demands of the cell. During all growth conditions a balance is maintained in the production of all ribosomal constituents. Coordinate expression of the ribosomal protein (rp) genes is primarily accomplished at the transcriptional level. Transcription activation of the majority of the rp-genes is mediated through common upstream activating sequences, so-called RPG boxes, which occur usually in a tandem at a distance of 200–500 bp from the start codon. These RPG-boxes represent binding sites for a transcriptional activator, called TUF or RAP. The concentration of TUF parallels the cellular growth rate and evidence exists that the response of rp-genes upon nutritional changes is mediated by this factor. Recent findings indicate that TUF/RAP also activates other gene families involved in cellular growth rate. Furthermore, this multifunctional protein also binds to the mating-type silencer and telomeres in yeast.

Some other rp-genes (e.g. those encoding S33 and L45) do not contain an RPG-box. They appear to be activated by another multifunctional protein, called ABF1 or SUF, by binding to another nucleotide motif. This multifunctional protein also activates other gene families, and in addition binds to the mating type silencer and ARS-elements.

Introduction

The rate of ribosome formation in yeast is perfectly geared to the cellular growth rate, and therefore accurately reflects the need for protein biosynthetic capacity [reviewed in ref. 1].

Coordinate control of rp-gene expression primarily occurs at the transcriptional level [reviewed in refs. 2, 3]. In addition, 'ribosome assembly-mediated' control is supposed to represent the ultimate level of control in the supply of r-proteins: excess, unassembled r-proteins are rapidly degraded [2, 3]. To account for the equimolar synthesis of rp-mRNAs the respective gene promoters are assumed to have evolved in a concerted fashion. At steady state growth conditions, rp-genes can be considered as real house-keeping genes displaying a balanced, constitutive level of transcription and giving rise to nearly equal amounts of mRNA [2, 3]. Upon alterations in nutritional conditions, for instance when glucose is added to an ethanol-based culturing medium, rp-gene expression is coordinately and rapidly adjusted to the new cellular demands (see Fig. 1).

In recent years studies in many laboratories have identified the functional sequence elements involved in the regulation of rp-gene expression and the transacting protein factors, binding to these elements, as will be described in this review.

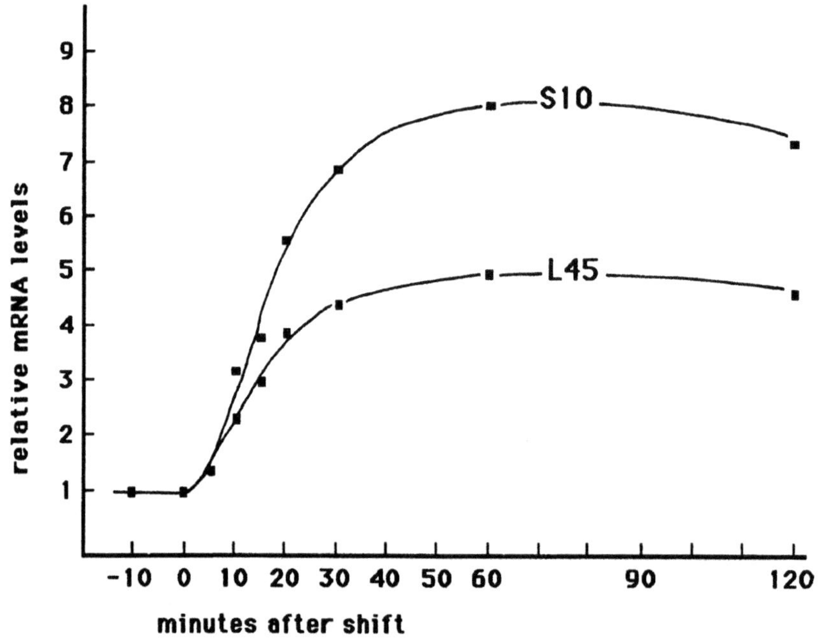

Fig. 1. Coordinate response of rp-gene expression to a nutritional upshift. Yeast cells were grown on 2% ethanol as a carbon source, then glucose was added to a final concentration of 2%. From cells harvested at several time points after the shift RNA was isolated and probed with gene-specific oligonucleotides as indicated. The autoradiograms were scanned and the intensity of the hybridization signals was expressed relative to that of the actin mRNA as an internal standard. L45 was taken as an example of an ABF1-controlled rp-gene; S10 is a RAP1-controlled rp-gene.

Transcriptional activation of yeast rp-genes

Sequence comparison of the upstream regions of many yeast rp-genes revealed the presence of a conserved nucleotide element, the RPG-box, having the consensus sequence motif ACACCCATACATTT [4]. RPG-boxes occur at a distance of 250–450 bp from the initiation codon, mostly in a tandem arrangement and in both possible orientations. For several rp-genes a deletion analysis of the extended promoter region demonstrated that RPG-boxes act as upstream activation sites (UAS_{rpg}; [2, 3]). In most cases, removal of RPG-boxes turned out to abolish transcription completely, whereas occasionally a residual transcription activity was measured, possibly due to the presence of a T-stretch downstream of the boxes [5]. T-stretches often occur in rp-gene promoters, as well as in the promoter regions of many other highly expressed genes in yeast, and contribute to the strength of the UAS (see furtheron). Synthetic oligonucleotides harbouring the consensus RPG-box sequence can act as heterologous UAS-elements replacing the authentic activating sequences [6].

By band shift assays and footprinting, RPG-boxes were identified as recognition sequences for an

Table 1. TUF = SBF − E = RAP1 = GRF1 = TBA

to be designated	consensus responsive site
RAP1	ACACCCATACATTT

functions*:
transcriptional activator for genes encoding r-proteins S10, S16A, S24, S31, L16, L17A, L25, L29, L34, L44, L44′, L46, rp 28, rp 51, rp 59; translation factors; subunits RNA polymerases; glycolytic enzymes PGK, ENO1, PYK1, PDC1, ADH1, TPI, TDH1; membrane-bound ATPase.
transcriptional repressor for silent mating type loci
telomere binding
attachment chromosomes to nuclear scaffold

* References cited in the text.

abundant yeast protein, initially designated TUF [7, 8], hereafter referred to as RAP1 (see below; Table 1). RAP1 has a molecular weight of 92.5 kD as deduced from the coding sequence of the isolated gene [9]. RAP1 is an acidic protein showing an unusually high content of Asn residues but otherwise not containing characteristic functional domains. Partial proteolysis generated a fragment of 50 kD that is still able to bind to the RAP1 responsive site [10]. In the RAP1-RPG-box complex all G-residues of the binding site are in close contact with the protein as could be concluded from a methylation interference study [11]. Analysis of single site mutations revealed that mainly the core sequence YCYR at position 4–7 is indispensable for transcription activity *in vivo* [11].

For this group of rp-genes the cis/trans combination of RPG-box/RAP1 is essential for efficient constitutive transcription activation. The question whether this primary protein-DNA complex is sufficient to mediate a nutritional (upshift) effect, is not easy to answer. The use of synthetic oligomers as heterologous UAS-elements however suggest that this transcriptional control occurs directly or indirectly, through RAP1 [12].

Not all rp-genes in yeast harbour RPG-boxes. Upstream of the genes encoding S33 and L3, for instance, a binding site for a second abundant yeast factor can be found, initially called SUF [13] or TAF [14], hereafter referred to as ABF1 (to be discussed below; Table 2). In the promoter regions of these genes the ABF1-responsive site is located at approx. 160–200 bp from the ATG. Deletion analysis demonstrated that the ABF1-site forms part of the upstream activation sequence of these genes [13, 14]. The use of synthetic oligomers encompassing the ABF1-responsive site indicated that this box, contrary to the RPG-box, does not represent a strong enhancer. It is likely therefore that the action of ABF1 as a transcriptional activator depends upon additional factors.

The ABF1-gene has been isolated and sequenced [15–17] and appears to encode a protein with a predicted molecular weight of 81.7 kD. A transcription activating domain has not been identified so far, but the protein does contain a potential DNA-binding motif, viz. the atypical metal-binding 'finger'-structure $CX_7HX_3HX_4CX_4C$. Comparison of the primary structures of ABF1 and RAP1 revealed a striking homology [17]. Except for the putative DNA-binding domain of ABF1, the protein shows about 40% similarity to RAP1, suggesting an evolutionary relationship between both factors but possibly also reflecting a functional equivalence.

Methylation interference studies of the ABF1-responsive site revealed a dual binding motif, RTCRYNNNNNACG, suggesting this factor to be bound to one side of the DNA-helix [18]. In agreement with this finding a single ACG → AGG mutation prevents binding of the protein *in vitro* and abolishes transcription *in vivo* (Doorenbosch, M.M., Mager, W.H. and Planta, R.J., unpublished data). Due to the split feature of this motif the

Table 2. SUF = TAF = SBF − B = ABF1 = GF1 = BAF1

to be designated	consensus responsive site
ABF1	RTCRYNNNNNACG

functions*:
transcriptional activator for genes encoding r-proteins S33, L2, L3, L45; translation factors; transcription factors; subunits RNA polymerases; enzymes involved in general metabolism, glycolysis, growth control; nuclear encoded mitochondrial proteins
transcriptional repressor for silent mating type loci
binding to ARS-associated element

* References cited in the text.

Table 3. RBP1 = REB1 = Y = GRF2

to be designated	consensus responsive site
REB1	YNNYYACCCG

functions*:
transcriptional activator for genes encoding RAP1; actin; enzymes involved in general metabolism (TRP, PYK, GAL10)
binding to enhancer and promoter of rRNA operon
binding to telomeres

* References cited in the text.

putative identity of SUF and TAF and ABF1 initially remained uncovered. Competition retardation analyses making use of oligonucleotides, however, proved these factors to be identical [18].

Upstream of the gene encoding the acidic ribosomal protein L45 a binding sequence for a third abundant yeast factor, CP1 (see Table 3), is situated (Kraakman, L.S., Mager, W.H. and Planta, R.J., unpublished results). Deletion or site-directed mutagenesis of this element, however, had no effect on the transcription activity of the plasmid-encoded L45-gene. However, we cannot exclude the possibility that this site does play a functional role in its normal chromosomal context. Downstream of this CP1 site a binding site for ABF1 was found, though displaying a deviating motif RTARYNNNNNACG. This ABF-1-responsive site turned out to be the major UAS-element of the L45-gene. A similar result was recently obtained for the rp-genes encoding L2 [19].

Since RAP1 and ABF1, identified as transcriptional activators of rp-genes, are abundant proteins in the yeast cell, it is not surprising that these factors are involved in the transcriptional activation of other yeast genes as well. Moreover, these proteins were found to fulfil very different functions in the yeast cell.

RAP1 and ABF1 belong to a group of multifunctional yeast proteins

RAP1 and ABF1 are not exclusively involved in the transcription activation of yeast rp-genes. The upstream region of many other genes, including those for glycolytic enzymes, contain RPG-boxes (see Table 1). In addition, in the 5′-flanking sequence of many genes an ABF1-responsive site is located (summarized in Table 2), though at present for only a limited number of these genes its transcription activating function has been established. The extended promoter regions of several genes, e.g. of PGK, even harbour both a RAP1 and an ABF1 site [20, 21]. In particular the genes that are under RAP1-control can be considered as housekeeping genes, the expression of which is supposed to be regulated in a cellular growth rate-dependent fashion. Therefore, we have estimated the level of RAP1- and ABF1-proteins able to bind to their responsive sites *in vitro,* isolated from cells grown on ethanol and glucose as a carbon source, respectively. The results revealed a considerable difference in cellular concentration of these factors dependent on the nutritional conditions. On the other hand, Northern analyses revealed only slight differences in the amount of the corresponding mRNAs (Kraakman, L.S., Hoekstra, R., Mager, W.H. and Planta, R.J., unpublished results). These findings suggest that either these factors are post-translationally modified to modulate their DNA-binding efficiency (see futheron) or, less likely, that expression of the respective genes is subject to a translational control.

Surprisingly, the factors shown to serve as transcriptional activators of rp-genes and many other genes in yeast, can also play other functional roles. First, the silencer regions of the silent mating type loci HML and HMR, which serve to keep these loci transcriptionally quiescent, contain binding sites for RAP1, ABF1 and an ARS-binding factor [22–24]. The silencing function of these multiple protein-DNA complexes, in addition, requires the action of the SIR-gene products [25].

Secondly, RAP1 binds to telomeres which contain the repeated sequence $(C_{1-3}A)_n$, thus occasionally representing RPG-boxes [23, 24, 26]. In addition, RAP1 is supposed to be involved in the attachment of chromosomal DNA to the nuclear scaffold [27]. It is presently not known whether a correlation exists between RAP1-transcription activation sites and chromosome-attachment sites.

ABF1, in addition to being a transcriptional activator or silencer, was identified as a protein that

Table 4. CP1 = GFII = CBP1

to be designated	consensus responsive site
CP1	RTCACGTG

functions*:
binding to centromere region I
binding upstream of genes encoding L45, GAL2, MET

* References cited in the text.

binds to nucleotide elements associated with ARS, autonomously replicating sequence [23].

A third abundant yeast factor, CP1, was found to associate with the conserved centromere region I but additional CP1-sites occur frequently in the yeast genome [28, 29; see Table 4]. No transcription activating role for this protein could be established so far (see above).

Finally, another abundant factor, REB1, was identified as a protein binding to the enhancer – and promoter – region of the yeast rRNA operon [30, 31]. Recently, it was found that the same factor binds to upstream sequences of several yeast genes, e.g. of the RAP1-gene, as well as to telomeres [32, 33; see Table 3]. REB1 has been proposed to mediate nucleosome positioning in promoter regions [32]. The various designations of these four abundant multifunctional DNA-binding proteins in yeast are summarized in Tables 1–4. Since RAP1 and ABF1 are involved in such diverse cellular processes as transcriptional regulation, replication, telomere function etc. the question arises how their actual functional role is determined. In both factors the transcription activating domain could not yet be identified, suggesting that this function may depend on the interaction with other factor(s). In addition, the complicated interplay of factors involved in transcriptional silencing and also the complicated architecture of some UASs (e.g. that of the PGK-gene) suggest that the nucleotide context of the protein binding sites and/or the interaction of RAP1 and ABF1 with additional factors will define their mode of action.

Evidence for auxiliary factors involved in yeast (rp-)gene transcription

Transcription activation of yeast rp-genes appears to be mediated by (at least) two distinct abundant DNA binding factors, RAP1 and ABF1. Therefore, again the question arises how the transcription regulation through these two general factors is coordinated. Are these factors modified in a similar and perhaps rp-gene specific fashion? Or do common additional factors, interacting with adjacent cis-acting nucleotide elements, e.g. the T-stretch mentioned before, or through protein-protein contacts, play a part? Several lines of evidence suggest that indeed additional activating factors may be involved in RAP1 and ABF1-mediated transcription activation. By UV cross-linking to the ABF1-site in front of the L3-gene, proteins with a similar size were found as with an RPG-box containing DNA fragment [14]. Furthermore, GCR1-deficiency was found to have a similar effect on ADH1 gene transcription as deletion of its RPG-box [34]. GCR1 has been identified as a positive transcriptional regulator of glycolytic enzymes. In gcr^--strains also the cellular level of rp59-mRNA was significantly reduced [34]. These results support the idea that RAP1 may activate the transcription of glycolytic genes – and perhaps also rp-genes – through the interaction with the GCR1 gene product.

Additional proof for the interaction of RAP1 and ABF1 with other factors is provided by the analysis of the PGK-gene. In the extended promoter of the PGK-gene, in addition to an ABF1-site and an RPG-box, several CTTCC-blocks occur [21, 22]. For these different cis-acting elements a synergistic interaction has been proposed, since efficient transcription of the PGK-gene only took place having the ABF1-site and the CTTCC blocks in conjunction with the RPG-box [20].

With respect to the ABF1-mediated transcription activation of rp-genes evidence for additional factors came forward from our studies of the S33- and L45-genes. After deletion of the ABF1-site located upstream of the S33-gene, the resulting promoter fragment still gave rise to complex formation in band shift assays and, in addition, showed a residual transcription activity [12]. In this respect, differences were observed for yeast cells, grown in media with different carbon sources. A factor predominantly present in glucose-grown cells was designated GDUF and a factor only observed in S100 extracts from ethanol-grown cells was named EDUF [13]. The binding site for GDUF was determined by footprint analysis and a subsequent methylation interference study suggested that the following bases are involved in protein binding GAAAATGAA (Doorenbosch, M.M., Mager, W.H. and Planta, R.J. unpublished data).

Making use of an oligonucleotide encompassing the GDUF-responsive site in band shift assays of yeast protein fractions, it appeared that more than one protein is able to bind to this site, probably in a mutually exclusive fashion. Also downstream of the ABF1-site in the L45-gene promoter a GDUF-site was found. This nucleotide element turned out to have a weak activating function in this promoter region, too. Experiments are in progress to determine the possible synergistic action of the ABF1- and GDUF-site in these promoters, preferably in their chromosomal context. Recently for the DED1-gene (an essential yeast gene with unknown function) the synergistic action of ABF1 and factors interacting with an adjacent T-rich sequence was proposed [35]. Interestingly, in this T-rich sequence a potential GDUF-site occurs. Computer searches so far elucidated only a limited number of genes harbouring the combination of an ABF1-site and a GDUF-site (Kraakman, L.S., Mager, W.H. and Planta, R.J., result not shown). We hypothesize that a protein factor interacting with the GDUF-site is needed for efficient transcriptional control of those rp-genes that are activated by ABF1.

Can protein modification play a part in transcriptional regulation?

An alternative way to explain the coordinate transcription regulation of rp-genes through RAP1 or ABF1 is by assuming a similar modification, for instance by phosphorylation. We have some preliminary evidence that RAP1 may be phosphorylated and that this modification increases its binding affinity to the RPG-box *in vitro*. Furthermore we used mutants, temperature-sensitive in cAMP-synthesis, to study the possible involvement of protein phosphorylation in transcriptional control of rp-genes. We observed that the characteristic nutritional upshift response of rp-mRNA levels does not occur after shifting these ts-mutant cells to the restrictive temperature (Kraakman, L.S., Zerp, S., Mager, W.H. and Planta, R.J., unpublished results). This effect was observed both for RAP1- and ABF1 controlled rp-genes and suggests that cAMP-dependent phosphorylation, directly or indirectly, mediates at least partially the control of rp-gene transcription.

Conclusion and perspectives

Transcription activation of rp-genes in yeast is mediated through two distinct abundant proteins, RAP1 and ABF1. These factors are multifunctional DNA binding proteins playing a part in diverse cellular processes, related to the cellular growth rate.

The promoter regions of rp-genes display differences with respect to the number, orientation, distance to the gene, and relative binding affinity of the responsive sites for RAP1 and ABF1. Yet the constitutive transcription of rp-genes leads to nearly equal amounts of mRNA. We propose that through the cooperative action of these primary complexes with auxiliary cis-elements (T-stretches, GDUF sites) and additional factors equivalent transcription initiation complexes are formed. It will be interesting to see whether the nuclear matrix is involved in the assembly of such complexes. Most probably, the coordinate control of rp-gene transcription upon alterations in cellular growth conditions requires the interaction with regulatory factors and/or changes in the phosphorylation state of the complex. Detailed mutational analyses of the primary transactivators, RAP1 and ABF1, are needed to unravel the putative involvement of such additional factors or modification.

Acknowledgements

The authors are indebted to their coworkers in the Biochemical Department for their research contributions and stimulating discussions. We thank Mrs. P.G. Brink for expert preparing of the typescript.

References

1. Planta RJ, Mager WH, Leer RJ, Woudt LP, Raué HA,

El-Baradi TTAL: In: B Hardesty and G Kramer (eds) Structure, function and genetics of ribosomes. Springer Verlag, New York, 1986, pp 699–718
2. Mager WH: Biochim Biophys Acta 949: 1–15, 1988
3. Planta RJ, Mager WH: In: MF Tuite, M Picard and M Bolotin-Fukuhara (eds) Genetics of translation. NATO ASI series H, vol 14, Springer-Verlag, 1988, pp 117–129
4. Leer RJ, van Raamsdonk-Duin MMC, Mager WH, Planta RJ: Curr. Genet 9: 273–277, 1985
5. Rotenberg HO, Woolford JL: Mol Cell Biol 6: 674–687, 1986
6. Woudt LP, Mager WH, Nieuwint RTM, Wassenaar GM, van der Kuyl AC, Murre JJ, Hoekman MFM, Brockhoff PGM, Planta RJ: Nucl Acids Res 15: 6037–6048, 1987
7. Huet J, Cottrelle P, Cool M, Vignais M-L, Thiele D, Marck C, Buhler JM, Sentenac A, Fromageot P: The EMBO J 4: 3539–3547, 1985
8. Vignais M-L, Woudt LP, Wassenaar GM, Mager WH, Sentenac A, Planta RJ: The EMBO J 6: 1451–1457, 1987
9. Shore D: Cell 51: 721–732, 1987
10. Huet J, Sentenac A: Proc Natl Acad Sci 84: 3648–3652, 1987
11. Nieuwint RTM, Mager WH, Maurer CTC, Planta RJ: Curr Genet 15: 247–251, 1989
12. Herruer MH, Mager WH, Woudt LP, Nieuwint RTM, Wassenaar GM, Groeneveld P, Planta RJ: Nucl Acids Res 15: 10133–10144, 1987
13. Herruer MH, Mager WH, Doorenbosch MM, Wessels PLM, Wassenaar GM, Planta RJ: Nucl Acids Res 17: 7427–7439, 1989
14. Hamil KG, Nam HG, Fried HM: Mol Cell Biol 8: 4328–4311, 1988
15. Rhode PR, Sweder KS, Oegema KF, Campbell JL: Genes and Developm 3: 1926–1939, 1989
16. Halfter H, Kavety B, Vandekerckhove J, Kiefer F, Gallwitz D: The EMBO J 8: 4265–4272, 1989
17. Diffley JFX, Stillman B: Science 246: 1034–1038, 1989
18. Dorsman JC, Doorenbosch MM, Maurer CTC, de Winde JH, Mager WH, Planta RJ, Grivell LA: Nucl Acids Res 17: 4917–4923, 1989
19. Della Seta F, Giafré S-A, Marck C, Santoro B, Presutti C, Sentenac A, Bozzoni I: Mol Cell Biol 10: 2437–2441, 1990
20. Chambers A, Tsang JSH, Stanway CA, Kingsman AJ, Kingsman SH: Mol Cell Biol 9: 5516–5524, 1989
21. Stanway CA, Chambers A, Kingsman AJ, Kingsman SH: Nucl Acids Res 17: 9205–9218, 1989
22. Brand AH, Micklen G, Nasmyth K: Cell 51: 709–719, 1987
23. Buchman AR, Kimmerly WJ, Rine J, Kornberg RD: Mol Cell Biol 8: 210–225, 1988
24. Buchman AR, Lue NF, Kornberg RD: Mol Cell Biol 8: 5086–5099, 1988
25. Shore D, Stillman DJ, Brand AH, Nasmyth K: The EMBO J 6: 461–467, 1987
26. Berman J, Tachinaba CY, Tye BK: Proc Natl Acad Sci USA 83: 3713–3717, 1986
27. Hofmann JF, Laroche T, Brand GH, Gasser SM: Cell 57: 725–737, 1989
28. Bram RJ, Kornberg RD: Mol Cell Biol 7: 403–409, 1987
29. Cai M, Davis RA: Mol Cell Biol 9: 2544–2550, 1989
30. Kulkens T, van Heerikhuizen H, Klootwijk J, Oliemans J, Planta RJ: Curr Genet 16: 351–359, 1989
31. Morrow BE, Johnson SP, Warner JF: J Biol Chem 264: 9061–9068, 1989
32. Chasman DI, Lue NF, Buchmann AR, LaPointe JW, Lorch Y, Kornberg RD: Genes and Developm 4: 503–514, 1990
33. Wang H, Nicholson PR, Stillman DJ: Mol Cell Biol 10: 1743–1753, 1990
34. Santangelo GM, Tornow J: Mol Cell Biol 10: 859–862, 1990
35. Buchman AR, Kornberg RD: Mol Cell Biol 10: 887–897, 1990

Address for offprints: R.J. Planta, Biochemisch Laboratorium, Vrije Universiteit, de Boelelaan 1083, 1081 HV Amsterdam, The Netherlands

Molecular and Cellular Biochemistry **104**: 189–194, 1991.

Regulation of ribosome biogenesis in differentiated rat myotubes

Peter Zahradka[1], Dawn E. Larson and Bruce H. Sells
Department of Molecular Biology and Genetics; [1] Department of Chemistry and Biochemistry, University of Guelph, Guelph, Ontario, Canada

Key words: differentiation, ribosome biogenesis, rRNA transcription, rat myoblasts

Abstract

The rate of ribosome biogenesis is closely coupled with cell proliferation, representing a unique model system for studying gene regulation. Terminal differentiation of rat L6 myoblasts, an example of a rapidly proliferating population of cells being converted into a non-dividing syncytial population, results in an 80% decline in the rate of ribosome accumulation. Ribosome production during myogenesis is regulated by a down-shift in the rate of rRNA accumulation, controlled at the level of transcription by specific trans-acting factors. The synthesis of both ribosomal proteins and 5S rRNA remains unchanged in myotubes, however, resulting in an over-production of these precursors. The excess molecules are rapidly degraded, preventing the accumulation of a static pool of ribosome components.

Introduction

In eucaryotic cells, ribosome biogenesis requires the equimolar accumulation of its various components: 18S, 28S, 5.8S and 5S rRNAs, and approximately 80 r-proteins. The individual components are then assembled into large and small ribosomal subunits in the nucleolus and subsequently transported to the cytoplasm. Since the rate of ribosome production varies with the cell growth rate [see 1–3], organisms have adopted diverse strategies to regulate production of ribosomal precursor molecules in response to i) variations in nutrient levels, ii) the presence of growth factors and hormones, and iii) different stages of development.

The development of muscle tissue during embryonic growth is one example in which ribosome production is altered [4]. Skeletal muscle develops from precursor cells, called satellite cells or myoblasts, which are capable of rapid proliferation and differ both biochemically and morphologically from multi-nucleated myofibres [4]. Differentiation is characterized by a cessation of growth, fusion of cell membranes, synthesis of muscle-specific proteins and a reduction in RNA production [5]. Difficulties in studying muscle development *in situ* and the heterogeneity of primary cultures, which contain both fibroblasts and other non-muscle cells, have led to the establishment of homogeneous myoblast cultures capable of both continuous growth and differentiation. The first permanent muscle cell line, L6, established by Yaffe from skeletal muscle tissue of newborn rats [6] has been extensively investigated for over 20 years.

Regulation of rRNA synthesis following differentiation of rat L6 myoblasts

While rat L6 myoblasts can be maintained in continuous culture, a shift in the growth conditions induces their terminal differentiation which is complete within 3–4 days. This process closely resembles the formation of myofibres during muscle development: cell multiplication ceases and the myoblasts fuse into multi-nucleated myotubes which synthesize muscle-specific proteins [6]. Either depletion of a specific growth factor(s) or the accumulation of a differentiation factor produced by the myoblasts is apparently the primary determinant for differentiation [7, 8].

Although the rates of mRNA and protein production are constant during myotube formation, a shift takes place in the population of proteins being synthesized [5]. A decrease in total RNA synthesis also occurs following myotube formation, suggesting a drop in rRNA production [9, 10]. Since the total RNA content per nucleus of rat L6 myoblasts and myotubes remains constant, no reduction in ribosome content is apparent following differentiation [11]. Furthermore, the fixed amount of rRNA in the two populations implies a close correlation between rRNA synthesis and cell growth. Since no change in the rate of turnover of cytoplasmic rRNA was noted, a down-shift in the rate of rRNA synthesis must occur.

A comparison of ^{3}H-uridine incorporation into the 45S rRNA precursor *in vivo* demonstrated that the rate of gene transcription in myotubes was approximately 5-fold less than observed in myoblasts [11]. Neither degradation of the precursor rRNA nor changes in the cellular UTP pools occurred. In addition, processing of the precursor to 18S and 28S rRNA was apparently unaffected by differentiation. These observations, indicating that rRNA synthesis is significantly reduced following myotube formation, have been confirmed and extended by studies from this laboratory [12, 13].

A comparison of rRNA synthesis by isolated nuclei demonstrated that rRNA gene transcription decreased by 80% in differentiated myotubes (Table 1). To identify the factors involved in regulating rRNA synthesis in myotubes, a cell-free system specific for rRNA gene transcription was established [13]. The development of *in vitro* transcription assays has led to the identification of specific DNA sequences and transcription factors involved in regulating gene expression following changes in growth. Although a number of trans-acting factors which participate in the formation of the transcription complex have been described, only two, designated TFIC and TFID, appear to be essential for transcription *in vitro* [14, 15].

In a cell-free system, myotube extracts display a significantly reduced capacity to transcribe the rat rRNA gene, mimicking the observations made *in vivo* [13]. This difference in transcriptional capacity occurs without a change in the activity of RNA polymerase I, implying the involvement of transcription factors in the regulation process. TFID is a promoter-binding protein essential for formation of the transcription initiation complex [14]. TFIC, on the other hand, is a factor that determines whether RNA polymerase I is competent for rRNA gene transcription [15]. Although the molecular identity of TFIC has not been determined, it enables RNA polymerase I to interact with TFID and thus coordinates its binding to the transcription initiation site. Analysis of these factors in myoblasts and myotube was accomplished by adding TFIC and TFID, prepared by phosphocellulose column chromatography of nuclear extracts from murine lymphosarcoma L1210 cells, to each extract and measuring relative transcript formation (Persson, Zahradka, Rothblum and Sells, unpublished observations). This investigation demonstrated that TFIC was limiting in myotube extracts and probably the primary target for regulation. Similar observations documented in other systems have demonstrated that control of rRNA synthesis involves modulation of TFIC [15–18]. Although control of rRNA synthesis in differentiated myotubes appears to be exerted through TFIC, it should be noted that TFID levels were reduced considerably

Table 1. RNA synthesis by nuclei isolated from proliferating myoblasts and differentiated myotubes. Transcription of ribosomal precursor genes by myoblast and myotube nuclei was assessed by hybridization of ^{32}P-labelled RNA synthesized *in vitro* with plasmids containing r-protein L32, 45S rRNA precursor and 5S rRNA gene sequences. Results are presented as a ratio for comparison of gene activity in the two cell populations. The transcriptional activity of the r-protein L32 gene was set to 1.0 [25].

Gene sequences	Relative transcription (myotube/myoblast)
r-protein L32	1.00
rRNA	0.30
5S rRNA	1.09

in myotube extracts, but the amount present remained sufficient to saturate the template and thus was not limiting for transcription (Persson, Zahradka, Rothblum and Sells, unpublished observations).

R-protein synthesis during myogenesis

A change in the demand for ribosomes may not only be reflected by alterations in the synthesis of rRNA, but also in the production of r-protein and 5S rRNA. Although transcriptional control of r-protein synthesis occurs in yeast in response to changes in growth conditions [19–21], r-protein production in most mammalian systems is usually regulated by other mechanisms. A down-shift in r-protein production, nevertheless, has been reported during differentiation of mouse MM14 myoblasts which results, in part, from a decline in the rate of gene transcription, as well as a decrease in the translational efficiency of r-protein mRNAs [22].

R-protein synthesis in the rat L6 myoblast cell line, however, is not altered by myogenesis [12, 23]. Consequently, the coordinated synthesis of r-protein and rRNA observed in myoblasts is lost following terminal differentiation. While the rate of r-protein synthesis continues unchanged after rRNA production decrease, excess r-proteins, however, do not accumulate. R-proteins, with half-lives decreasing from > 25 h in myoblasts to 30 min in myotubes, are therefore rapidly degraded. Regulation of r-protein production at the post-translational level is supported by the observation of similar rates of r-protein mRNA synthesis by i) nuclei isolated from myoblasts and myotubes [12] and ii) cell-free extracts in an *in vitro* transcription assay [13]. Since both the gene transcription rate and mRNA stability (Table 2) are unchanged, degradation of excess r-proteins in myotubes implies that a specific mechanism is required for reducing r-protein levels without affecting the stability of other proteins.

Although other myoblast systems are available, only the mouse MM14 myoblast cell line has been studied with respect to ribosome biogenesis. Following MM14 myoblast differentiation, both rRNA and r-protein gene transcription are decreased following terminal differentiation [22, 24]. While the differences noted between rat L6 and mouse MM14 myoblasts may reflect their developmental lineage, the MM14 myoblast system can be used for examining the regulation of r-protein gene expression and its coupling with rRNA.

Table 2. Half-lives of ribosomal protein mRNAs and 5S rRNA in rat L6 myoblasts and myotubes. The half-lives ($t_{1/2}$) of r-protein mRNAs in myoblasts and myotubes were measured by determining the relative level of each mRNA by Northern blot analysis following inhibition of synthesis with actinomycin D [12]. For $t_{1/2}$ of 5S rRNA, myoblasts and myotubes were pulse-labelled with ^{3}H-uridine and the amount of radioactive 5S rRNA/total 5S rRNA was determined after incubation with actinomycin D, preventing re-incorporation of label, during the chase period [25].

RNA		$t_{1/2}$ (hours)	
		myoblasts	myotubes
r-protein mRNA	S16	11.1 ± 0.4	9.5 ± 0.6
	L18	12.3 ± 0.5	12.4 ± 1.2
	L32	9.6 ± 0.6	10.0 ± 1.2
5S rRNA		6.2	1.1

5S rRNA synthesis in differentiated myotubes

The genes for 5S rRNA, unlike those for the 45S rRNA precursor, are located outside the nucleolus. The newly synthesized 5S rRNA must therefore be transported to the nucleolus where, in combination with r-proteins, it is assembled into the large ribosomal subunit prior to its exit from the nucleus. The levels of 5S rRNA appear not to be controlled as closely as rRNA and r-protein, with an excess of up to 20% present in many cells. An excess was also observed in both L6 myoblasts and myotubes as indicated by the 5S rRNA/rRNA ratio which remained unchanged following differentiation [25].

As with rRNA and r-protein, accumulation of 5S rRNA must be reduced in myotubes. Since the rates of 5S rRNA synthesis by nuclei from myoblasts and myotubes are similar, the level of this

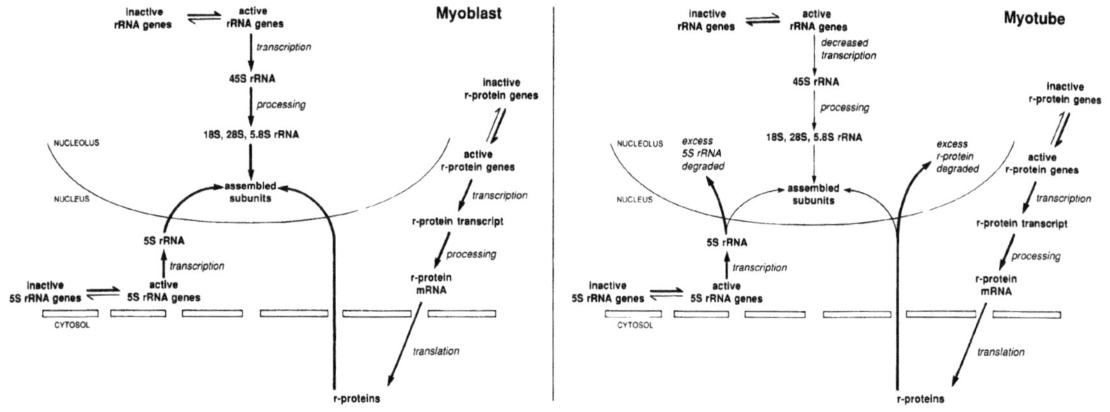

Fig. 1. Ribosome biogenesis in myoblasts and myotubes.

ribosomal precursor is apparently not regulated at the level of transcription (Table 1). Furthermore, the transcriptional capacity of S100 extracts prepared from myoblasts is comparable to myotube extracts, indicating neither transcription initiation nor elongation was altered as a result of the differentiation process [25]. Transcriptional control of 5S rRNA production has been observed in several systems [25–28]. Suppression of 5S rRNA gene transcription in serum-starved cells, for example, may relate to changes such as starvation, in which high levels of protein synthesis do not occur [26] in contrast to the high protein synthetic rate in myotubes. Other genes transcribed by RNA polmerase III may be controlled by this mechanism following differentiation.

Once 5S rRNA is synthesized, it is incorporated into the 60S ribosomal subunit. If excess 5S rRNA is produced but does not accumulate, decreased stability of the product may account for this paradox. A comparison of 5S rRNA turnover in the two populations demonstrated a significant decrease in its stability in myotubes (Table 2). The regulation of 5S rRNA levels thus involves degradation of the excess product synthesized in myotubes. Since excess 5S rRNA and r-proteins are degraded in myotubes, there is an uncoupling of the balanced production of ribosomal precursor molecules observed in myoblasts.

Ribosome biogenesis and muscle tissue

Studies of gene regulation during muscle development are more easily accomplished with cell culture systems. How closely these cell lines reflect the events in muscle tissue, however, is difficult to determine. Differences exist between the available cell lines, suggesting that their establishment, often requiring treatment with mutagens, result in altered characteristics. In addition, muscle satellite cells, especially from embryonic sources, are heterogeneous and may develop into smooth, cardiac or skeletal muscle [29]. To determine whether the mechanisms regulating ribosome biogenesis during terminal differentiation of rat L6 myoblasts reflect those occurring during muscle development *in situ*, a cell-free transcription system using nuclear extracts prepared from rat skeletal muscle tissue was developed [30]. Many similarities in gene expression were found for L6 myotubes and muscle, including the pattern of r-protein and rRNA gene transcription [30, 31]. Of note, transcription of the rRNA and r-protein genes are uncoupled, suggesting that post-transcriptional degradation of r-proteins may also occur in muscle tissue. Although this analysis was limited to *in vitro* studies, the results suggest that regulation of ribosome biogenesis involves similar mechanisms in both L6 myotubes and skeletal muscle tissue.

Summary

Ribosome biogenesis involves synthesis of three classes of precursor molecules and their assembly into a functional unit. Since the shift from a population of proliferating myoblasts to non-dividing myotubes requires a substantial reduction in ribosome production, the rat L6 myoblast cell line has proven itself useful as a suitable model system for investigating those mechanism(s) involved in regulating the synthesis and assembly of ribosomal precursor molecules. While synthesis of ribosomal components is coordinated in myoblasts, their balanced production is lost in myotubes. Following differentiation, rRNA synthesis decreases, whereas r-protein gene transcription remains constant and the r-protein mRNAs are translated. The r-proteins, synthesized in excess, are then degraded. Similarly, 5S rRNA gene transcription continues unchanged while the half-life of newly synthesized 5S rRNA is reduced, implying the degradation of 5S rRNA molecules not incorporated into ribosomes is enhanced. As summarized in Fig. 1, rRNA transcription appears to be the primary event controlling ribosome biogenesis in myotubes. This mechanism may also function to control ribosome levels in muscle tissue. Since a drop in rRNA transcription apparently results from changes in TFIC, as described in other systems, a common regulatory step may operate in eucaryotic cells in response to changes in ribosome demand. More information, both from additional myoblast cell lines and other regulated systems, is necessary before a consensus can be reached concerning the mechanisms regulating ribosome biogenesis.

Acknowledgements

Studies from this laboratory were supported by grants from the Medical Research Council (Canada) and the Natural Sciences and Engineering Research Council.

References

1. Becker H, Stanners CP, Kudlov JE: Control of macromolecular synthesis in proliferating and resting syrian hamster cells in monolayer culture. J Cell Physiol 77: 43–50, 1971
2. Bowman LH, Emerson Jr CP: Post-transcriptional regulation of ribosome accumulation during myoblast differentiation. Cell 10: 587–596, 1977
3. Dabeva MD, Dudov KP: Transcriptional control of ribosome production in regenerating rat liver. Biochem J 208: 101–108, 1982
4. Fischman DA: The synthesis and assembly of myofibrils in embryonic muscle. Curr Topics Develop Biol 5: 235–280, 1970
5. Merlie JP, Buckingham ME, Whalen RG: Molecular aspects of myogenesis. Curr Topics Develop Biol 11: 61–114, 1977
6. Yaffe D: Retention of differentiation potentialities during prolonged cultivation of myogenic cells. Proc Natl Acad Sci USA 61: 477–483, 1968
7. Wright WE, Sassoon DA, Lin VK: Myogenin, a factor regulating myogenesis, has a domain homologous to myoD. Cell 56: 607–617, 1989
8. Olson EN, Sternberg E, Hu JS, Spizz G, Wilcox C: Regulation of myogenic differentiation by type β transforming growth factor. J Cell Biol 103: 1799–1805, 1986
9. Yaffe D, Fuchs S: Autoradiographic study of the incorporation of uridine-^{3}H during myogenesis in tissue culture. Develop Biol 15: 33–50, 1967
10. Clissold P, Cole RJ: Regulation of ribosomal RNA synthesis during mammalian myogenesis in culture. Expt Cell Res 80: 159–169, 1973
11. Krauter KS, Soeiro R, Nadal-Ginard B: Transcriptional regulation of ribosomal RNA accumulation during L_6E_9 myoblast differentiation. J Mol Biol 134: 727–741, 1979
12. Jacobs FA, Bird RC, Sells BH: Differentiation of rat myoblasts. Regulation of turnover of ribosomal proteins and their mRNAs. Eur J Biochem 150: 255–263, 1985
13. Zahradka P, Sells BH: Transcription factors mediate rRNA synthesis during myogenesis. Eur J Biochem 171: 37–43, 1988
14. Tower J, Culotta VC, Sollner-Webb B: Factors and nucleotide sequences that direct ribosomal DNA transcription and their relationship to the stable transcription complex. Mol Cell Biol 6: 3451–3462, 1986
15. Tower J, Sollner-Webb B: Transcription of mouse rDNA is regulated by an activated subform of RNA polymerase I. Cell 50: 873–883, 1987
16. Buttgereit D, Pflugfelder G, Grummt I: Growth-dependent regulation of rRNA synthesis is mediated by a transcription initiation factor (TIF-IA). Nucl Acids Res 13: 8165–8180, 1985
17. Cavanaugh AH, Thompson Jr EA: Hormonal regulation of transcription of rDNA: glucocorticoid effects upon initiation and elongation *in vitro*. Nucl Acids Res 13: 3357–3369, 1985

18. Bateman E, Paule MR: Regulation of eukaryotic ribosomal RNA transcription by RNA polymerase modification. Cell 47: 445–450, 1986
19. Kim CH, Warner JR: Mild temperature shock alters the transcription of a discrete class of *Saccharomyces cerevisiae* genes. Mol Cell Biol 3: 457–465, 1983
20. Donovan DM, Pearson NJ: Transcriptional regulation of ribosomal proteins during a nutritional upshift in *Saccharomyces cerevisiae.* Mol Cell Biol 6: 2429–2435, 1986
21. Herruer MH, Mager WH, Woudt LP, Nieuwint RTM, Wassenaar GM, Groeneveld P, Planta RJ: Transcriptional control of yeast ribosomal protein synthesis during carbon-source upshift. Nucl Acids Res 15: 10133–10144, 1987
22. Agrawal MG, Bowman LH: Transcriptional and translational regulation of ribosomal protein formation during mouse myoblast differentiation. J Biol Chem 262: 4868–4875, 1987
23. Krauter KS, Soeiro R, Nadal-Ginard B: Unco-ordinate regulation of ribosomal RNA and ribosomal protein synthesis during L_6E_9 myoblast differentiation. J Mol Biol 142: 145–159, 1980
24. Bowman LH: rDNA transcription and pre-rRNA processing during the differentiation of a myoblast cell line. Develop Biol 119: 152–163, 1987
25. Zahradka P, Larson DE, Sells BH: Post-transcriptional regulation of 5S rRNA synthesis following myoblast differentiation. Eur J Biochem 184: 261–266, 1989
26. Tower J, Sollner-Webb B: Polymerase III transcription factor B activity is reduced in extracts of growth-restricted cells. Mol Cell Biol 8: 1001–1005, 1988
27. Fradkin LG, Yoshinaga SK, Berk AJ, Dasgupta A: Inhibition of host cell RNA polymerase III-mediated transcription by poliovirus: Inactivation of specific transcription factors. Mol Cell Biol 7: 3880–3887, 1987
28. White RL, Stott D, Righby PWJ: Regulation of RNA polymerase III transcription in response to F9 embryonal carcinoma stem cell differentiation. Cell 59: 1081–1092, 1989
29. Cossu G, Molinaro M: Cell heterogeneity in the myogenic lineage. Curr Topics Develop Biol 23: 185–208, 1987
30. Zahradka P, Larson DE, Sells BH: RNA polymerase II-directed gene transcription by rat skeletal muscle nuclear extracts. Expt Cell Res 185: 8–20, 1989
31. Larson DE, Zahradka P, Stein G, Stein J, Sells BH: Down-regulation of histone H3 and H4 gene transcription in differentiated L6 myotubes. Biochim Biophys Acta 1009: 177–183, 1989

Address for offprints: B.H. Sells, Dean's Office, Biological Sciencxe, University of Guelph, Guelph, Ontario, Canada N1G 2W1

Molecular and Cellular Biochemistry **104**: 195–200, 1991.

Glucocorticoid regulation of rRNA synthesis

Pramod B. Mahajan and E. Aubrey Thompson
Department of Human Biological Chemistry and Genetics, University of Texas Medical Branch, Galveston, Texas, USA

Key words: rRNA synthesis, transcription *in vitro,* glucocorticoids, dexamethasone, transcription factors

Abstract

Glucocorticoids inhibit transcription of the genes encoding rRNA (rDNA) in P1798 lymphoma cells. This is due to a decrease in the amount or activity of a transcription factor, called TFIC. TFIC has been purified to apparent homogeneity and the properties of this protein have been investigated in detail. TFIC is tightly associated with RNA polymerase I. The data indicate that TFIC is a *bona fide* initiation factor that is required for formation of the first phosphodiester bond of nascent pre-rRNA. Extracts from dexamethasone-treated cells are devoid of this factor and cannot form initiated complexes *in vitro.*

Introduction

Glucocorticoids inhibit the synthesis of mature rRNA in P1798 cells, a murine lymphoma line of thymic origin [1, 2]. Greater than 95% inhibition is observed within 24 hr after addition of 0.1 μM dexamethasone. Inhibition of rRNA synthesis is primarily, if not exclusively, due to inhibition of transcription of those genes that encode the primary transcript of rDNA (45S pre-rRNA). A large number of mechanisms could be proposed to account for this phenomenon. The mouse rRNA promoter contains a potential glucocorticoid receptor binding site (GRE) [3] and it is formally possible that the receptor binds to the promoter and directly represses transcription. Alternatively, glucocorticoids might induce the expression of a repressor. Indirect modification of the chromatin could be involved. Glucocorticoids could induce a nuclease that rapidly degraded nascent rRNA. Finally, glucocorticoids could inhibit the expression of an essential rDNA transcription factor. In order to differentiate among these alternative hypotheses, it was necessary to develop a system for analysis of transcription of rDNA *in vitro* [4–6]. This paper describes the use of such an assay system to elucidate certain aspects of the molecular mechanism that accounts for regulation of rRNA synthesis in malignant thymocytes.

Materials and methods

P1798 cells were maintained in culture, as described elsewhere [1]. Mid-log phase cells were used for all experiments. S100 extracts were prepared [6] and used to transcribe the cloned mouse rDNA promoter [7]. The details of heparin- and KCl-start transcription experiments are given elsewhere [8–11], as are the procedures for purification of TFIC [11].

Results

The data shown in Fig. 1 illustrate glucocorticoid regulation of rRNA synthesis in P1798 cells. Labeling of mature 28S rRNA is reduced by >95% within 24 hr after addition of the steroid, as shown in Fig. 1A. Labeling of mature (i.e., ribosomal) 5S

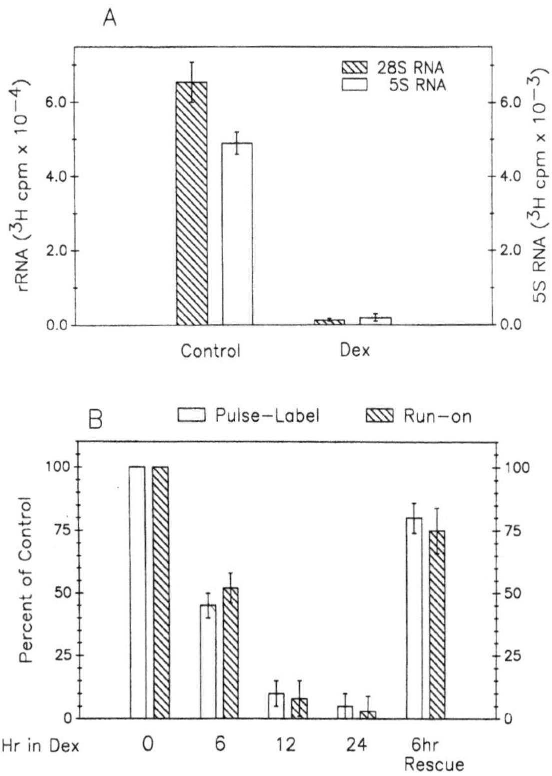

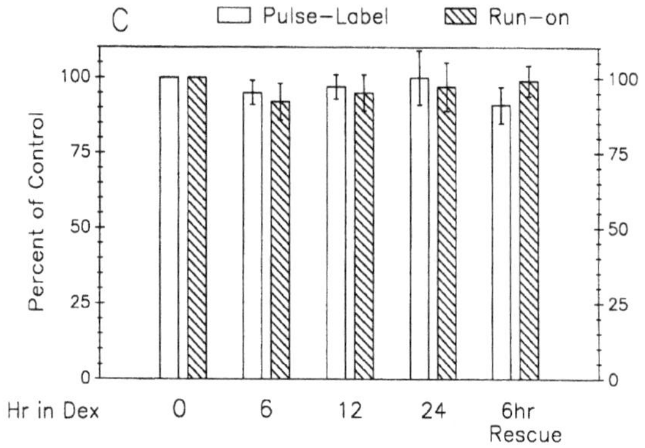

Fig. 1. Glucocorticoid effects upon the synthesis of 5S and rRNA species. Panel 1A shows the results of an experiment in which P1798 cells were treated for 24 hr with 0.1 μM dexamethasone (Dex). Thereafter, the cells were labeled for 3 hr with 50 μCi/ml [^{3}H]uridine. Labeled RNA was extracted and 28S and 5S RNA were resolved by agarose gel electrophoresis of 5 μg of total RNA under denaturing conditions. RNAs were identified by co-migration with authentic standards. These were cut from agarose gels and quantified by liquid scintillation counting. Panel 1B shows pulse-labeling and nuclear run-on analysis of the synthesis of 45S pre-rRNA. Pulse labeling was carried out for 15 min and the products were analyzed by solution hybridization [1]. Nuclear run-on was assayed under standard conditions [13]. Panel 1C shows pulse labeling and nuclear run-on transcription of 5S RNA genes.

RNA is reduced to a similar extent. Inhibition of 28S rRNA synthesis is accompanied by inhibition of pulse labeling of pre-rRNA and is associated with reversible inhibition of nuclear run-on transcription of rDNA (Fig. 1B). Neither pulse labeling nor run-on transcription of the genes encoding 5S RNA are inhibited (Fig. 1C), suggesting that newly synthesized 5S RNA is rapidly degraded so that there is no accumulation of mature 5S RNA. The mechanism that accounts for inhibition of transcription of rDNA has been studied in detail, using the technique of run-off transcription of cloned mouse rDNA [4–6]. Fig. 2 contains representative data from experiments in which extracts from control and dexamethasone-treated cells were assayed for ability to form the 292 nucleotide rDNA transcript *in vitro*. Extracts from control cells transcribe rDNA, as evidenced by more or less linear accumulation of the product for 20–30 min (circles). The rate of product accumulation decreases subsequently. Extracts from dexamethasone-treated cells are virtually inactive in this respect (triangles). Control and dexamethasone-treated extracts transcribe the gene encoding 5S RNA with more or less equal efficiency [6]. A mixing experiment was carried out, in addition. Equal aliquots of control and treated extract were incubated with template, under appropriate conditions [6], mixed, and transcription was measured (Fig. 2, squares). One notes that the initial velocity of transcription in the mixture was significantly greater than that observed in the control, alone. However, the total amount of product formed was additive (i.e., equal to that formed in the control extract).

The data shown in Fig. 2 reveal several important clues concerning the mechanism that accounts for inhibition of transcription by glucocorticoids. Extracts of this sort are very inefficient in forming nucleosomes *in vitro*. Consequently, the observation that the transcriptional activity of these extracts recapitulates that observed in cells tends to discount the hypothesis that glucocorticoid inhibition is attributable to modification of the chromatin structure. Furthermore, the observation that the activity of such extracts is additive is not consistent with the hypothesis that glucocorticoid-treated

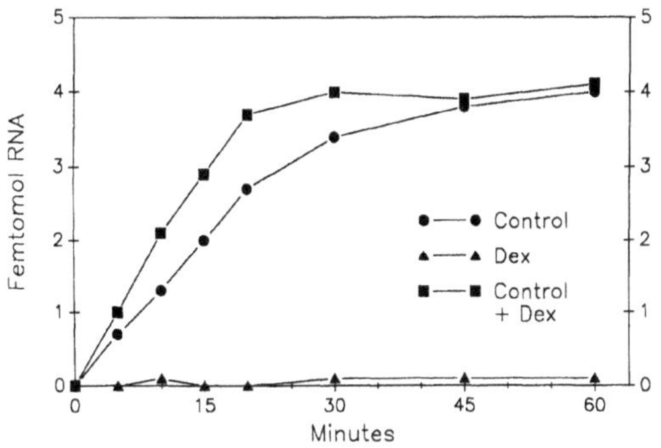

Fig. 2. Transcription of mouse rDNA *in vitro*. The cloned mouse rRNA promoter was transcribed *in vitro* and the products were quantified as described in Materials and methods.

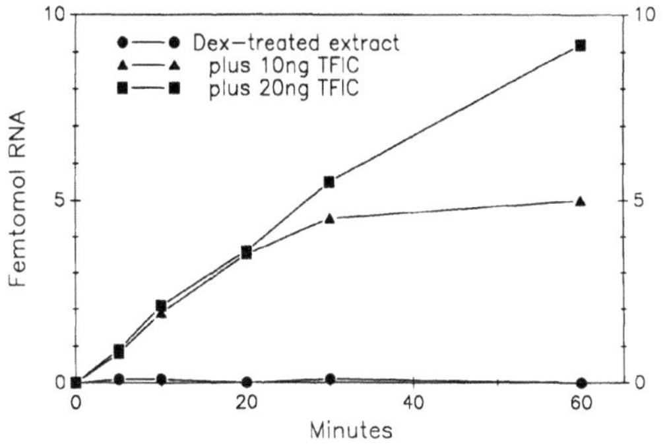

Fig. 3. Transcription of rDNA by extracts from dexamethasone-treated cells. An extract was prepared from cells that had been treated with dexamethasone for 24 hr. This extract was assayed for ability to transcribe rDNA *in vitro*. Reconstitution of transcription was affected by addition of highly purified TFIC.

cells contain a repressor of transcription. By the same argument, these data tend to rule out the hypothesis that apparent differences in transcription are due to activation of an enzyme that degraded the template or the products of transcription.

The simplest explanation of the data shown in Fig. 2 holds that extracts from glucocorticoid-treated cells are depleted of a substance that is required for transcription of rDNA. This hypothesis has been confirmed by purification and characterization of the active principle, a transcription factor isolated from control extracts by virtue of its ability to reconstitute transcription in extracts from hormone-treated cells [7, 11]. This factor is called TFIC, according to the convention proposed by Mishima *et al.* [12]. TFIC activity co-purifies with three polypeptides (Mr 55, 50, & 42K). These are tightly associated with RNA polymerase I and, in so far as can be determined by gel-mobility assays, are devoid of sequence-specific DNA-binding activity. The data suggest that TFIC interacts with core RNA polymerase I to give rise to an 'active' form of the enzyme that is capable of recognizing the rDNA pre-initiation complex.

Extracts from glucocorticoid-treated P1798 cells are depleted of TFIC activity and cannot transcribe rDNA. The ability of such extracts to transcribe rDNA can be reconstituted by addition of TFIC, as shown in Fig. 3. Extracts from hormone-treated cells cannot synthesize the 292 nucleotide run-off transcript of rDNA (circles). Addition of 10 ng of TFIC reconstitutes transcription (triangles). Addition of 20 ng of the factor does not affect the initial velocity of transcription but does cause an increase in the total amount of product formed within a 60 min incubation period.

The data shown in Fig. 3 are consistent with the hypothesis that TFIC is an initiation factor. This hypothesis was confirmed by measuring single round transcription *in vitro* [8–10]. The basic features of this assay are illustrated by the data shown in Fig. 4, which indicate that formation of heparin- or KCl-resistant transcription complexes requires pre-incubation with the first two nucleotides of nascent rRNA. When control extracts are incubated with ATP and CTP, the first phosphodiester bond is formed, giving rise to an initiated complex which is resistant to heparin or KCl. If such complexes are fortified by addition of all four nucleoside triphosphates, an elongation complex is formed that gives rise to one full length transcript per initiated complex. By measuring the number of such transcripts formed in an heparin- or KCl-start experiment, one may estimate the number of initiation complexes formed under a given set of circumstances. The data in Fig. 4 indicate that a representative extract from control cells is capable of forming about 0.1 fmol of initiated complex upon the rRNA promoter.

Extracts from dexamethasone-treated P1798 cells do not form initiated complexes, as indicated by the heparin- and KCl-start transcription data

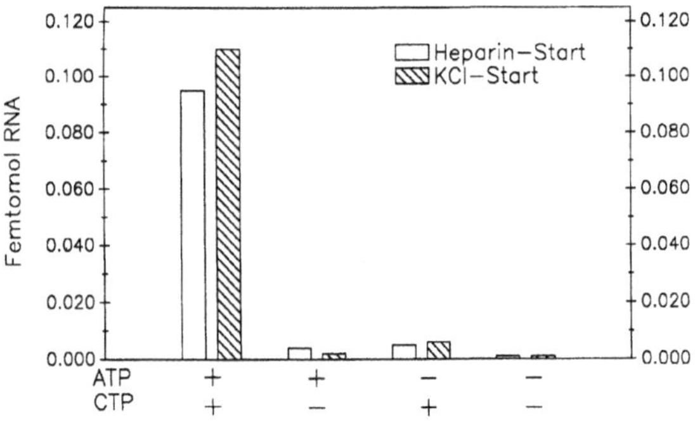

Fig. 4. Heparin- and KCl-start transcription in control extracts. An extract from control cells was assayed for single round transcription of rDNA, as described elsewhere [8].

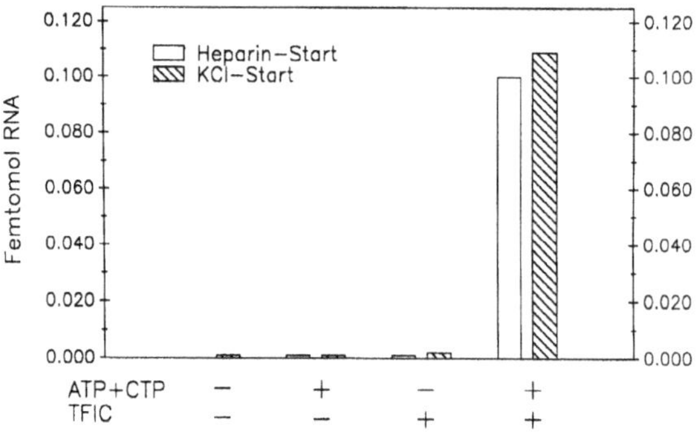

Fig. 5. Heparin- and KCl-start transcription in extracts from dexamethasone-treated cells. Single round transcription was measured in an extract prepared from cells that had been exposed to dexamethasone for 24 hr. Transcription was reconstituted by addition of 10 ng of highly purified TFIC. The conditions employed for transcription and quantification of the products are described elsewhere [8].

shown in Fig. 5. However, the ability to form such complexes can be restored by addition of TFIC. Approximately 0.1 fmol of initiated complexes was formed by addition of 10 ng of TFIC. Addition of greater amounts of the factor does not increase the number of initiated complexes that can be formed (not shown).

Discussion

Glucocorticoid inhibition of ribosome synthesis in P1798 cells is associated with inhibition of transcription of rDNA. Transcription of 5S RNA genes is not inhibited, but the nascent 5S RNA is degraded, presumably due to defective ribosome assembly occasioned by decreased synthesis of mature rRNAs and ribosomal proteins [13]. The data indicate that inhibition of transcription of rDNA, in and of itself, is sufficient to account for inhibition of synthesis of mature rRNAs. However, it is known that glucocorticoids inhibit transcription of U3 snRNA genes in P1798 cells [14], and processing of the primary transcript may also be impaired.

The mechanism that accounts for inhibition of transcription by glucocorticoids appears to be novel. Glucocorticoids cause a decrease in the amount or activity of an essential transcription factor, TFIC. This effect appears to be relatively specific for glucocorticoids, as the amount of TFIC activity does not decrease in P1798 cells that have ceased to proliferate in response to depletion of serum growth factors (A. Thompson, unpublished observation). The mechanism that accounts for decreased TFIC activity is unknown. Preliminary data suggest that extracts from dexamethasone-treated cells contain TFIC polypeptides, although TFIC activity cannot be detected. This suggests that an inactivation mechanism may be involved. It is known, however, that RNA polymerase I from hormone-treated cells is capable of specific initiation if supplemented with TFIC [6, 7, 11]. This suggests that modification of TFIC, rather than the polymerase, is a likely mechanism to account for failure of initiation of transcription.

A large number of studies have been carried out to elucidate the biochemical role of TFIC in transcription of rDNA. A few salient observations have been recapitulated in this paper, with the objective of emphasizing those unanswered questions that prove to be of interest with respect to this protein. Of particular interest is the suggestion that TFIC may play a stoichiometric role in synthesis of pre-rRNA. An indication of this may be obtained by consideration of the results of the mixing experiment shown in Fig. 2. One notes that the initial velocity in mixtures is approximately twice the sum of that observed in control and hormone-treated extracts. It is known that extracts from glucocorticoid-treated cells are capable of forming pre-initiation complexes. Consequently, the concentration of such complexes in a mixing experiment would be about twice that predicted in those reactions in which the activity of the individual extracts was measured. The data suggest that, in the presence of excess TFIC, the initial velocity of transcription is determined by the number of pre-initiation complexes that are formed by a given extract.

The single round transcription experiments indicate that the number of pre-initiation complexes formed in a given control extract is quite small, on the order of 0.1 fmol/50 μg of extract protein. Nevertheless, the initial velocity of transcription is 0.1–0.2 fmol/min/50 μg protein and 2–3 fmol of RNA can be synthesized within 10–15 min. This is consistent with the conclusion that a substantial degree of re-initiation occurs in these extracts. We propose that the extent of re-initiation is determined by the amount of TFIC. This paper contains two observations that form the basis of this proposition. Initially, one notes whereas the initial velocity increases in mixing experiments (Fig. 2), the total amount of product formed does not. Furthermore, in reconstitution experiments (Fig. 3), the amount of product formed, but not the initial velocity, is determined by the amount of TFIC added. These observations have been confirmed in greater detail by experiments published elsewhere [9, 11].

Based upon these data, we propose the following working hypothesis. Initiation by RNA polymerase I requires association between the enzyme and TFIC. It is not known if TFIC is necessary for binding of the polymerase to the pre-initiation complex. An alternative hypothesis would hold that TFIC is not required for binding but is necessary for melting of the DNA and formation of the first phosphodiester bond of the nascent transcript. Glucocorticoids may regulate the interaction between TFIC and the polymerase. Alternatively, the activity of the factor may be attenuated. We propose that TFIC functions in a stoichiometric manner *in vitro,* such that the number of nascent transcripts formed is directly related to the number of TFIC molecules available. This is an attractive mechanism that could account for coordinate regulation of translation and rRNA synthesis. However, no data preclude the possibility that TFIC dissociates from the polymerase subsequent to initiation and is diluted to the extent that its physicochemical activity (if not its biochemical activity) is lost due to mass action. Detailed kinetic analysis of the process of initiation indicates that the initiated complex is not kinetically equivalent to the elongation complex [9], suggesting that the polymerase undergoes a significant change in properties during promoter escape. It is tempting to speculate that association with TFIC is required for promoter recognition and dissociation of TFIC may be involved in promoter escape. Experiments are currently in progress to address these, and other, questions with respect to the biochemistry of TFIC and the role of steroids in regulating the activity of this protein.

References

1. Cavanaugh AH, Thompson EA: Hormonal regulation of transcription of rDNA I: inhibition of transcription during glucocorticoid-mediated inhibition of proliferation. J Biol Chem 258: 9768–9773, 1983
2. Thompson EA: Glucocorticoid inhibition of gene expression and cellular proliferation. Cancer Res 49: 2259–2265, 1989
3. Beato M: Steroid receptors as gene regulators. In Advances in Gene Technology: Molecular Biology of the Endocrine System (Puett *et al.*, eds) ICSU Short Reports Vol 4, ICSU Press, pp. 250–253, 1986
4. Grummt I: Specific transcription of mouse ribosomal DNA in a cell-free system that mimics control *in vivo.* Proc Natl Acad Sci USA 78: 727–731, 1981
5. Miller KG, Sollner-Webb B: Transcription of mouse rRNA genes by RNa polymerase I: *In vitro* and *in vivo* initiating and processing sites. Cell 27: 165–174, 1981
6. Cavanaugh AH, Gokal PK, Lawther RP, Thompson EA: Glucocorticoid inhibition of transcription of the DNA encoding rRNA (rDNA) in lymphosarcoma P1798 cells. Proc Natl Acad Sci USA 81: 718–721, 1984
7. Cavanaugh AH, Thompson EA: Hormonal regulation of transcription of rDNA III: Glucocorticoid effects upon initiation and elongation *in vitro.* Nucleic Acids Res 13: 3357–3369, 1985
8. Gokal PK, Mahajan PB, Thompson EA: Hormonal regulation of transcription of rDNA VI: Formation of initiated complexes by RNA polymerase I *in vitro.* J Biol Chem 265: 16234–16243, 1990
9. Mahajan PB, Gokal PK, Thompson EA: Hormonal regulation of transcription of rDNA VII: The role of TFIC in formation of initiated complexes. J Biol Chem 265: 16244–16247, 1990

10. Cavanaugh AH, Thompson EA: Hormonal regulation of transcription of rDNA IV: initiation of transcription *in vitro*. J Biol Chem 261: 12738–12744, 1986
11. Mahajan PB, Thompson EA: Hormonal regulation of transcription of rDNA V: Purification and characterization of the hormone-regulated transcription factor TFIC. J Biol Chem 265: 16225–16233, 1990
12. Mishima Y, Financsek L, Kominami R, Muramatsu M: Fractionation and reconstitution of factors required for accurate transcription of mammalian ribosomal RNA genes: identification of a species-dependent factor. Nucleic Acids Res 10: 6659–6670, 1982
13. Meyuhas O, Thompson EA, Perry RP: Glucocorticoids selectively inhibit the translation of ribosomal protein mRNA in P1798 lymphosarcoma cells. Moll Cell Biol 7: 2691–2699, 1987
14. Meyuhas O, Baldin V, Bouche G, Amalric F: Glucocorticoids repress ribosome biosynthesis in lymphosarcoma cells by affecting gene expression at the level of transcription, posttranscription and translation. Biochim Biophys Acta 1049: 38–44, 1990

Address for offprints: A. Thompson, Department of Human Biological Chemistry and Genetics, University of Texas Medical Branch, Galveston, TX 77550, USA

Molecular and Cellular Biochemistry **104**: 201–207, 1991.

Serum, insulin and phorbol esters stimulate rRNA and tRNA gene expression in both dividing and nondividing *Drosophila* cells

Heather W. Weber, Shawn Vallett, Lorna Neilson, Mark Grotke, Yesu Chao, Mark Brudnak, Alan San Juan and Maria Pellegrini
Molecular Biology Section, University of Southern California, Los Angeles, California, USA

Key words: ribosomal RNA, tRNA, phorbol ester, male accessory gland, RNA polymerase I

Abstract

The expression of genes that code for the large ribosomal RNAs (rRNAs) and tRNAs can be regulated by calcium, serum, insulin and a tumor-promoting phorbol ester, TPA. These effectors can rapidly alter rRNA and tRNA synthesis in dividing and nondividing *Drosophila* cells. In an *in vitro* assay system of the nondividing cells of the male accessory glands, calcium, insulin and TPA were shown to increase both rRNA and tRNA synthesis. Exposure of actively dividing *Drosophila* culture cells to differing serum concentrations or TPA also altered rRNA and tRNA synthesis. Nuclear run-on assays demonstrate that the exposure of these cells to increased serum concentrations coordinately alters RNA polymerase I loading on both 18S and 28S rDNA. These data indicate that calcium, growth factors and a tumor-promoter each can signal changes in ribosomal and tRNA gene expression.

Introduction

Protein synthesis is an essential function in all cells. The structural RNAs of the ribosome along with tRNAs are key components of the protein synthesis machinery. We have obtained data that show the coordinate regulation of the synthesis of rRNAs and tRNAs in the *Drosophila* system. Reagents such as serum, and the tumor-promoting phorbol ester, TPA, rapidly induce transcription of rRNA and tRNA genes in *Drosophila* cells. Although mechanisms for the induction of RNA Polymerase II gene expression by serum and phorbol esters have been well-studied, regulation of RNA Pol I and Pol III gene expression by these effectors has not been examined in any system.

We now show that various reagents which mimic *in vivo* situations can either up- or down-regulate ribosome synthesis in *Drosophila* cells. This regulation includes the control of rRNA and tRNA transcription in both mitotically active and terminally differentiated (nonmitotic) cells. Our results show that 5S and tRNA levels are increased in concert with the 18S and 28S rRNAs when *Drosophila* cells are treated with serum or TPA.

Materials and methods

Accessory gland isolation and labeling

Wild type *D. melanogaster* (Oregon-R) were grown at 25° C on a standard cornmeal-Karo-yeast agar medium supplemented with live yeast. After eclosion, the flies were separated according to sex. The experimental males were 7 to 10 days old. Accessory glands from 10 virgin males were dissected free of testicular and gut tissue and incubated for 1 h at 4° C in MOPS buffer [10 mM MOPS, pH 7.0, 80 mM NaCl, 10 mM KCl, and 0.2 mM $MgCl_2$, and varying concentrations of $CaCl_2$ [1] plus 10 μCi [^{32}P]-orthophosphate (8 mCi/ml, Amersham Corp.)]. Similarly when dissected glands were exposed to insulin or calcium channel antagonists, they were incubated continuously for 1 h in 0.1 mM $CaCl_2$ MOPS buffer containing these reagents and [^{32}P]-orthophosphate.

When the glands were treated with the tumor-promoting phorbol ester, TPA, (12-tetradecanoyl-phorbol-13-acetate), the time of exposure to the drug was 15 minutes. The reagent-containing media was then removed and the glands washed once and resuspended in MOPS buffer containing radioactive label and the incubation continued for 1 h. The control glands were treated in a similar fashion only using the control media for all procedures.

Isolation and analysis of RNA from accessory glands

Total RNA was extracted using the hot phenol method of Pelham [2]. The RNA was lyophilized, resuspended in water, denatured with glyoxal [3] and electrophoresed on a horizontal 1.5% agarose gel [4] at room temperature. For analysis of tRNA species the ^{32}P-labeled RNA was electrophoresed on vertical 8% polyacrylamide gels containing 8.3 M urea [5]. The gels were stained with 0.5 μg/ml of ethidium bromide, and photographed which allowed the quantitation of the recovery of RNA. The gels were then dried and the labeled RNA visualized by autoradiography. The autoradiographs were scanned on an LKB densitometer and integrated areas corresponding to rRNA and tRNA species were summed.

Maintenance of cell cultures

Schneider's line-2 tissue culture cells [6] were maintained in *Drosophila* Schneider's medium supplemented with 0.1% yeastolate, 0.25% bactopeptone, 1% antibiotic-antimycotic and 2% or 15% fetal bovine serum. To adapt Schneider's cells to growth in media containing 2% serum, the serum content of the medium was gradually reduced from 15% to 2% during which time the cells were passed continuously. Cell viability was monitored by trypan blue staining. Gradual reduction of serum content did not result in significant cell death.

Labeling of culture cells

Cells from exponentially growing cultures were harvested, washed and aliquoted for resuspension in media containing 100 μCi/ml [^{32}P]-orthophosphate (8 mCi/ml, Amersham Corp.), and indicated amounts of serum and then incubated for 1 h at room temperature. TPA treatment was identical to that described for accessory gland cells.

Total cellular RNA was isolated using the guanidinium monothiocyanate/LiCl method of Cathala *et al.* [7]. RNA samples were lyophilized, resuspended in a small volume of distilled water, denatured by glyoxal treatment and electrophoresed as described above. Following autoradiography of the dried gels, the films were scanned by densitometry and the integrated areas corresponding to the rRNA bands were summed.

Nuclear isolation

Nuclei were prepared from cells using the method of Schibler *et al.* [8] as modified by Tautz and Dover [9]. Following treatment, all manipulations were carried out on ice.

Cell suspensions were pelleted by low speed centrifugation (500 × g, 10 min, 4° C) and the pellet was gently resuspended by the addition of 0.3 M sucrose in lysis buffer A (60 mM KCl, 15 mM NaCl, 0.15 mM spermine, 0.5 mM spermidine, 14 mM 2-mercaptoethanol, 0.5 mM EGTA, 2 mM EDTA, 15 mM HEPES [pH 7.0], 5 mM $MgCl_2$). Cells were lysed with 10% NP40 for 5 min on ice and then immediately layered over 20% sucrose pads in lysis buffer A. The crude nuclear pellet was recovered by centrifugation (2500 × g, 30 min, 4° C). Endogenous RNA was digested by incubating the crude nuclear pellet in lysis buffer A plus 10 μg/ml DNase-free RNAse at 4°C. RNAse treated nuclei were further purified by centrifugation through 2 successive 20% sucrose pads in lysis buffer A. Clean nuclei were obtained by resuspension of the nuclear pellets in 0.3 M sucrose in lysis buffer B (same as A except containing 0.1 mM EGTA, 0.1 mM EDTA) followed by ultracentrifugation (36,000 × g, 60 min, 4° C) through 2 M sucrose pads in lysis buffer B. The nuclei were resuspended in nuclei suspension buffer (20 mM Tris-HCl[pH 7.9], 75 mM NaCl, 0.5 mM EDTA, 0.85 mM DTT, 0.125 mM PMSF, 50% [v/v] glycerol) at approximately 1×10^6 nuclei/μL. Aliquots from these suspensions were removed prior to freezing at − 70° C and used for quantitation. Nuclei were counted using 0.4% crystal violet in 0.1 M citric acid. Preparations were checked for evidence of nuclease activity, elongation rate, size of nascent transcripts, and alpha-amanitin sensitivity.

Nuclear run-on transcription

Transcription reactions were carried out in a reaction buffer containing 60 mM Tris-HCl [pH 7.9], 62.5 mM NaCl, 0.45 mM EDTA, 0.625 mM DTT, 0.112 mM PMSF, 350 mM $(NH_4)_2SO_4$, 1 mg/ml heparin sulfate, 4 mM $MnCl_2$, 5 mM $MgCl_2$, 0.6 μM alpha-[^{32}P] UTP (800 Ci/mmole), 1 mM each of ATP, CTP, GTP, 30% glycerol, 10 mM creatine phosphate, 130 U/ml RNasin, 100 μg/ml alpha-amanitin, and 5×10^7 nuclei. Transcription was terminated by vortexing in 10 volumes (2 ml) of RNA lysis buffer (4 M guanidine isothiocyanate, 25 mM sodium citrate (pH 7.0), 0.5% sarkosyl, 0.1 mM 2-mercaptoethanol). Total RNA was isolated from nuclei as described by Chomczynski *et al.* [10].

Two plasmids, designated pAB1915 and pAB2103 (a gift from B. Kohorn) were used in the nuclear run-on experiments. Digestion of pAB1915 with EcoR1 results in a restriction fragment of 1.4 Kb which contains 1.03 Kb fragment of 18S rDNA. pAB2103 was digested with Hind III to give a 0.90 Kb fragment of 28S rDNA. Equivalent amounts of rDNA sequences were electrophoresed in each lane of 1.5% agarose (BRL Ultrapure) gel. The amount of rDNA in each lane was determined to be in excess of ^{32}P-labeled nuclear run on transcripts.

Hybridization of rDNA specific nuclear run-on transcripts was visualized by autoradiography. Multiple exposures of the Southern blots were obtained and scanned using a LKB Scanning Densitometer. Changes in transcription in response to treatment with serum and TPA were determined by calculating the change in hybridization signal of 18S and 28S sequences relative to the level of hybridization observed in nuclei from cells maintained in 2% serum.

Results

RNA synthesis in the nondividing cells of the male accessory glands

The male accessory glands of *D. melanogaster* produce and secrete proteins that are found in the seminal fluid. These proteins are essential for reproductive success. Following mating there is a rapid increase in protein synthesis in the accessory glands [11]. This stimulation of secretory protein synthesis was accompanied by a stimulation of rRNA transcription, as well as an increase in ribosomal protein accumulation [12]. We observed a parallel increase in labeling of 5S and tRNAs. However, levels of several mRNAs were unaffected by mating.

Cells of the male accessory glands are nondividing as established by electron microscopy studies and the lack of labeling of gland nuclei with [^{3}H]-thymidine [12]. We have also developed an *in vitro* system in which we observe similar alterations in protein synthesis as those seen in our *in vivo* studies [13, 14]. We found that added calcium, TPA and juvenile hormone III each increased protein synthesis in the glands.

Using this system calcium and other reagents were tested for their affects on rRNA and tRNA synthesis in male accessory glands *in vitro*. With no added calcium in the media synthesis of RNA is about half that observed with 0.1 mM added calcium (Fig. 1A). Increasing the calcium concentration

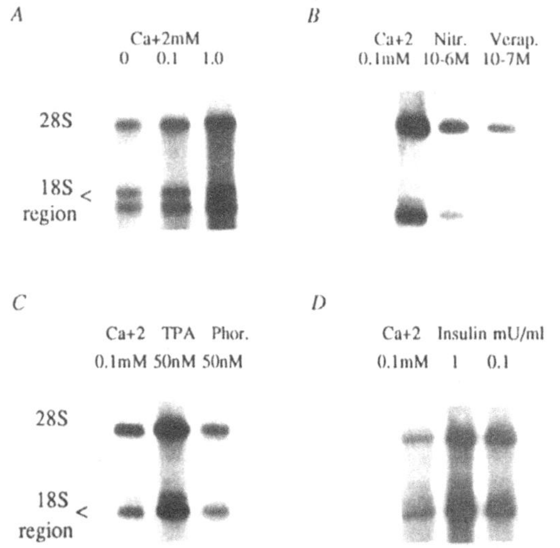

Fig. 1. Autoradiographs of agarose gels of ^{32}P-labeled rRNA from glands treated with (A) different calcium concentrations, (B) calcium channel antagonists (C) TPA and phorbol, (D) different insulin concentrations.

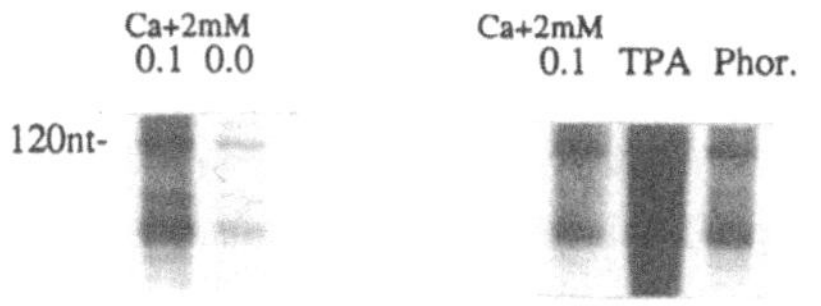

Fig. 2. Autoradiographs of the acrylamide gels of ^{32}P-labeled tRNA from glands treated with different calcium concentrations and with phorbol or TPA.

from 0.1 mM to 1.0 mM increased the labeled RNA 2-fold. These data confirm that rRNA synthesis, as well as protein synthesis, are modulated by the level of calcium in the incubation media. The specific calcium channel antagonists, nitrendipine and verapamil, significantly inhibited gland RNA synthesis in our *in vitro* system. The maximum inhibition of RNA synthesis was seen at 10^{-6} M nitrendipine and 10^{-7} M verapamil (Fig. 1B). These observations strongly implicate calcium as a major regulator of rRNA synthesis in male accessory glands.

Experiments using the tumor-promoting phorbol ester, TPA, which acts to activate protein kinase C, demonstrate a stimulation of rRNA synthesis when the glands are exposed for 15 min. The maximum stimulation of rRNA synthesis in the glands was seen at 50 nM TPA (Fig. 1C). Exposure of the glands to the kinase C inactive compound, phorbol, at 50 nM did not effect rRNA synthesis. These data support the argument that protein kinase C may mediate the affects of external calcium on rRNA synthesis. However, involvement of other calcium-dependent protein kinases is also a possibility. It is interesting to note that Soprano and Baserga [15] reported that in human/mouse hybrid cells TPA exposure activated 'silent' rRNA genes.

Incubation of the nonmitotic accessory glands with insulin for 1 h, leads to a significant increase in rRNA synthesis (Fig. 1D). Thus treatment of the nonmitotic cells of the accessory glands with varying calcium concentrations, with a tumor-promoting phorbol ester or insulin all result in increased rRNA synthesis. We have further used our *in vitro* system to investigate whether these effectors also increase tRNA synthesis. Removal of calcium from the external media caused a reduction in the labeled tRNA species (Fig. 2). Further, a 15 min exposure to TPA caused a significant increase in the tRNA labeling. When the accessory glands were treated with phorbol, which does not activate protein kinase C, there is no change in tRNA labeling. Finally, insulin treatment of the accessory glands also causes a significant increase in tRNA labeling (not shown). Thus variations in calcium concentration, treatment with TPA or insulin each cause coordinate changes in both rRNA and tRNA synthesis in the nondividing cells of the male accessory gland.

RNA synthesis in the mitotically active Drosophila *Schneider's cultured cells*

We next examined rRNA and tRNA synthesis in *Drosophila* tissue culture cells which are mitotically active. *Drosophila* Schneider's line 2 cells were grown in Schneider's medium containing 2% or 15% fetal bovine serum. Cells grown in 2% serum double every 50 hr. while cells in 15% serum double every 24 hr. We observed a significant difference in ^{32}P-labeled rRNA from cells grown in the presence of 15% serum as compared to cells grown in the presence of 2% serum. If cells growing in 2% serum are shifted to a medium with 15% serum, there is a significant increase in labeled rRNA within 1 h (Fig. 3).

The effect of the tumor-promoting phorbol ester, TPA, on rRNA synthesis in Schneider's cells was also determined. If the cells are exposed to TPA for 15 min, then washed, resuspended in the same media without TPA and labeled with [^{32}P]-phosphate for 1 h, the TPA stimulates rRNA synthesis even at a concentration of 0.5 nM. These data and other data not shown here demonstrated that a concentration of 5 nM TPA maximally stimulates rRNA synthesis. Cells treated with the inactive compound, phorbol (5 nM) showed no stimulation in rRNA synthesis and perhaps a slight inhibition.

The experiments presented here show that changes in rRNA accumulation result from the treatment of glands or tissue culture cells with a variety of agents. We re-examined these effectors for their influence specifically on rRNA transcription using nuclear run-on assays (Fig. 3). Run-on RNAs complementary to both 18S and 28S rDNA were assayed. The increase of RNA polymerase loading as judged by increased labeling of run-on rRNA reflects the increases observed in rRNA accumulation in cells maintained in 15% serum and following serum stimulation. These data are summarized in Table 1.

In summary, the data show that several conditions can affect the level of rRNA and tRNA accumulation in mitotically active and terminally differentiated (nonmitotic) *Drosophila* cells. This regulation could be occurring by similar signaling pathways involving calcium and protein kinase C. The changes in rRNA levels are paralleled by changes in transcription as measured by nuclear run-on experiments.

Discussion

Several experiments including those presented here have begun to delineate those factors which affect changes in rRNA and tRNA gene expression. In mammalian cultured cells and, in *Acanthamoeba,* starvation or inhibition of protein synthesis by treatment with cycloheximide causes a significant down-regulation of rRNA and tRNA gene transcription [16] and [17]. Tower and Sollner-Webb [18] reported that in growth-restricted cells the activity of RNA Pol I was reduced causing lowered rDNA transcription, while a decrease in the activity of RNA Pol III transcription factor, TFIIIB, was responsible for lowered 5S and tRNA synthesis. During rat myogenesis rDNA activity may be regulated by factors other than Pol I [19]. Rat heart cells in culture showed accelerated rates of rRNA synthesis when contraction began [20]. Transcriptionally silent mouse rRNA genes are activated by TPA in human-mouse cell hybrids [15]. The experiments presented here, in the *Drosophila* system, show that in both actively dividing and terminally differentiated (nondividing) cells rRNA

A

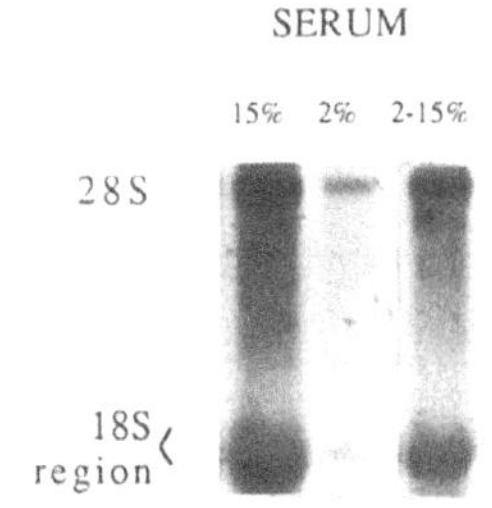

Fig. 3. (Top Panel) Autoradiographs of agarose gels of ^{32}P-labeled RNA from Schneider's cells maintained in 2% or 15% serum or stimulated with added serum for 1 h. (Bottom Panel) Autoradiographs of the hybridization of nuclear run-on labeled RNA to 18S and 28S rDNA. The nuclear run-on RNA was derived from Schneider's cells grown continuously in 2% serum, 15% serum or serum stimulated for 1 h.

B

and tRNA synthesis is substantially increased by serum and tumor-promoting phorbol esters. In addition, several cases have been examined in which viral infection can either up-regulate or down-regulate transcription by Pol I or III (for example, [21]).

It is currently thought that tumor-promoting phorbol esters exert their effect by altering the pattern of gene expression in cells through the activation of protein kinase C [22]. The induction of certain viral and cellular RNA Pol II transcribed genes by TPA has been characterized and includes metallothionien II_A, collagenase and SV40 genes [23]. The inducibility of these genes occurs via a common DNA-sequence element recognized by the AP-1 or AP-1-related transcription factors.

The calcium messenger system can be mediated by at least 2 mechanisms, one involving protein kinase C and the other involving calmodulin [24]. Only genes transcribed by Pol II have been examined in detail for TPA-inducibility. An effect of phorbol esters on *Drosophila* gene expression had not been demonstrated. Our studies are therefore the first to demonstrate inducibility of the transcription of tRNAs and rRNAs by TPA in *Drosophila.*

Our *in vivo* results which demonstrate the regulation of rRNA and tRNA synthesis by serum and phorbol esters have been extended to show these same regulatory effects in *in vitro* assays. We have constructed a plasmid with upstream and downstream sequences around the rDNA transcriptional start site attached to one of two distinct T7 phage reporter sequences [25]. This plasmid gives stable T7 transcripts whose levels are increased to the same extent as the endogenous rRNA levels when cells are treated with serum and TPA. It is clear that all the essential sequences for the proper regulation of the rDNA response to TPA and serum are present between − 305 and + 680 nt on either side of the start site of rRNA transcription. It is important to note that there are sequence elements in the essential regulatory DNA that show some similarity to the sequences which have been identified as consensus elements for serum and TPA responsiveness in mammalian genes transcribed by RNA Pol II. In addition, we have recently prepared nuclear extracts from *Drosophila* cells that accurately and reproducibly transcribe Pol I and Pol III templates *in vitro* [26]. Extracts from TPA and serum-stimulated cells show a marked increase in transcriptional activity *in vitro.*

Drosophila rRNA and tRNA genes are rapidly regulated by reagents such as calcium, serum, insulin and TPA. These effectors act on rRNA at least in part at the level of transcription and these regulatory effects can be reproduced in *in vitro* assays. Together these data suggest that growth factors, calcium and compounds that activate protein kinase C can signal coordinate effects on rRNA and tRNA synthesis.

Table 1. Control levels of rRNA and tRNA following a 1 h labeling or hybridization of nuclear run-on RNA to rDNA sequences are set at 1.0. Treated cell levels are relative to the control. For cells maintained at 15% serum, the numbers given are relative to the control for cells grown on 2% serum. Numbers in parentheses are the relative values as compared to control cells grown on 15% serum.

	rRNA levels	Run-on		tRNA levels
		18S	28S	
2% Serum				
Control	1.0	1.0	1.0	1.0
TPA	6.2	2.3	2.3	6.5
2% → 15% Serum 1 h	7.5	2.6	2.6	6.5
15% Serum				
Control	12.0	2.8	2.8	10.1
TPA	96.0 (8.0)	6.4 (2.3)	6.4 (2.3)	75.0 (7.5)

Acknowledgements

We would like to thank Linda Reppas for preparation of the manuscript, Jane Padilla for preparation of figures, Dr. Bruce Kohorn for the gift of rDNA plasmids and Dr. Deborah Johnson for helpful discussion. This work was supported by the USC Faculty Research and Innovation Fund and NIH grant GM30443.

References

1. Mitchell HK, Chen PS, Lipps LS, Moller G: Separation of *Drosophila* RNAs on acrylamide gels in formamide. Insect Biochem 8: 29–35, 1978
2. Pelham HRB: Techniques for eukaryotic protein synthesis *in vitro*. In: Techniques in the Life Sciences, Vol. B4/1, Elsevier Scientific, County Clare, Ireland. 1982, pp 1–18
3. McMaster GK, Carmichael GC: Analysis of single- and double-stranded nucleic acids on polyacrylamide and agarose gels by using glyoxal and acridine orange. Proc Natl Acad Sci USA 74: 4835–4838, 1977
4. Maniatis T, Fritsch EF, Sambrook J: Molecular cloning: A laboratory manual. Cold Spring Harbor Laboratory, Cold Spring Harbor, N.Y. 1982, pp 149–172
5. Burke DJ, Soll D: Functional analysis of fractionated *Drosophila* KC cell tRNA gene transcription components. J Biol Chem 260: 816–823, 1985
6. Schneider I: Cell lines derived from late embryonic stages of *Drosophila melanogaster*. J Embryol Exp Morph 27: 353–365, 1972
7. Cathala G, Savouret J-F, Mendez B, West BL, Karin M, Martial J, Baxter B: Laboratory methods. A method for isolation of intact translationally active ribonucleic acid. DNA 2: 329–335, 1983
8. Schibler V, Hagenbuchle O, Wellauer P, Pillet A: Two promoters of different strengths control the transcription of the mouse alpha-amylase gene amy-1[a] in the parotid gland of the liver. Cell 33: 501–508, 1983
9. Tautz D, Dover G: Transcription of the tandem array of ribosomal DNA in *Drosophila melanogaster* does not terminate at any fixed point. EMBO J 5: 1267–1273, 1986
10. Chomczynski P, Sacchi N: Single-step method of RNA isolation by acid guanidinium thiocyanate-phenol-chloroform extraction. Anal Biochem 162: 156–159, 1987
11. Chen PS: The functional morphology and biochemistry of insect male accessory glands and their secretions. Annu Rev Entomol 29: 233–255, 1984
12. Schmidt T, Chen PS, Pellegrini M: The induction of ribosome biosynthesis in a nonmitotic secretory tissue. J Biol Chem 260: 7645–7650, 1985a
13. Schmidt T, Stumm-Zollinger E, Chen PS: Protein metabolism of *Drosophila melanogaster* male accessory glands – III. Ins Biochem 15: 391–401, 1985b
14. Yamamoto K, Chadarevian A, Pellegrini M: Juvenile hormone action mediated in male accessory glands of *Drosophila* by calcium and kinase C. Science, 239: 916–919, 1988
15. Soprano KJ, Baserga R: Reactivation of ribosomal RNA genes in human mouse hybrid cells by 12-O-tetradecanoylphorbol-13-acetate. Proc Natl Acad Sci USA 77: 1566–1569, 1980
16. Bateman E, Iida C, Kownin P, Paule M: Footprinting of ribosomal RNA genes by transcription initiation factor and RNA Polymerase I. Proc Natl Acad Sci USA 82: 8044–8048, 1985
17. Sollner-Webb B, Tower J: Transcription of cloned eukaryotic ribosomal RNA genes. Ann Rev Biochem 55: 801–830, 1986
18. Tower J, Sollner-Webb B: Pol III Transcription factor B activity is reduced in extracts of growth restricted cells. Molec Cell Biol 8: 1001–1005, 1988
19. Zahradka P, Sells BH: Transcription factors mediate rRNA synthesis during myogenesis. Eur J Biochem 171: 37–43, 1988
20. McDermott PJ, Rothblum LI, Smith SD, Morgan HE: Accelerated rates of ribosomal RNA synthesis during growth of contracting heart cells in culture. J Biol Chem 264: 18220–18227, 1989
21. Hoeffler WK, Kovelman R, Roeder RG: Activation of transcription factor IIIC by the adenovirus EIA protein. Cell 53: 909–920, 1988
22. Nishizuka Y: The role of protein kinase C in cell surface transduction and tumour-promotion. Nature 308: 693–697, 1984
23. Karin M, Herrlich P: Cis- and trans-acting genetic elements responsible for induction of specific genes by tumor-promoters, serum factors, and stress. In: Genes and Signal Transduction in Multistage Carcinogenesis, N.H. Colburn, (ed) Marcel Dekker, Inc, New York, 1989, pp 415–440
24. Rasmussen H, Barrett PQ: Calcium messenger system: An integrated view. Physiol Rev 64: 938–984, 1984
25. Elion EA, Warner JR: Major promoter element of rRNA transcription in yeast lies 2 Kb upstream. Cell 39: 663–673, 1984
26. Parker CS, Topol J: A *Drosophila* RNA polymerase II transcription factor contains a promoter-region-specific DNA-binding activity. Cell 36: 357–369, 1984

Address for offprints: M. Pellegrini, Molecular Biology Section, University of Southern California, Los Angeles, CA 90089-1340, USA

Molecular and Cellular Biochemistry **104**: 209–211, 1991.

Index